Algebra $\frac{1}{2}$

An Incremental Development

Algebra $\frac{1}{2}$: An Incremental Development

Printed in the United States of America
ISBN: 0-939798-05-0
Fifth Printing, May 1985

Grassdale Publishers, Inc.
1002 Lincoln Green
Norman, Oklahoma 73069

Algebra $\frac{1}{2}$

An Incremental Development

JOHN H. SAXON, JR.

Rose State College

GRASSDALE PUBLISHERS, INC.

Preface

This book is designed to provide the guidance for a full year of practice in the fundamentals of arithmetic while basic abstractions of beginning algebra are introduced and practiced. Testing has revealed that many students at this point still have some difficulty in using fractions, mixed numbers and decimal numbers in the four basic operations. Some students have difficulty in only one area while others have difficulty in more than one area. Thus, the homework problem sets provide practice in every one of these skill areas for the entire year. The students who are deficient will be afforded the opportunity to learn, and the practice afforded the others will increase and solidify their abilities in these basic areas.

Special emphasis is given to reading numbers written in numerical form and to translating numbers in numerical form to the word form. Practice problems in these skills are provided in homework problem sets for over half the book. This translation is difficult and lack of emphasis and long-time practice at the proper point is the reason that many older students are without this skill.

Conceptualization of area, volume and perimeter is given considerable emphasis. These abstractions are introduced early and problems on at least one of these topics appear in every lesson beginning with Lesson 9. Since the words area and volume designate abstractions, students are encouraged to associate floor tiles with area and sugar cubes with volume. Reification of abstractions facilitates comprehension and leads to long-term retention.

Word problems are given special attention. Every problem set after Lesson 11 begins with three or four word problems. At first, the problems are straightforward problems that require only addition, subtraction, division, multiplication or finding the average. The problems are designed to give students practice in reading word problems and deciding what is given and what procedure is required so that the question asked may be answered. This ability to read and to translate is a skill that must be mastered before the solution of more difficult problems can be attempted. Problems about rate, time and distance are presented in Lesson 61 and later simple two-step word problems are introduced.

Major emphasis is placed on the fundamental problems of mathematics which are those that deal with the fractional parts of a number. This problem is introduced early and practiced continually until its companion, the percent problem, is introduced. Then both problems appear regularly until the end of the book. Both problems are approached conceptually and students are encouraged to draw diagrams of the completed problems. Some people believe that somehow, somewhere, a trick can be found that will allow the answers to meaningful percent problems to be obtained without understanding. Since no such trick exists, understanding must occur, and a picture of the problem has proved to be most helpful. This approach will be continued in my Algebra I book and in the Algebra II book. Students can no longer be permitted to finish mathematics without a total understanding of percent and the relationship of percent to the fractional part of a number problem.

Simple concepts in algebra are introduced early and practiced for the rest of the book. Variables are introduced in Lesson 34 where the numbers of arithmetic are used as replacements for variables in algebraic expressions. Simple equations are introduced in Lesson 39, and equations with fractions are introduced in Lesson 40. These problems allow practice in adding, subtracting and dividing fractions at the same time that the two basic rules for solving equations are being introduced. This early introduction of simple equations permits the introduction of ratio problems in Lesson 56 and elementary ratio word problems in Lesson 66. These problems are practiced in the problem sets until the end of the book.

Integral exponents and integral roots are introduced in Lesson 45 and are practiced gently thereafter. This early introduction of exponents and roots is necessitated by the long-term practice that many students need in order to understand these notations. Negative numbers are introduced in Lessons 70 and 71 and simplification problems that contain elementary combinations of positive and negative numbers will appear in the next forty problem sets. This will provide excellent preparation for the more complicated expressions that will be encountered in Algebra I.

A study of the homework problem sets will show how this book provides a comprehensive and complete review and continued practice of the skills of arithmetic while basic facets of more advanced topics are introduced. The early introduction of these topics will ease the transition to algebra and should insure higher success percentages for the students who attempt algebra. Thus, we see that this book can be used successfully as a pre-algebra book for the gifted and average students and can be used as the first year's course in a two year Algebra I sequence for other students. It can also be used as a general mathematics book for students who will not attempt algebra. The only difference will be in the age of the students when they use this book.

The book was written to provide continued practice in skills for an entire year. **To gain maximum advantage from the use of this book, it is necessary that all students work all the problems. The book does not have extra problems and was designed with the understanding that every problem would be worked.** In this book, the learning is spread out rather than being concentrated. **Teachers should avoid the temptation to furnish extra problems of the new kind at the expense of review problems. Testing has shown that this procedure will increase short-term understanding but will vitiate long-term retention.**

Again, I thank Frank Wang for checking the answers, for help in proofreading and for his other contributions. Also, I again thank Detia Roe, my graduate physicist typist, for her typing, for pointing out errors and for her moral support.

John Saxon Norman, Oklahoma

Contents

Basic

110 Algebra Lessons

Course

LESSON 1 Whole number place value · Reading and writing whole numbers

1.A whole number place value

We use the **Hindu-Arabic** system to write our numbers. This system is a base 10 system and has 10 different symbols. The symbols are called **digits**, and they are:

0, 1, 2, 3, 4, 5, 6, 7, 8, 9

When we write whole numbers, we can write the decimal point at the end of the number, or we can leave it off. Thus, both of these

427.   427

represent the same number. In the right-hand number, the decimal point is assumed to be after the 7. The value of a digit in a number depends on where the digit appears in the number. The 5 in the left-hand number below

415,623   701,586   731,235

has a value of 5000 because it is in the thousands' place. The value of the 5 in the center number is 500 because it is in the hundreds' place. The value of the 5 in the right-hand number is 5 because it is in the units' (ones') place. The first place to the left of the decimal point is the ones' place. We also call this place the **units' place.** The next place is the **tens' place.** The next place is the **hundreds' place.** The next place is the **thousands' place.** Each place to the left has one more zero.

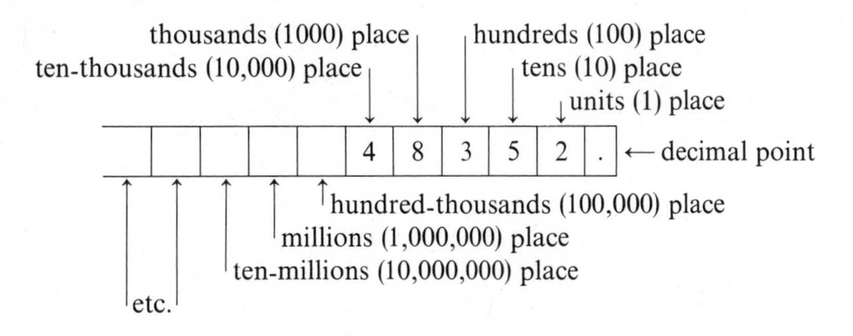

example 1.A.1 In the number 46,235:

(a) What is the value of the digit 5?
(b) What is the value of the digit 2?
(c) What is the value of the digit 4?

solution First we write the decimal point at the end.

46,235.

(a) The 5 is one place to the left of the decimal point. This is the units' place. This digit has a value of 5×1, or **5.**

(b) The 2 is three digits to the left of the decimal point. This is the hundreds' place. This digit has a value of $2 \times 100 =$ **200.**

(c) The 4 is five digits to the left of the decimal point. This is the ten-thousands' place. This digit has a value of $4 \times 10{,}000 =$ **40,000.**

1.B reading and writing whole numbers

We begin by noting that all numbers between 20 and 100 that do not end in zero are hyphenated words when we write them out.

23 is written twenty-three
35 is written thirty-five
42 is written forty-two
51 is written fifty-one
64 is written sixty-four
79 is written seventy-nine
86 is written eighty-six
98 is written ninety-eight

The word **and** is not used when we write whole numbers by using words.

501	is written	five hundred one
	not	five hundred and one
370	is written	three hundred seventy
	not	three hundred and seventy
422	is written	four hundred twenty-two
	not	four hundred and twenty-two

The hyphen is also used in whole numbers when the whole number is used as a modifier. The words

ten thousand

are not hyphenated. But when we use these words as a modifier, as when we say

ten-thousands' place

the words are hyphenated. Other examples of this rule are

hundred-millions' digit
ten-billions' place
hundred-thousands' place

Before we read whole numbers, we begin at the decimal point and move to the left, placing a comma after every three digits. These commas divide the digits into groups called **periods.**†

Etc,	Trillions			Billions			Millions			Thousands			Units (Ones)			·Decimal Point
	Hundreds	Tens	Ones	Hundreds	Tens	Ones	Hundreds	Tens	Ones	Hundreds	Tens	Ones	Hundreds	Tens	Ones	

† The word *period* has two meanings. One meaning is a dot. The other meaning is a space or a space in time such as a period of days.

Then we read the number from left to right beginning with the leftmost period. First read the number in the period and then read the name of the period. Then we move to the right and repeat the procedure.

example 1.B.1 Use words to write the number 51723642

solution We write the decimal point on the right end. Then we move to the left and place a comma after each group of three digits.

51,723,642.

The left period is the millions' period. We read it as

fifty-one million

The next period is the thousands' period. We read it as

seven hundred twenty-three thousand

The last period is the units' period. We do not say *units* at the end but read it as

six hundred forty-two.

Note that the words **fifty-one, twenty-three,** and **forty-two** are hyphenated words. Also note that we do not use the word *and* between the periods. Now we put the parts together and read the number as

fifty-one million, seven hundred twenty-three thousand, six hundred forty-two.

Note that we also use commas to separate the periods when we write the number in words.

example 1.B.2 Use digits to write the number fifty-one billion, twenty-seven thousand, five hundred twenty.

solution The first period is the billions' period. It contains the number fifty-one.

51, , , .

All the periods after the first period must have three digits. There are no millions so we use three 0s.

51,000, , .

There are twenty-seven thousands. **So we write 027 in the next period so the period will contain three digits.**

51,000,027, .

Now we finish by writing 520 in the last period.

51,000,027,520.

1.C addition

When we add numbers, we call each of the numbers **addends,** and we call the answer a **sum.**

523	addend
619	addend
512	addend
1654	sum

When we add whole numbers, we write the numbers so that the last digits of the numbers are aligned vertically.

example 1.C.1 Add 4 + 407 + 3526

solution We write the numbers so that the last digits are aligned vertically. Then we add.

$$\begin{array}{r} 4 \\ 407 \\ 3256 \\ \hline \mathbf{3667} \end{array}$$

problem set 1

1. In the number 5062973, what is the value of the digit (a) 6, (b) 9, (c) 3?

2. Write the six-digit number that has the digit 4 in the thousands' place with every other digit being 6.

3 Write the seven-digit number that has the digit 3 in the millions' place, the digit 7 in the hundreds' place, with every other digit being 6.

4. A number has eight digits. Every digit is 9 except the ten millions' digit which is 3, the ten-thousands' digit which is 5 and the units' digit which is 2. Use digits to write the number.

***5.** Use digits to write the number: fifty-one billion, twenty-seven thousand, five hundred twenty.

6. Use digits to write the number: five hundred seven billion, six hundred forty million, ninety thousand, forty-two.

7. Use digits to write the number: four hundred seven trillion, ninety million, seven hundred forty-two thousand, seventy-two.

8. Use digits to write the number nine hundred eighty million, four hundred seventy.

Use words to write the following numbers.

***9.** 51723642 **10.** 90807060 **11.** 32000000652

12. 3250000652

Add:

13. 416 + 200 + 33 + 499

14. 908 + 43 + 6 + 293 + 365

15. 5090 + 6241 + 198 + 17 + 849

16. 9321 + 47 + 183 + 18 + 1037

17. 20,831 + 8432 + 293 + 84 + 6953

18. 91,032 + 40,516

19. 501 + 22 + 619 + 512

20. 7 + 299 + 915 + 205

21. 92 + 301 + 1037 + 120

22. 9183 + 5176 + 5098

23. 24,756 + 9805 + 132

24. 70,918 + 93,265 + 40,527

25. 52,913 + 5047 + 23

26. 50,214 + 69,207 + 5143

27. 25,172 + 6,284 + 75

28. 47 + 321 + 205 + 4167

29. 9 + 3102 + 81 + 753

30. 81 + 2 + 506 + 398

31. 31 + 7 + 503 + 2815

LESSON 2 Rounding off whole numbers

2.A rounding off

Often we use the rounded-off form of a number. If the distance to the barn is 209 yards, the farmer would probably say that it is 200 yards to the barn. If so, the farmer has rounded off the number 209 to the nearest hundred.

When we round off a number, it helps to use a circle and an arrow. To demonstrate we will round off 24,374 to the nearest thousand. We will use three steps.

1. Circle the digit in the place to which we are rounding off and mark the digit to its right with an arrow.

$$2\,\textcircled{4},\overset{\downarrow}{3}\,7\,4$$

2. Change the arrow-marked digit and all digits to its right to zero.

$$2\,\textcircled{4},\overset{\downarrow}{0}\,0\,0$$

3. Leave the circled digit unchanged or increase it one unit as determined by the following rules:
   (a) If the arrow-marked digit was less than 5, do not change the circled digit.
   (b) If the arrow-marked digit was 5 or greater, increase the circled digit by one unit.†

Rule (a) applies in this problem so our answer is

24,000

example 2.A.1 Round off 471,326,502 to the nearest ten thousand.

solution First circle the ten-thousands' digit and mark the digit to its right with an arrow.

$$4\,7\,1,3\,\textcircled{2}\,\overset{\downarrow}{6},5\,0\,2$$

Next change the arrow-marked digit and all digits to its right to zero.

$$4\,7\,1,3\,\textcircled{2}\,\overset{\downarrow}{0},0\,0\,0.$$

Since the arrow-marked digit was greater than 5, we increase the circled digit from 2 to 3 and get

471,330,000.

example 2.A.2 Round off 83,752,914,625 to the nearest ten thousand.

solution First we circle the ten thousands digit and mark the digit to its right with an arrow.

$$8\,3,7\,5\,2,9\,\textcircled{1}\,\overset{\downarrow}{4},6\,2\,5.$$

† This is a simplified rule. Sometimes a more complicated rule is used when the arrowed-marked digit is 5 or is 5 followed by one or more zeros.

Now we change the arrow-marked digit and all digits to its right to zero.

$$8\,3,\,7\,5\,2,\,9\,\textcircled{1}\,\overset{\downarrow}{0},\,0\,0\,0.$$

Since the arrow-marked digit was less than 5, we leave the circled digit unchanged and get

83,752,910,000.

problem set 2

*1. Round off 24,374 to the nearest thousand.

*2. Round off 471,326,502 to the nearest ten thousand.

*3. Round off 83,752,914,625 to the nearest ten thousand.

4. Round off 4,185,270 to the nearest hundred.

5. Round off 83,721,525 to the nearest thousand.

6. Round off 415,237,842 to the nearest hundred thousand.

7. A number has nine digits. All the digits are 7 except the ten-thousands' digit, which is 2, and the tens' digit, which is 5. Write the number.

8. A number has seven digits. All the digits are 3 except the hundred-thousands' digit, which is 6; the thousands' digit, which is 4; and the hundreds' digit, which is 7. Write the number.

9. Use digits to write one hundred seven million, forty-seven thousand, twenty.

10. Use digits to write ninety-three billion, four hundred sixty-two million, forty-seven.

Use words to write the following numbers.

11. 731284006

12. 903721625

13. 9003001256

14. 7234000052

Add:

15. 432 + 189

16. 405 + 67 + 838

17. 893 + 699 + 26

18. 6104 + 20 + 8349

19. 9186 + 2243 + 1786

20. 9035 + 87 + 632 + 4268

21. 9317 + 4526 + 9015

22. 7316 + 4582 + 9143

23. 88,871 + 40,012 + 90,375

24. 78,524 + 91,325 + 70,026 + 91,358

25. 42,715 + 90,826 + 41,222 + 39,057

26. 37,251 + 81,432 + 90,256 + 21,312

27. 14 + 32 + 16 + 21 + 932 + 21

28. 1 + 2 + 21 + 12 + 122 + 1222

29. 33 + 333 + 313 + 1313

30. 4 + 314 + 134 + 13,245

LESSON 3 Austrian subtraction

3.A subtraction, the inverse of addition

Two operations are called inverse operations if one operation will "undo" the other operation. Addition and subtraction are inverse operations because addition will undo subtraction and subtraction will undo addition. To demonstrate we will begin with the number 5.

To this number we will add 4 and get a sum of 9.

$$\begin{array}{rl} 5 & \text{addend} \\ +\ 4 & \text{addend} \\ \hline 9 & \text{sum} \end{array}$$

Now if we subtract 4 from 9 we will "undo" the addition of 4, and we will be back at 5 again.

$$\begin{array}{rl} 9 & \text{minuend} \\ -\ 4 & \text{subtrahend} \\ \hline 5 & \text{difference} \end{array}$$

We call the number subtracted from the **minuend.** The number subtracted is called the **subtrahend,** and the answer is called the **difference.**

3.B Austrian subtraction

We often make mistakes when we have to borrow when we subtract. This is due partly to the untidy result of crossing out some numbers and writing in other numbers. The Austrian method of subtraction is much neater and causes fewer errors. If we wish to subtract 47 from 82, we normally borrow 1 from the 8 and replace the 8 with 7, as shown on the left.

$$\begin{array}{r} {\overset{7}{\not 8}}{}^{1}2 \\ -\ 4\ 7 \\ \hline 3\ 5 \end{array} \qquad \begin{array}{r} 8^{1}2 \\ -\ 4\ 7 \\ {\scriptstyle 1}\ \ \\ \hline 3\ 5 \end{array}$$

On the right we use the Austrian method. We still borrow 1, but **instead of crossing out the 8, we put a 1 below the 4.** We add this 1 to 4 and subtract from 8 to get 3. We get 3 either way, but the right-hand method is much neater.

example 3.B.1 Use the Austrian method to subtract.

$$\begin{array}{r} 462 \\ -\ 174 \\ \hline \end{array}$$

solution We borrow so that we can subtract the 4. When we borrow, we also place a 1 below the 7. Then we subtract 4 from 12 to get 8.

$$\begin{array}{r} 4\ 6^{1}2 \\ -\ 1\ 7\ 4 \\ {\scriptstyle 1}\ \ \\ \hline 8 \end{array}$$

Now we must borrow again. **Every time we borrow, we must record two 1s.**

$$\begin{array}{r} 4^{1}6^{1}2 \\ -\ 1\ 7\ 4 \\ {\scriptstyle 1}\ {\scriptstyle 1}\ \ \\ \hline \mathbf{2\ 8\ 8} \end{array}$$

Then we subtracted 7 plus 1 from 16 to get 8, and 1 + 1 from 4 to get 2. We finish by adding the last two lines to check.

$$\begin{array}{rl} 174 & \\ +\ 288 & \\ \hline 462 & \text{check} \end{array}$$

example 3.B.2 Use the Austrian method to subtract.

$$\begin{array}{r} 583 \\ -\ 284 \\ \hline \end{array}$$

solution Each time we borrow, we record two 1s.

$$\begin{array}{r} 5\ 8\ {}^{1}3 \\ -\ 2\ 8\ 4 \\ {}_{1}\ \ \ \\ \hline 9 \end{array}$$

Now we must borrow again.

$$\begin{array}{r} 5\ {}^{1}8\ {}^{1}3 \\ -\ 2\ 8\ 4 \\ {}_{1}\ {}_{1}\ \ \ \\ \hline \mathbf{2\ 9\ 9} \end{array}$$

We check by adding the last two lines.

$$\begin{array}{r} 284 \\ +\ 299 \\ \hline 583 \end{array} \quad \text{check}$$

There is nothing magical about the Austrian method of subtraction. It is not a better way to subtract—just a neater way to subtract. Some people prefer to use the Austrian method because it is neater. Other people prefer to subtract using the other method because it is more familiar. Both methods of subtraction result in the same answers.

problem set 3 Subtract. Add to check.

*1. $\begin{array}{r} 82 \\ -\ 47 \\ \hline \end{array}$ 2. $\begin{array}{r} 73 \\ -\ 46 \\ \hline \end{array}$ 3. $\begin{array}{r} 83 \\ -\ 47 \\ \hline \end{array}$

4. $\begin{array}{r} 92 \\ -\ 27 \\ \hline \end{array}$ *5. $\begin{array}{r} 462 \\ -\ 174 \\ \hline \end{array}$ 6. $\begin{array}{r} 853 \\ -\ 284 \\ \hline \end{array}$

7. $\begin{array}{r} 211 \\ -\ 122 \\ \hline \end{array}$ 8. $\begin{array}{r} 521 \\ -\ 245 \\ \hline \end{array}$

9. Round off 4,152,783 to the nearest ten thousand.

10. Round off 8,752,146 to the nearest thousand.

11. Round off 84,163,278 to the nearest million.

12. Round off 310,523 to the nearest thousand.

13. A number has six digits. All the digits are 2 except the thousands' digit, which is 5, and the units' digit, which is 3. What is the number?

14. A number has five digits. All of them are 7 except the thousands' digit, which is 0. What is the number?

15. Use digits to write fourteen million, seven hundred five thousand, fifty-two.

16. Use digits to write five hundred billion, four hundred sixty-five thousand, one hundred eighty-two.

Use words to write the following numbers.

17. 707070705 **18.** 5803125702

19. 9510282105 **20.** 42000002001

Add:

21. 423 + 718 + 935 + 218

22. 305 + 261 + 528 + 913

23. 295 + 486 + 588 + 714

24. 205 + 937 + 483 + 286

25. 913 + 405 + 709 + 203

26. 41,325 + 80,926 + 71,452 + 52,061

27. 90,125 + 30,627 + 95,132 + 40,061

28. 90,325 + 31,722 + 50,155 + 10,932

29. 13 + 216 + 2163 + 25,004

30. 12 + 1233 + 3012 + 51,623

LESSON 4 *Multiplication*

4.A multiplication

Multiplication is a shorthand notation we use to denote repeated addition. If we wish to add 7 eleven times, we could write

$$7 + 7 + 7 + 7 + 7 + 7 + 7 + 7 + 7 + 7 + 7$$

or we could write either

$$11 \times 7 \qquad \text{or} \qquad 7 \times 11$$

Thirty years ago when someone multiplied two numbers, one number was called the **multiplicand** and the other number was called the **multiplier.** The answer was called the **product.**

11	multiplicand
× 7	multiplier
77	product

We still call the answer the product, but we don't use the words multiplicand and multiplier much any more. Instead, we call both of the numbers **factors.**

11	factor
× 7	factor
77	product

We will use calculators in the next course to help us with multiplication and other arithmetic operations. This course is one-half arithmetic and one-half algebra so we will continue to practice multiplication, addition, and other arithmetic fundamentals in every lesson.

example 4.A.1 Multiply 421 by 335.

solution Either number may be placed on top. We will put 335 on top.

$$\begin{array}{r} 335 \\ \times \quad 421 \\ \hline 335 \\ 670 \\ 1340 \\ \hline \mathbf{141{,}035} \end{array}$$

problem set 4 Multiply:

*1. $\begin{array}{r} 335 \\ \times\ 421 \\ \hline \end{array}$

2. $\begin{array}{r} 778 \\ \times\ 563 \\ \hline \end{array}$

3. $\begin{array}{r} 904 \\ \times\ 725 \\ \hline \end{array}$

4. $\begin{array}{r} 528 \\ \times\ 806 \\ \hline \end{array}$

5. $\begin{array}{r} 234 \\ \times\ 526 \\ \hline \end{array}$

6. $\begin{array}{r} 215 \\ \times\ 404 \\ \hline \end{array}$

7. $\begin{array}{r} 923 \\ \times\ 315 \\ \hline \end{array}$

8. $\begin{array}{r} 406 \\ \times\ 207 \\ \hline \end{array}$

Subtract. Add to check.

9. $\begin{array}{r} 413 \\ -\ 277 \\ \hline \end{array}$

10. $\begin{array}{r} 724 \\ -\ 538 \\ \hline \end{array}$

11. $\begin{array}{r} 2165 \\ -\ 496 \\ \hline \end{array}$

12. $\begin{array}{r} 5205 \\ -\ 2403 \\ \hline \end{array}$

13. Round off 51,783,642 to the nearest hundred.

14. Round off 56,287,365 to the nearest thousand.

15. A number has five digits. All the digits are 7 except the digit in the tens' place, which is 2, and the digit in the ten-thousands' place, which is 6. Write the number.

16. A number has seven digits. The hundreds' digit is 4 and the hundred-thousands' digit is 5. All the rest of the digits are 3. Write the number.

17. Use digits to write two hundred seven million, seventy-five thousand, nine hundred three.

18. Use digits to write forty-one trillion, six hundred seventy-five million, nine hundred thirty-two thousand, four hundred seventy-five.

Use words to write the following numbers:

19. 4051632 20. 215320625004 21. 98061502

Add:

22. $51 + 6 + 97 + 52 + 78 + 44$

23. $93 + 24 + 97 + 81 + 56 + 21$

24. $9 + 51 + 86 + 13 + 21 + 52$

25. $316 + 32 + 152 + 937 + 214 + 602$

26. $42{,}351 + 18{,}216 + 90{,}527$

27. $93{,}215 + 82{,}906 + 90{,}517$

28. $502{,}154 + 200{,}316 + 905{,}213$

29. $5 + 217 + 63 + 582 + 7152$

30. $90{,}234 + 6235 + 4006 + 21{,}325$

LESSON 5 Division

5.A division

We remember that addition and subtraction are inverse operations because one of these operations will "undo" the other operation. **Multiplication and division are also inverse operations.** To demonstrate we begin with the number 5

$$5$$

and then multiply by 4 to get 20.

$$5 \times 4 = 20$$

Now if we divide 20 by 4, we will undo the multiplication by 4 and be back at 5 again.

$$20 \div 4 = 5$$

Here we show 3 different ways to designate division.

$$\text{(a)}\quad 20 \div 4 \qquad \text{(b)}\quad 4\overline{)20} \qquad \text{(c)}\quad \frac{20}{4}$$

All three notations indicate that 20 is to be divided by 4. The notations (a) and (b) are often used in arithmetic books but are not used much in algebra books. In algebra most authors prefer to use the fractional form (c). In this book we will use all three notations. The number we divide by is the **divisor,** and the number it goes into is the **dividend.** The answer to a division problem is called a **quotient.**

$$\begin{matrix} \text{dividend} & \longrightarrow \\ \text{divisor} & \longrightarrow \end{matrix} \frac{20}{4} = 5 \longleftarrow \text{quotient}$$

5.B division as a grouping process

It is sometimes helpful to think of division as a process of separating the dividend into a number of equal groups. For instance, if we wish to divide 12 by 4, we may write

$$\frac{12}{4}$$

The question we are asking is "Into how many groups of 4 can we divide 12 objects?" We can display the solution visually by using 12 dots and arranging them in groups of 4.

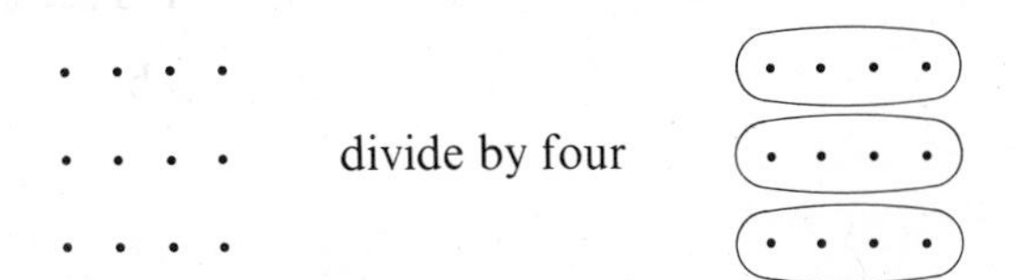

We see that 12 dots can be divided into three groups of 4, so we may say that

$$\frac{12}{4} = 3$$

In the same manner if we write

$$\frac{16}{5}$$

we are asking "Into how many groups of 5 can 16 be divided?" Again we use dots to permit a visual solution.

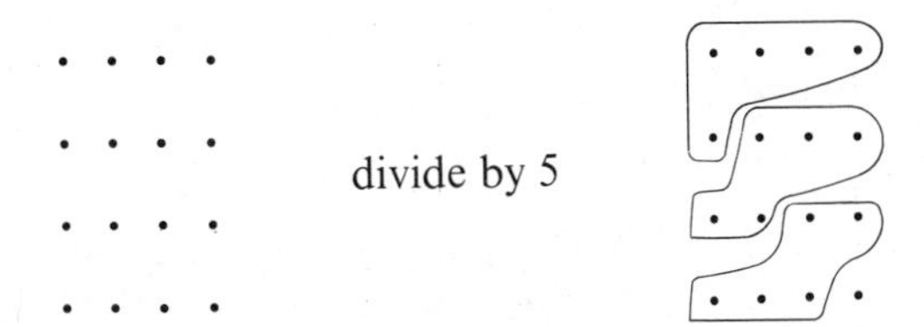

We find that 16 dots can be divided into three groups of 5 with one dot left over, and thus we say that the quotient is 3 with a remainder of 1, or

$$\frac{16}{5} = \mathbf{3,\ R1}$$

If we try to divide 17 objects into groups of three, we indicate our purpose by writing

$$\frac{17}{3}$$

and again we find that our division does not come out with a remainder of zero.

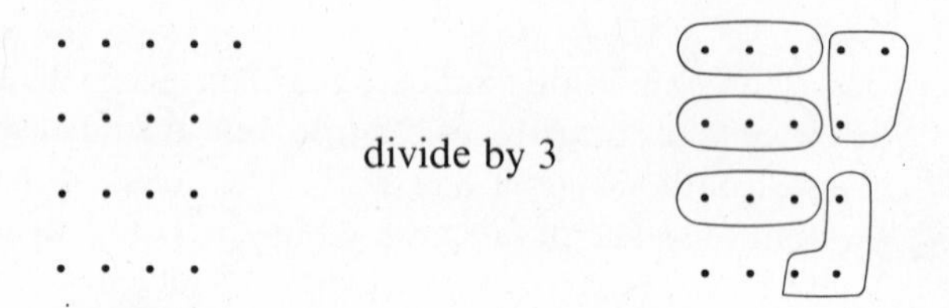

We get five groups of three dots and have two dots left over. Thus we say that 17 divided by 3 equals 5 with a remainder of 2.

$$17 \div 3 = \mathbf{5, R2}$$

We can see from the diagrams why some people use the words **goes into** instead of the word **divides.** They would say that 3 goes into 17 five times with a remainder of 2. The phrase *goes into* is more meaningful but the word *divides* is preferred by some people.

5.C two-digit divisors

The word **algorithm** means a way to do something. Many mathematicians like to use this word. When they speak of a division algorithm, they are using unusual words to say a way to do division. For the present, we will restrict our divisors to two digits.

example 5.C.1 Divide $\frac{251}{40}$.

solution We will use the common division algorithm.

$$\begin{array}{r|l} & \ \ \ 6 \\ \hline 40 & 251 \\ & 240 \\ \hline & \ \ 11 \text{ R} \end{array}$$

Thus $$\frac{251}{40} = \mathbf{6, R11}$$

example 5.C.2 Divide: $2183 \div 47$

solution We use the same algorithm.

$$\begin{array}{r|l} & \ \ \ 46 \\ \hline 47 & 2183 \\ & 188 \\ \hline & \ \ 303 \\ & \ \ 282 \\ \hline & \ \ \ 21 \text{ R} \end{array}$$

Thus $$\frac{2183}{47} = \mathbf{46, R21}$$

problem set 5

Divide:

*1. $\frac{251}{40}$ *2. $2183 \div 47$ 3. $43\overline{)50{,}217}$

4. $\frac{9300}{7}$ 5. $4165 \div 23$ 6. $21\overline{)30{,}215}$

Multiply:

7. 285×321 8. 506×275 9. 512×632

10. 773×667 11. 513×295 12. 806×932

Subtract. Add to check.

13. $943 - 277$ 14. $205 - 179$ 15. $3182 - 2257$

16. $2227 - 1349$

17. Round off 716,487,250 to the nearest ten million.

18. Round off 716,487,250 to the nearest ten thousand.

19. A number has four digits. All the digits are 3 except for the thousands' digit, which is 7. What is the number?

20. A number has seven digits. All are 3 except for the hundreds' digit and the thousands' digit. Both of these are 9. What is the number?

21. Use digits to write forty-seven million, fourteen.

22. Use digits to write fourteen billion, forty-two thousand, seven hundred fifty-five.

Use words to write the following numbers.

23. 5000021 24. 75400700215 25. 39002

Add:

26. $73 + 816 + 92 + 47 + 321 + 5432$ 27. $92 + 184 + 3182 + 915 + 21$

28. $408{,}627 + 915{,}634 + 589{,}062$ 29. $957{,}125 + 826{,}015 + 902{,}121$

30. $321 + 51 + 80{,}642 + 5123$

LESSON 6 ***Decimal numbers***

6.A decimal numbers

We have noted that whole numbers have a decimal point just after the last digit in the number. Sometimes the decimal point is written down. Many times it is not written but is understood to be there. Thus the two notations

615 615.

both designate the number six hundred fifteen. In the number on the left we did not write the decimal point, but in the number on the right we did write it.

In some numbers the decimal point is not at the end of the number. We often call these numbers **decimal numbers.** Some people call these numbers **decimal fractions** because they can be written as whole numbers divided by 10 or 100 or 1000 or 10,000 etc, as shown here.

$$61.23 = \frac{6123}{100}$$

The first place to the right of the decimal point in a decimal fraction is the $\frac{1}{10}$ or tenths' place. The next place is the $\frac{1}{100}$ or hundredths' place. The next place is the $\frac{1}{1000}$ or thousandths' place, etc. The value of a digit is the digit times the place value. Thus, the 6 in

.0006724

has a value of 6 times $\frac{1}{10,000}$, or 6 ten-thousandths, because it is in the ten-thousandths' place.

etc.	millions place	hundred-thousands place	ten-thousands place	thousands place	hundreds place	tens place	units place	decimal point	tenths place	hundredths place	thousandths place	ten-thousandths place	hundred-thousandths place	millionths place	etc.
	1,000,000	100,000	10,000	1000	100	10	1	.	$\frac{1}{10}$	$\frac{1}{100}$	$\frac{1}{1000}$	$\frac{1}{10,000}$	$\frac{1}{100,000}$	$\frac{1}{1,000,000}$	

6.B reading decimal numbers

Although it is not necessary, we can use commas to the right of the decimal point to help us read decimal fractions. We begin on the right end of the number and move left, placing a comma after every three digits.

example 6.B.1 Place the commas in: (a) .0000416 (b) .003102 (c) .10705014

solution **In each case we begin on the right end and move left.**

(a) .0,000,416 (b) .003,102 (c) .10,705,014

If the number has digits on both sides of the decimal point, we use the above procedure to the right of the decimal point and begin again at the decimal point.

example 6.B.2 Place the commas in: (a) 4165283.61805 (b) 7324.0062582

solution We begin on the right end and move left. Then we begin again at the decimal point and move to the left again.

(a) 4,165,283.61,805 (b) 7,324.0,062,582

To read a decimal number, we begin on the left end and read according to the following procedure:

(a) The digits to the left of the decimal point are read in the same way as they are read in whole numbers.
(b) **The decimal point is read as *and*.**
(c) **The digits to the right of the decimal point are read as if they formed a whole number and this reading is followed by naming the *place* of the last digit in the number.**

Commas are placed between the words in the same locations that commas appear in the numbers.

We will demonstrate by reading several decimal fractions.

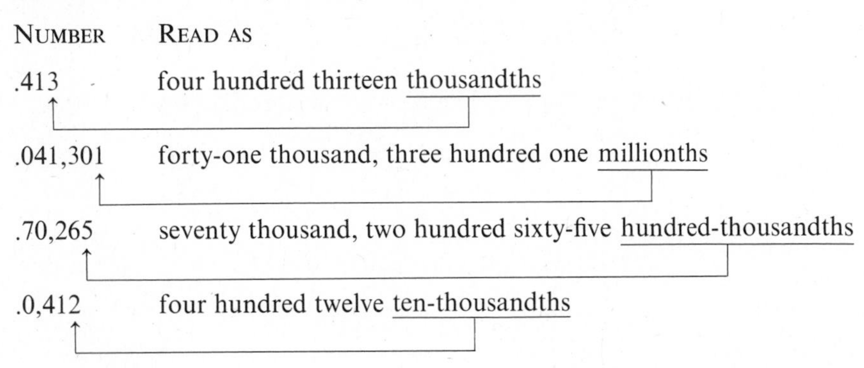

NUMBER	READ AS
.413	four hundred thirteen thousandths
.041,301	forty-one thousand, three hundred one millionths
.70,265	seventy thousand, two hundred sixty-five hundred-thousandths
.0,412	four hundred twelve ten-thousandths

We will finish by reading two decimal numbers that have nonzero digits on both sides of the decimal point.

example 6.B.3 Read the numbers: (a) 4165.0162 (b) 7108000.21578

solution First we insert the commas and then we read the nuumbers.

(a) 4,165.0,162 Four thousand, one hundred sixty-five **and** one hundred sixty-two ten-thousandths

(b) 7,108,000.21,578 Seven million, one hundred eight thousand **and** twenty-one thousand, five hundred seventy-eight hundred-thousandths

6.C adding and subtracting decimal numbers

We add and subtract decimal numbers just as we do whole numbers. When we add and subtract decimal numbers we must remember to write the numbers so that the decimal points are aligned one above the other.

example 6.C.1 Simplify: (a) 6.231 + .044 (b) 6.231 − .044

solution To begin, we write the numbers down with the decimal points one above the other. Then we add or subtract as indicated.

(a)
$$\begin{array}{r} 6.231 \\ +\ .044 \\ \hline \mathbf{6.275} \end{array}$$

(b)
$$\begin{array}{r} 6.2{}^{1}3{}^{1}1 \\ -\ .0\ 4\ 4 \\ {\scriptstyle 1\ 1} \\ \hline \mathbf{6.1\ 8\ 7} \end{array}$$

6.D multiplying decimal numbers

We do not align the decimal points when we multiply decimal numbers. We multiply decimal numbers just as we multiply whole numbers. Then we add the decimal places in both numbers to find the position of the decimal point in the product.

example 6.D.1 Multiply 4.12 by 63.2.

solution We multiply just as if the decimal points were not present.

```
      4.12
  ×   63.2
  --------
       824
      1236
     2472
  --------
    260384
```

In the top number, the decimal point is two places from the end. In the next number, the decimal point is one place from the end. Two plus one equals three so the decimal point in the product is placed three places from the end.

260.384

problem set 6

Multiply:

*1. 4.12 by 63.2 | 2. .513 by 4.23 | 3. 41.06 by .0005

4. .0732 by 1.63 | 5. 4.16 by .305

Subtract. Add to check.

6. 3.14 − .123 | 7. 5.32 − .163 | 8. 7.04 − .0062

9. 14.03 − .0132 | 10. 941.2 − 14.23

Divide:

11. $\frac{3624}{23}$ | 12. 1275 ÷ 17 | 13. $51\overline{)41,362}$

14. $27\overline{)2198}$ | 15. 2546 ÷ 41 | 16. $\frac{92,438}{51}$

17. Round off 5,143,782 to the nearest thousand.

18. Round off 90,521,765 to the nearest ten thousand.

19. A number has nine digits. All the digits are 7 except the millions' digit, which is 3; the ten-thousands' digit, which is 5; and the tens' digit, which is 9. What is the number?

20. A number has seven digits. All the digits are 6 except the hundred-thousands' digit, which is 2, and the thousands' digit, which is 4. What is the number?

*21. Use digits to write forty-one thousand, three hundred one millionths.

*22. Use digits to write four hundred twelve ten-thousandths.

Use words to write the following numbers.

*23. 4165.0162 | *24. 7108000.21578 | 25. 504327.001510512

Add:

26. .073 + 2.146 + .932 + .725

27. 32.142 + 3.0695 + 21.04 + 893.2006

28. 51,372.413 + 921.02 + .0425 + 42.6352

29. .005 + 21.62 + 9.035 + 5165.2

30. 70.02 + .0013 + 9.062 + .142

LESSON 7 Multiplying and dividing by powers of 10

7.A powers of 10

When we use 10 as a factor two times, the product is 100.

$$10 \times 10 = 100 \qquad \text{second power of 10}$$

When we use 10 as a factor three times, the product is 1000.

$$10 \times 10 \times 10 = 1000 \qquad \text{third power of 10}$$

When we use 10 as a factor four times, the product is 10,000.

$$10 \times 10 \times 10 \times 10 = 10{,}000 \qquad \text{fourth power of 10}$$

From this, we can see that the number of zeros in the product equals the number of times 10 is used as a factor. This number is called the power of 10. Thus, we see that the number

$$100{,}000{,}000$$

has eight 0s and must be the eighth power of 10. This is the product we get if 10 is used as a factor eight times.

$$10 \times 10 \times 10 \times 10 \times 10 \times 10 \times 10 \times 10 = 100{,}000{,}000 \qquad \text{eighth power of 10}$$

When we multiply a number by a power of 10, all we do is move the decimal point to the right. The number of places moved equals the number of zeros in the power of 10.

example 7.A.1 Multiply 47,162.314 by 100.

solution This time we will do the multiplication.

$$\begin{array}{r} 47{,}162.314 \\ \times \qquad 100 \\ \hline \mathbf{4{,}716{,}231.400} \end{array}$$

All that happened was that the decimal point was moved two places to the right.

example 7.A.2 Multiply .031652 by 10,000.

solution We see that 10,000 has four zeros. Thus, if we multiply by 10,000, all we will do is move the decimal point four places to the right. This time we will not multiply but will just write down the answer.

316.52

7.B dividing by powers of 10

When we divide a number by a power of 10, we move the decimal point to the left. The number of places moved equals the number of zeros in the power of 10.

example 7.B.1 Divide 41.32 by 1000.

solution We will do the division this time.

$$\begin{array}{r} .04132 \\ 1000\overline{)41.32} \\ \underline{40\ 00} \\ 1\ 320 \\ \underline{1\ 000} \\ 3200 \\ \underline{3000} \\ 2000 \\ \underline{2000} \end{array}$$

All that happened was that the decimal point was moved three places to the left.

example 7.B.2 Divide 48.512 by 10,000.

solution This time we will just write the answer. Dividing by the fourth power of 10 will move the decimal point four places to the left so the answer is

.0048512

problem set 7

Simplify mentally:

*1. 47,162.314 × 100

*2. .031652 × 10,000

*3. 41.32 ÷ 1000

*4. $\frac{48.512}{10{,}000}$

Multiply:

5. .0632 × 1.42

6. 1.413 × 216

7. .00413 × .312

8. 914.23 × .0132

Subtract. Add to check.

9. 1.416 − .0168

10. 23.41 − 2.666

11. 38.04 − 1.687

12. 10.421 − 1.687

13. 42.73 − 4.86

Divide:

14. $\frac{4832}{17}$

15. 9016 ÷ 23

16. $41\overline{)74{,}316}$

17. $\frac{90{,}327}{43}$

18. 42,153 ÷ 19

19. Round off 91,648,573 to the nearest hundred thousand.

20. Round off 84,165,812 to the nearest thousand.

21. A number has eight digits. All of them are 8 except the ten-thousands' digit, which is 3, and the units' digit, which is 7. What is the number?

22. Use digits to write four thousand, seven and nine thousand seven hundred forty-two hundred-thousandths.

23. Use digits to write seven hundred two and nine hundred forty-two hundred-thousandths.

Use words to write the following numbers.

24. 14372.015264

25. 9056213.00057328

Add:

26. .00152 + .31564 + 3.9123 + 2.406

27. 94.061 + 432.916 + 5.143 + 632.642

28. 91,642.82 + 4,130.957 + .088 + 423.725

29. 3.164 + 75.236 + 4328.914 + 508.21

30. 3.0624 + 783.91 + 9053.216

LESSON 8 *Rounding off decimal numbers · Dividing decimal numbers*

8.A rounding off decimal numbers

We find that the circle and arrow also can be used to explain how we round off decimal numbers. The procedure is the same as the procedure we use for whole numbers.

example 8.A.1 Round off 212.0165725 to the nearest ten-thousandth.

solution First we circle the digit in the ten-thousandths' place. Then we mark the digit to its right with an arrow.

$$2\,1\,2.\,0\,1\,6\,\textcircled{5}\,\overset{\downarrow}{7}\,2\,5$$

Next we change the arrow-marked digit and the digits to its right to zero.

$$2\,1\,2.\,0\,1\,6\,\textcircled{5}\,0\,0\,0$$

The arrow-marked digit was greater than 5 so we increase the circled digit by 1,

$$2\,1\,2.\,0\,1\,6\,\textcircled{6}\,0\,0\,0$$

and since terminal zeros to the right of the decimal point have no value, we can omit them.

212.0166

example 8.A.2 Round off 4057.2138362 to two decimal places.

solution We begin by circling the second digit to the right of the decimal point and marking the next digit to the right with an arrow.

$$4\,0\,5\,7\,.\,2\,\textcircled{1}\,\overset{\downarrow}{3}\,8\,3\,6\,2$$

Now we change the arrow-marked digit and the digits to its right to 0.

4 0 5 7 . 2 ① 0 0 0 0 0

Since the arrow-marked digit was less than 5, we do not change the circled digit. Also we discard the terminal zeros and get

4057.21

8.B dividing decimal numbers

To divide by a decimal number, we first move the decimal point in the divisor to the right as necessary to make the divisor a whole number. Then the decimal point in the dividend is moved to the right the same number of places. The decimal point in the answer is placed just above the decimal point in the dividend.

example 8.B.1 Divide .004415 by .032.

solution First we record the numbers.

$$.032 \overline{)\,.004415}$$

Next we move the decimal point in the divisor to the right so that the divisor is a whole number. We use a caret ($\wedge$) to mark the location of the old decimal point.

$$_{\wedge}032. \overline{)\,.004415}$$

Now we must move the decimal point in the dividend the same number of places and write the decimal point for the answer just above it.

$$\begin{array}{r} . \\ _{\wedge}032 \overline{)\,_{\wedge}004.415} \end{array}$$

This result is untidy so we recopy it, omitting the carets and the extra 0s. Then we divide.

$$\begin{array}{r} \mathbf{.137} \\ 32 \overline{)\,4.415} \\ 3\;2 \\ \hline 1\;21 \\ 96 \\ \hline 255 \\ 224 \\ \hline 31 \end{array}$$

This does not come out even. We decide to round off the answer to two decimal places, and we get

.14

problem set 8

Simplify mentally:

1. $\dfrac{41{,}362.68}{100}$

2. 305.2165×100

3. $9315.21 \div 1000$

4. $32.1652 \times 10{,}000$

Multiply:

5. $.00526 \times 3.14$

6. 2.315×413

7. $.00312 \times .642$

8. $313.65 \times .0147$

Subtract. Add to check.

9. 2.042 − 1.306 **10.** 42.31 − 37.44 **11.** 392.163 − 4.077

12. 3.2421 − 1.363 **13.** 47.23 − 21.44

Divide. Round off answers to two decimal places.

***14.** .004415 ÷ .032 **15.** $\frac{14.045}{.014}$ **16.** .0020 ÷ .013

17. $\frac{321.4}{.071}$ **18.** 3.22 ÷ .0022

19. Round off 42.12345678 to the nearest millionth.

20. Round off 31.6372052 to two decimal places.

21. Use digits to write the number forty-seven billion, sixty-seven thousand and four hundred seventeen hundred-thousandths.

22. Use digits to write the number one thousand three and four thousand seven hundred forty-two hundred-thousandths.

Use words to write the following numbers.

23. .00006184 **24.** 4000062.0130023 **25.** 75004213.0001652

Add:

26. 32.0165 + 40.2144 + 4.3785

27. 4152.32 + 62.41 + 930.22 + 507.11

28. 7852.165 + 7186.132 + 9185.624

29. 42.16 + .0032 + 3.165 + 305.321

30. .0016 + 32.1005 + 9.0312 + .00324

LESSON 9 Perimeter

9.A perimeter

The word **perimeter** comes from the Greek prefix *peri-*, which means "around" and the Greek word *metron*, which means "measure." Thus, perimeter means the measure around or the distance around.

example 9.A.1 Find the perimeter of the following figure. The dimensions are in feet.

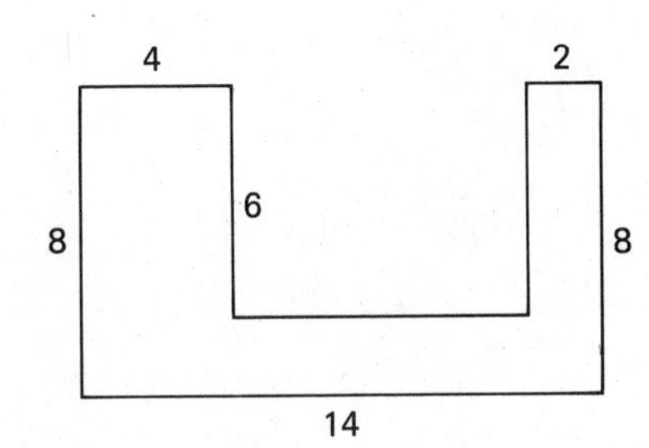

solution Several lengths are not given. We will assume all corners are square so we can determine the missing lengths. The dip on top goes down 6 feet so it must go up 6 feet. Finally, since it is 14 feet across the bottom, the missing length on top must be 8 feet.

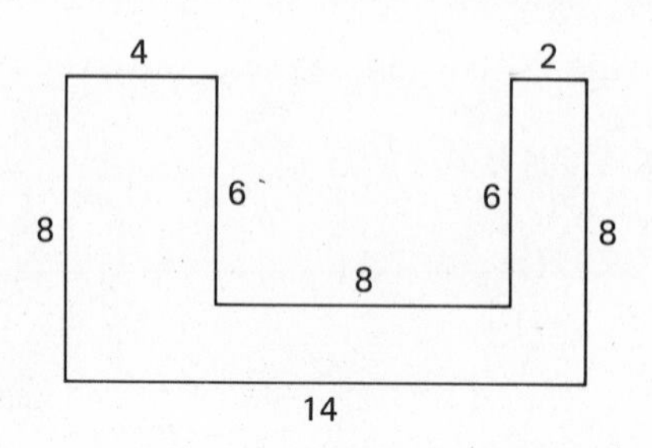

The perimeter is the distance around or the sum of these lengths.

$$8 + 4 + 6 + 8 + 6 + 2 + 8 + 14 = \mathbf{56\ feet}$$

problem set 9 Find the perimeter of these two figures. The dimensions are in inches.

1.

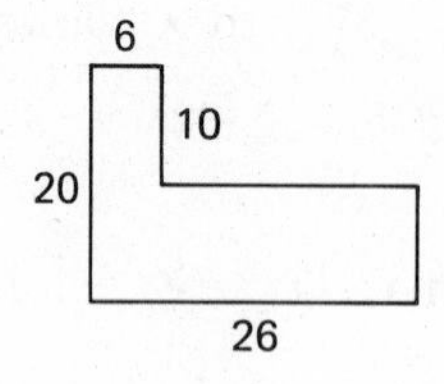

2.

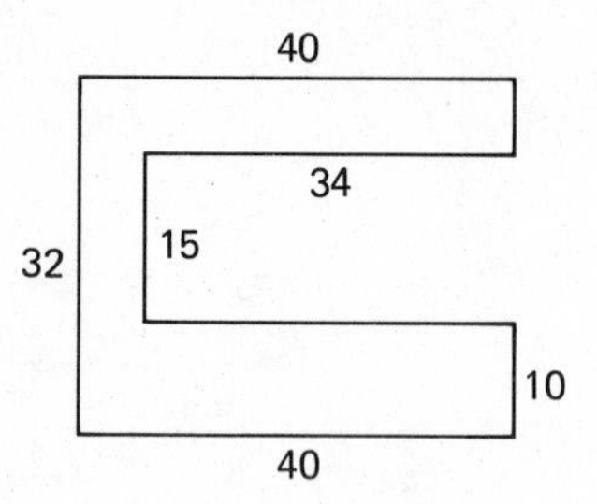

Simplify mentally.

3. $\dfrac{3164.215}{100}$

4. 3164.215×100

5. $\dfrac{417{,}365.20}{1000}$

6. $2.1532 \times 10{,}000$

Multiply:

7. $.0316 \times 2.4$

8. $2.862 \times .013$

9. $.08421 \times .22$

10. 8.123×3.13

Subtract. Add to check.

11. $3.065 - 1.423$

12. $43.24 - .000613$

13. $3.065 - .2121$

14. $2.092 - 1.166$

15. $438.2 - 2.0162$

Divide. Round off answers to two decimal places.

16. $.002215 \div .042$

17. $\dfrac{16.032}{.024}$

18. $.0040 \div .024$

19. $\dfrac{416.5}{.073}$

20. $3.44 \div .0066$

21. Round off 61.373737842 to the nearest ten-millionth.

22. Round off 433.6851472 to five decimal places.

23. Use digits to write the number seven hundred forty-two million, five hundred thirty-seven and ten thousand nine hundred forty-eight millionths.

24. Use digits to write the number one thousand seven hundred forty-eight ten-millionths.

Use words to write the following numbers:

25. .00128647 **26.** 27000316.08156 **27.** 51786.00785

Add:

28. 904.682 + 513.976 + 214.685 **29.** 4293.015 + 2172.062 + 5091.799

30. .51232 + 9.0572 + 72 + 82.3258

LESSON 10 Divisibility

10.A divisibility

The number 2 will divide into the number 30 and will have no remainder.

$$\frac{30}{2} = 15$$

Thus we say that 2 is a divisor of 30 and also that 30 is divisible by 2. Other divisors of 30 are 3, 5, 6, 10, and 15.

$$\frac{30}{3} = 10 \qquad \frac{30}{5} = 6 \qquad \frac{30}{6} = 5 \qquad \frac{30}{10} = 3 \qquad \frac{30}{15} = 2$$

To be called a divisor a number must be a whole number. If we list all the divisors of 30, we get

1, 2, 3, 5, 6, 10, and 30

It is often helpful to know if a whole number is divisible by 2, 3, 5, or 10. There are rules that we can use to find out. There are also rules to test for divisibility by 4 and 8 and a few other numbers. These rules are not used as often as the rules for divisibility by 2, 3, 5, and 10; and thus we will concentrate on the rules for these four numbers.

The rules are:

1. A whole number is divisible by 2 if its last digit is either 0, 2, 4, 6 or 8.
2. A whole number is divisible by 10 if its last digit is 0.
3. A whole number is divisible by 5 if its last digit is either 5 or 0.
4. A whole number is divisible by 3 if the sum of its digits is divisible by 3.

example 10.A.1 Which of these numbers are divisible by 3? (a) 99 (b) 1239 (c) 1561

solution We will add the digits in each number and divide the sum by 3.

(a) 99 $9 + 9 = 18$ $\frac{18}{3} = 6$

(b) 1239 $1 + 2 + 3 + 9 = 15$ $\frac{15}{3} = 5$

(c) 1561 $1 + 5 + 6 + 1 = 13$ $\frac{13}{3} =$ not divisible

The numbers (a) **99** and (b) **1239** are divisible by 3. The number 1561 is not divisible by 3 because 13 is not divisible by 3.

example 10.A.2 Which of these numbers are divisible (a) by 2? (b) by 5? (c) by 10?

4, 32, 75, 99, 4165, 4020

solution (a) Any even number is divisible by 2 so 2 is a divisor of **4**, **32**, and **4020** because the last digit of these numbers is an even digit.

(b) The numbers that are divisible by 5 have either a 5 or 0 as their last digit. These numbers are

75, **4165**, and **4020**

(c) To be divisible by 10 the last digit must be 0. Thus **4020** is the only one that is divisible by 10.

problem set 10

1. Use the rules for divisibility to tell which of the numbers 235, 300, 4888, 9132, and 72,654 are divisible (a) by 2 (b) by 3 (c) by 5 (d) by 10.

Find the perimeters of these figures. The dimensions are in meters.

2.

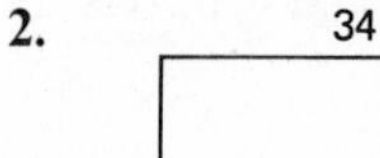
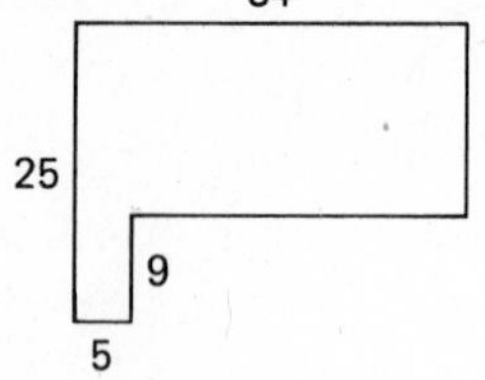

3.

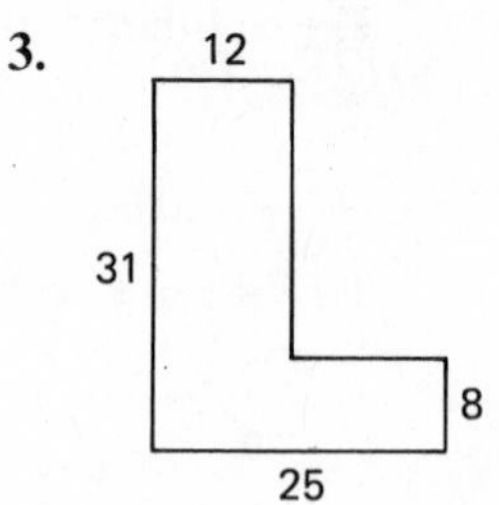

Simplify mentally:

4. 47,123 ÷ 1000
5. 40.265 × 1000
6. $\frac{.00143}{100}$

Multiply:

7. .0352 × 2.24
8. 305 × 2.42
9. 3.062 × 410
10. 3.06 × 4.18

Subtract. Add to check.

11. 4.016 − 3.217
12. 23.21 − .0034
13. 2.049 − .2343
14. 5.063 − 1.184
15. 421.6 − 24.8

Divide. Round off the answers to two places.

16. .003326 ÷ .021
17. $\frac{18.034}{.047}$
18. .0030 ÷ .031
19. $\frac{428.3}{.067}$
20. 2.77 ÷ .0055

21. Round off 4283.52162 to the nearest hundred.

22. Round off 478.64385 to three decimal places.

23. Use digits to write the number seven hundred sixty-two million, four hundred forty-two and twelve thousand, seven hundred ninety-two hundred-thousandths.

24. Use digits to write the number fourteen thousand, seven hundred two millionths.

Use words to write the following numbers.

25. .000120123
26. 472.058003
27. 5000162.0008

Add:

28. 903.625 + 819.423 + 459.685 + 918.416

29. 723.528 + 804.526 + 613.912 + 844.504

30. .00032 + 416.52 + 3.006 + 215.006

LESSON 11 Word problems about sums and differences

11.A word problems about sums

Many word problems require that we add the numbers in the problem to find the answer. These problems often contain the words **in all** or contain the word **sum.**

example 11.A.1 One hundred forty goblins wore red suits and 1360 wore green suits. How many goblins were there in all?

solution The words **in all** tell us to put all the goblins in one group. If we do, we add the numbers to find the number in this group.

1360	green
140	red
1500	in all

example 11.A.2 On the first jump, the frog jumped 18.0310 inches. On the next jump, the frog jumped 23.2105 inches. What was the sum of the two jumps?

solution The word **sum** tells us to add. We want to find the total distance the frog jumped.

18.0310	1st jump
23.2105	2nd jump
41.2415	**inches (sum)**

Some problems that require addition do not use the words in all or the word sum. These problems give us the same thought by using other words.

example 11.A.3 The first distance was .00184 inches. The second distance was 1.15030 inches. What was the total of the two distances?

solution The word **total** also tells us to add.

.00184 inches
1.15030 inches
1.15214 inches (total)

11.B word problems about differences

Many problems concern themselves with the difference of two numbers. We must subtract to find the answers to these problems. These problems often contain statements such as

how much larger
how much smaller
what was the difference

When we see these phrases, we know we are being asked to find a difference and must subtract to find the answer.

example 11.B.1 When the king measured it, he found the distance to be one hundred forty-two thousand and seven hundred seventeen thousandths centimeters. When the queen measured it, her answer was one hundred forty-two thousand and three hundred thirteen thousandths centimeters. Whose measurement was greater and by how much?

solution The words **how much greater** tell us that this is a difference problem. We must subtract to find the answer. We put the larger number on top.

$$\begin{array}{rl} 142{,}000.717 & \text{king} \\ -\ 142{,}000.313 & \text{queen} \\ \hline .404 & \text{difference} \end{array}$$

Thus, the king's measurement was greater by **.404 centimeters.**

problem set 11

*1. One hundred forty goblins wore red suits and 1360 wore green suits. How many goblins were there in all?

*2. On the first jump the frog jumped 18.0310 inches. On the next jump the frog jumped 23.2105 inches. What was the sum of the two jumps?

*3. When the king measured it, he found the distance to be one hundred forty-two thousand and seven hundred seventeen thousandths centimeters. When the queen measured it, her answer was one hundred forty-two thousand and three hundred thirteen thousandths centimeters. Whose measurement was greater and by how much?

4. Use the rules for divisibility to tell which of the following numbers are divisible (a) by 2, (b) by 3, (c) by 5, and (d) by 10: 6132, 6325, 9130, 6111, 6130.

Find the perimeters of these figures. The dimensions are in inches.

5.

31
14
25
25

6.

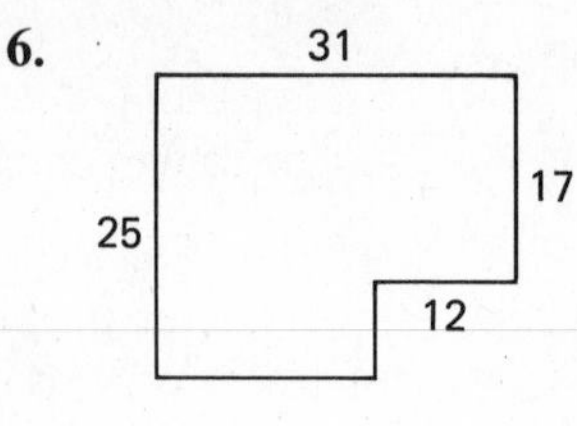

Simplify mentally:

7. 132,116 ÷ 10,000

8. 136.131 × 1000

9. $\frac{123.631}{1000}$

Multiply:

10. 162 × 2.25

11. 1.811 × 20.1

12. 6.61 × 3.16

13. 30.5 × 1.62

Subtract. Add to check.

14. 6.316 − 3.124

15. 61.81 − .0012

16. 129.631 − 2.48

17. 2.11 − 1.031

18. 510.510 − 96.9969

Divide. Round off answers to two decimal places.

19. $\frac{411.23}{61}$

20. .0016 ÷ .011

21. $\frac{11.031}{3.1}$ **22.** $\frac{.123}{.013}$

23. Round off 223.092870 to the nearest hundredth.

24. Round off 1621.32161 to the nearest hundred.

25. Use digits to write the number one trillion, six hundred twenty-five billion, two hundred fifty thousand, twenty-five and one hundred twenty-three thousandths.

Use words to write the following numbers.

26. 123.9621 **27.** 223092870

Add.

28. 1135.62 + 32.61 + 311.82 + 99.01

29. 1132.8 + 6251.6 + 312.1 + 11.3

30. 1.2 + .00315 + 61.312

LESSON 12 Prime numbers and composite numbers • Products of primes

12.A prime numbers and composite numbers

The number 6 can be composed by multiplying the two whole numbers 3 and 2.

$$3 \times 2 = 6$$

We say that 6 is a **composite number** because it can be composed by multiplying two other whole numbers. The number 21 is also a composite number. We can compose 21 by multiplying 3 and 7.

$$3 \times 7 = 21$$

The number 11 can be composed in only one way. That way is to multiply 11 by 1.

$$11 \times 1 = 11$$

There are many other numbers whose only factors are the numbers themselves and the number 1. Some are

$$5 \times 1 = 5 \qquad 23 \times 1 = 23 \qquad 31 \times 1 = 31$$

There is no other way to compose 5, 23, or 31 by multiplying. We call these numbers **prime numbers.**

PRIME NUMBERS

A prime number is a whole number greater than 1 whose only whole number divisors are 1 and the number itself.

example 12.A.1 Write the whole numbers 1 through 40 and circle the prime numbers. Do not circle 1, because 1 is not a prime number.

solution 1 (2) (3) 4 (5) 6 (7) 8 9 10 (11) 12 (13) 14 15 16 (17) 18 (19) 20 21 22 (23) 24 25 26 27 28 (29) 30 (31) 32 33 34 35 36 (37) 38 39 40

12.B products of primes

Sometimes it is necessary to write a composite number as a product of prime numbers. We can write 12 as a product of prime numbers as

$$2 \times 2 \times 3$$

Here we found the prime factors of 12 by inspection. However, if we wish to write large numbers such as 84 or 1260 as products of prime numbers, it is nice to have a procedure to follow. A procedure often used is to divide the given number by the prime numbers 2, 3, 5, etc., until the prime factors are found. To find the prime factors of 84, we begin by dividing by 2.

$$\frac{84}{2} = 42$$

Now 42 can be divided by 2 so we divide again.

$$\frac{42}{2} = 21$$

and 21 can be divided by 3

$$\frac{21}{3} = 7$$

So we can write 84 as a product of prime numbers as

$$2 \times 2 \times 3 \times 7$$

Many people use the following division format to find the prime factors of a number.

$$\begin{array}{r|l} 2 & 84 \\ \hline 2 & 42 \\ \hline 3 & 21 \\ \hline & 7 \end{array} \qquad \text{so } 84 = 2 \times 2 \times 3 \times 7$$

We did the same divisions as before, but this time we used a more compact format.

example 12.B.1 Write 1260 as a product of prime numbers.

solution We will work the problem twice to demonstrate both formats.

$$\frac{1260}{2} = 630 \qquad \frac{630}{2} = 315 \qquad \frac{315}{5} = 63 \qquad \frac{63}{3} = 21 \qquad \frac{21}{3} = 7$$

Now we use the other format.

$$\begin{array}{r|l} 2 & 1260 \\ \hline 2 & 630 \\ \hline 5 & 315 \\ \hline 3 & 63 \\ \hline 3 & 21 \\ \hline & 7 \end{array}$$

Both formats yield the same answer.

$$1260 = \mathbf{2 \times 2 \times 3 \times 3 \times 5 \times 7}$$

problem set 12

1. The circumference of the great wheel was twelve thousand and forty-one thousandths inches. The circumference of the lesser wheel was only one thousand twenty-one and two hundred-thousandths inches. How much larger was the great wheel?

2. When the smoke cleared, the judges found that on the first try Roger had covered fourteen million, seven hundred sixty-two and seventy-five ten-thousandths' units. The second try was only eight hundred forty-two thousand, fifteen and seven thousandths' units. What was the sum of both tries?

3. What is the sum of the prime numbers that are greater than 3 and less than 28?

4. What is the sum of the 8 smallest prime numbers?

5. Use the divisibility rules to determine which of the following numbers are divisible (a) by 2, (b) by 3, (c) by 5, and (d) by 10: 625; 302; 9172; 3132; 62,120.

Find the perimeters of the figures. The dimensions are in feet.

6.

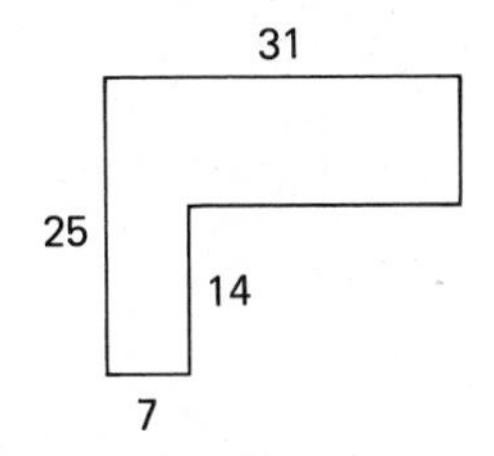

7.

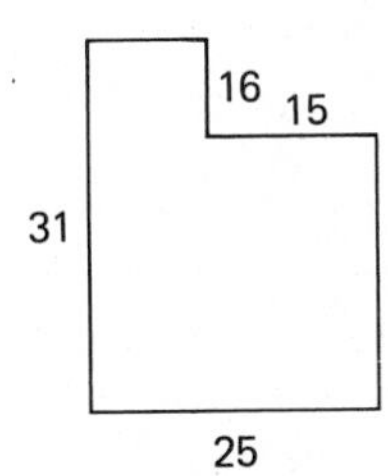

Simplify mentally:

8. $91,865 \div 100$ **9.** 36.8211×1000

Multiply:

10. $.0316 \times 72.1$ **11.** 913×6.19 **12.** 3.16×71.3

13. Write the following numbers as products of primes: *(a) 84 *(b) 1260 (c) 5400

Subtract. Add to check.

14. $31.625 - 12.110$ **15.** $1.6251 - .0132$ **16.** $162.123 - .0123$

17. $1.329 - .999$ **18.** $1362.13 - 321.63$

Divide and round off the answers to two places.

19. $18.621 \div 6.1$ **20.** $\frac{32.631}{.03}$

21. $\frac{3012.3}{12}$ **22.** $\frac{621}{3.1}$

23. Round off 4692.83215 to the nearest hundred.

24. Round off 4113.62185 to two decimal places.

25. Use digits to write the number nine hundred sixty-one billion, three hundred thirteen million, twenty-five.

Use words to write the following numbers.

26. .001621 **27.** 16.0562 **28.** 6231562.01

Add:

29. $931.62 + 621.73 + 631.81 + 713.13$ **30.** $.0031 + 612.13 + .721 + 16.11$

LESSON 13 *Three-digit divisors • Multiplication word problems*

13.A three-digit divisors

Division problems whose divisors contain three or more digits are just as easy as the problems in which the divisors have only two digits. They just appear to be more difficult.

example 13.A.1 Divide .41623 by .0215.

solution The decimal point in .0215 must be moved four places to make this a whole number. Thus, the decimal point in .41623 is also moved four places.

```
        19.359
215 ) 4162.300
      215
      ----
      2012
      1935
      ----
        773
        645
        ----
        1280
        1075
        ----
         2050
         1935
         ----
```

We round off to two decimal places and get **19.36.**

13.B multiplication word problems

Some word problems require that numbers be multiplied to find the answer. These word problems sometimes contain the word **product.** Many of them contain the word **times** used in a phrase such as 5 times as many.

example 13.B.1 The second game score was 25 times the score of the first game. If Homer had scored 14,025 points in the first game, how many points did he score in the second game?

solution The word **times** tells us to multiply.

$$25 \times 14{,}025 = \mathbf{350{,}625}$$

example 13.B.2 Roger scored 26,142 points in the first game. He scored 7 times this many points in the second game. How many points did he score in all?

solution The word **times** tells us to multiply and the words **in all** tells to add.

	26,142	first game
26,142 × 7 =	182,994	second game
	209,136	in all

He scored **209,136** points in all.

problem set 13

*1. The second game score was 25 times the score of the first game. If Homer had scored 14,025 points in the first game, how many did he score in the second game?

*2. Roger scored 26,142 points in the first game. He scored 7 times this many points in the second game. How many points did he score in all?

3. The first one measured fourteen million, seven hundred forty-two thousand and seventeen hundred-thousandths. The second one measured only eight hundred thousand and forty-two millionths. By how much was the first one larger?

4. Find the sum of the prime numbers that are between 12 and 42.

5. Use the rules for divisibility to determine which of the following numbers are divisible (a) by 2, (b) by 5, (c) by 3, and (d) by 10: 1020, 125, 130, 1332, 185, 132.

Find the perimeters of these figures. Dimensions are in centimeters.

6.

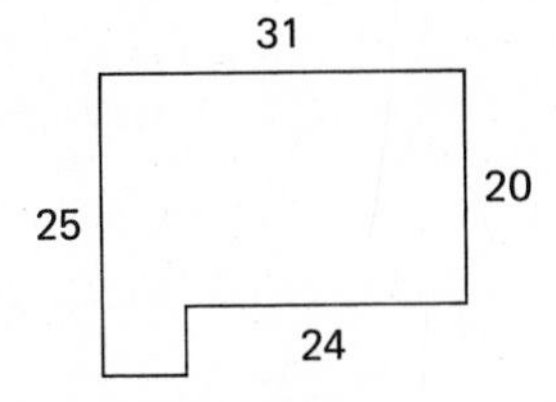

7.

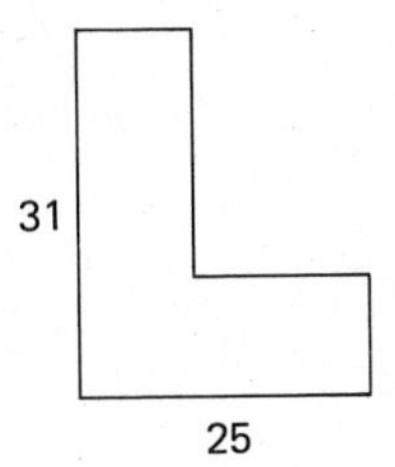

Simplify mentally:

8. 31,621 ÷ 1000 9. 311.836152 × 10,000

Multiply:

10. .0121 × 62.1 11. 621 × 8.11 12. 2.28 × 22.4

13. Write each of the following numbers as the product of primes:
(a) 360 (b) 720 (c) 1440

Subtract. Add to check.

14. 16.162 − 12.373 15. 1.6132 − .1316 16. 17.132 − 1.693

17. 1932.016 − .918 18. 852.161 − 161.321

Divide and round off the answers to two decimal places:

*19. $\frac{.41623}{.0215}$ 20. 19.312 ÷ .061

21. $\frac{311.12}{1.2}$ 22. $\frac{621}{.31}$

23. Round off 1231.62567 to three decimal places.

24. Round off 6,469,693,230 to the nearest ten million.

25. Use digits to write the number three hundred twenty-one million, six hundred seventeen thousand, two hundred twelve and two hundred thirty-one thousandths.

Use words to write the following numbers.

26. 161.016 27. 613.162 28. 111111112

Add:

29. 621.81 + 31.62 + 62.11 + 12.61 30. .1123 + 61.831 + .817 + 1.13

LESSON 14 Fractions · Reducing to lowest terms · Division word problems

14.A fractions

When we write two numbers vertically and draw a line between them, we have written a fraction. Thus 3 over 4, written as follows,

$$\begin{array}{lcl} \text{numerator} \longrightarrow & 3 & \\ & - & \longleftarrow \text{fraction line} \\ \text{denominator} \longrightarrow & 4 & \end{array}$$

is a fraction. We call the top number the **numerator** of the fraction. The bottom number is called the **denominator** of the fraction. We call the line between the numbers a **fraction line.**

Fractions are used to designate parts of a whole. The bottom of the fraction tells us how many parts there are in all. The top of the fraction tells us how many of these parts we are considering. Here we show what we mean when we write $\frac{1}{4}$, $\frac{2}{4}$, $\frac{3}{4}$, and $\frac{4}{4}$.

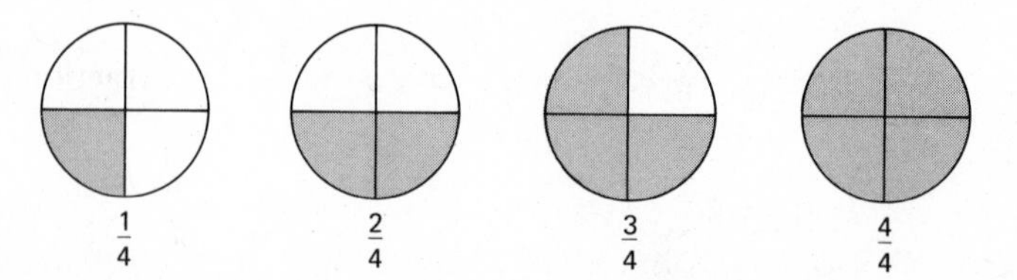

If two fractions have the same value, we say that the fractions are **equivalent fractions.** The two fractions

$$\frac{1}{2} \quad \text{and} \quad \frac{2}{4}$$

are equivalent fractions, for they have the same value. We can show this by drawing a picture that represents each of these fractions. First, we draw two circles. We divide the first circle into two equal parts, or halves. Then we divide the second circle into four equal parts, or fourths.

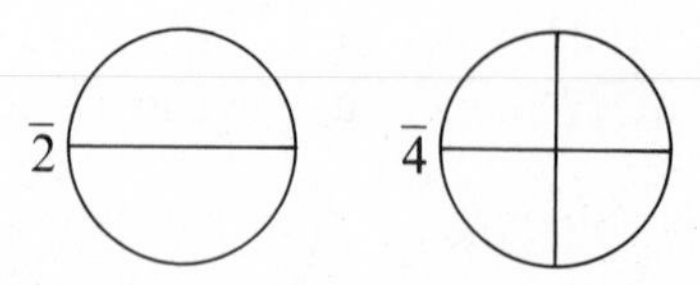

On the left we will shade one of the halves to represent $\frac{1}{2}$, and on the right we will shade in two of the fourths to represent $\frac{2}{4}$.

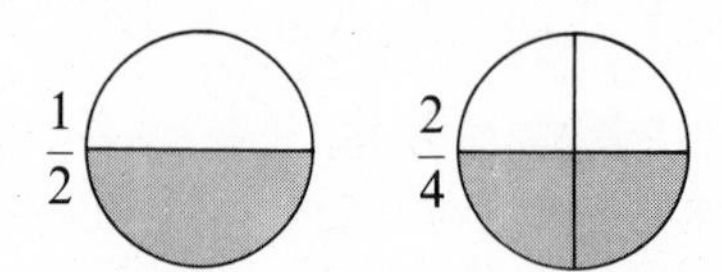

We see that we have shaded equal parts of each whole. We say that $\frac{1}{2}$ and $\frac{2}{4}$ are equivalent fractions because they represent equal parts of the whole and thus have the same value.

14.B reducing to lowest terms

We have only one rule that we use for fractions. It can be stated by using (a) multiplication or (b) division.

(a) **The top and bottom of a fraction can be multiplied by the same number (except zero) without changing the value of the fraction.**
(b) **The top and bottom of a fraction can be divided by the same number (except zero) without changing the value of the fraction.**

We use this rule to reduce fractions to lowest terms. To reduce a fraction to lowest terms, we will divide the top and the bottom by the same prime numbers until no more common prime factors can be found.

example 14.B.1 Reduce $\frac{30}{45}$.

solution As the first step, we divide the top and the bottom of the fraction by 5.

$$\frac{\overset{6}{\cancel{30}}}{\underset{9}{\cancel{45}}}$$

Next we divide the top and bottom of the fraction by 3.

$$\frac{\overset{\overset{2}{\cancel{6}}}{\cancel{30}}}{\underset{\underset{3}{\cancel{9}}}{\cancel{45}}} = \mathbf{\frac{2}{3}}$$

This fraction is an equivalent fraction to the original fraction. It is now in lowest terms because 2 and 3 do not have a common factor.

14.C division word problems

Some word problems indicate that things should be divided into groups. The number of groups can be found by dividing the total number by the number in a group.

example 14.C.1 Ramirez had 5282 taco shells. If each box would hold 12 taco shells, how many completely full boxes could he ship?

solution The problem tells us that the taco shells will be placed in groups of 12. Since division is a grouping process, we can find the answer by dividing.

$$\frac{5282}{12} = \textbf{440, R2}$$

Thus, he will ship 440 boxes, each containing 12 taco shells, and will have 2 taco shells left over.

example 14.C.2 The fairy queen found that 14 dryads could sit on one toadstool. If 780 dryads were coming to the convocation, how many toadstools would she have to provide?

solution She wants to put the dryads into groups of 14. Since division will tell us how many groups, we will divide.

$$\frac{780}{14} = \textbf{55, R10}$$

Thus she will need 56 toadstools. Fifty-five toadstools will have 14 dryads and one toadstool will have only 10 dryads.

problem set 14

*1. Ramirez had 5282 taco shells. If each box would hold 12 taco shells, how many completely full boxes can he ship?

*2. The fairy queen found that 14 dryads could sit on one toadstool. If 780 dryads were coming to the convocation, how many toadstools would she have to provide?

3. The first microbe had a measure of one hundred sixty-two ten-millionths. The measure of the second microbe was four hundred twenty-three hundred-millionths. By how much was the first microbe larger?

4. In the flang-flung contest, Jesse Lee flang his fourteen million, six hundred forty-two units. When Jethroe tried, he flung his thirty-two million, fifteen thousand, thirty-two units. How much farther did Jethroe flung than Jesse Lee flang?

5. Use the rules for divisibility to determine which of these numbers are divisible (a) by 10, (b) by 5, (c) by 2, and (d) by 3: 120, 135, 122, 1332, 1620.

Find the perimeters of these figures. Dimensions are in kilometers.

6.

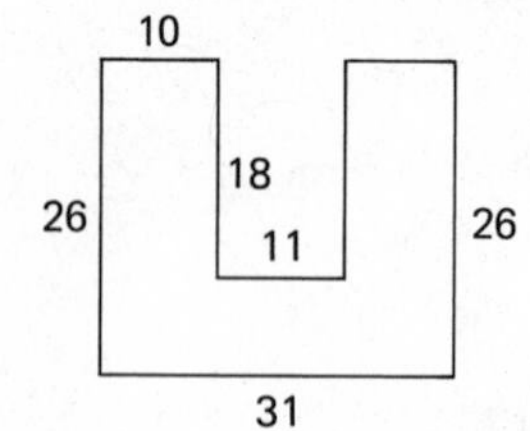

7.

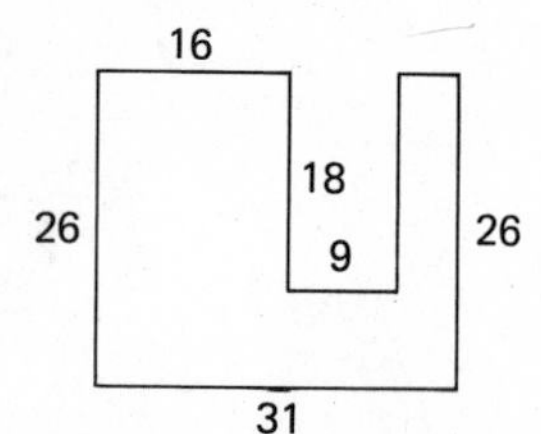

Reduce the following fractions to lowest terms.

8. *(a) $\frac{30}{45}$ (b) $\frac{30}{75}$ (c) $\frac{80}{220}$

9. (a) $\frac{70}{245}$ (b) $\frac{48}{112}$ (c) $\frac{80}{720}$

Simplify mentally:

10. 12,361 ÷ 1000

11. 11.36121 × 1000

Multiply:

12. 16.21 × 11.3

13. 113 × 1.28

14. 2.14 × 11.6

15. Write each of the following as a product of primes:
(a) 1800 (b) 900 (c) 450

Subtract. Add to check.

16. 131.61 − 11.87

17. 181.811 − 6.329

18. 1361.13 − 13.782

19. 1114.13 − 61.17

Divide and round off the answer to two places:

20. $\frac{612.13}{.603}$

21. 913.62 ÷ .025

22. 6111.12 ÷ 7.5

23. $\frac{611.21}{1.2}$

24. Round off 1612.316289 to four decimal places.

25. Round off 20,056,049,013 to the nearest hundred million.

Use words to write the following numbers:

26. 1231.161

27. 11123.121

28. 1612.12

Add:

29. 126.81
31.71
36.18
1137.78

30. .1189 + 123.1862 + .119 + 1.2

LESSON 15 Fractions to decimals

15.A fractions and decimals

Fractions and decimals (decimal fractions) can be used to represent the same parts of a whole.

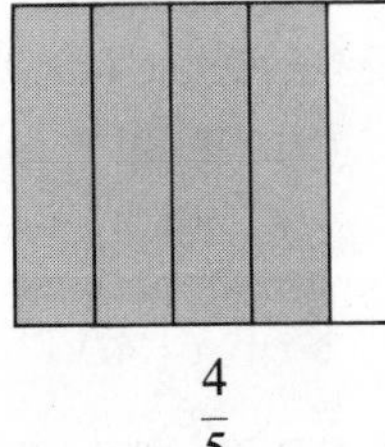

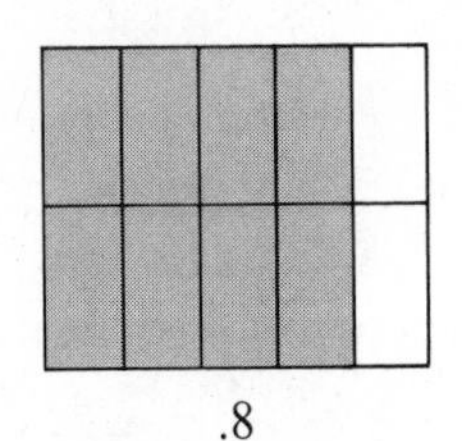

.8

On the left we have divided the whole figure into 5 parts and shaded in 4 of them to represent $\frac{4}{5}$. On the right we have divided the same figure into 10 parts and shaded in 8 of them to represent .8, which is eight-tenths. We see that the shaded areas are equal. This is a picture that shows us that

$$\frac{4}{5} \quad \text{equals} \quad .8$$

Every fraction can be written as a decimal number. To write a fraction of whole numbers as a decimal number, we divide the bottom number into the top number. **The answer will always be a terminating decimal number or will be a decimal number that repeats in a definite pattern. Sometimes it is easy to see whether the digits terminate or repeat. Other times there are a large number of repeating digits in the pattern, and the pattern is not easy to see.** When this happens, we will just round off to a convenient number of digits. The important thing is to remember that the repeating pattern is there even if we don't see it. We show several examples here.

(a) $\frac{1}{8} = .125$

(b) $\frac{3}{7} = .428571428571428571\cdots$

(c) $\frac{21}{99} = .2121212121\cdots$

(d) $\frac{5}{6} = .8333333333\cdots$

(e) $\frac{56,156}{99,000} = .5672323232323\cdots$

We can use a bar over the repeating digits to designate a repeating pattern. Here we use bars as necessary to write the above answers again.

(b) $.\overline{428571}$

(c) $.\overline{21}$

(d) $.8\overline{3}$

(e) $.567\overline{23}$

15.B rounding off repeaters

To round off repeating decimal numbers, it is often necessary to write out the repeating pattern several times.

example 15.B.1 Round off $42.\overline{617}$ to the nearest one-millionth.

solution We begin by writing the repeating digits until the millionths' place is passed.

$$42.617617617617 \cdots$$

Then we circle the digit in the millionths' place and mark the digit to its right with an arrow.

$$42.61761\,\textcircled{7}\,\overset{\downarrow}{6}17617 \cdots$$

Then we change the arrow-marked digit and the digits to its right to zero.

$$42.61761\,\textcircled{7}\,\overset{\downarrow}{0}000 \cdots$$

Since the arrow-marked digit was greater than 5, we increase the circled digit by 1 and get

$$42.6176180000 \cdots$$

Terminal zeros to the right of the decimal point have no value. Thus we can omit these zeros and write the answer as

42.617618

example 15.B.2 Round off $718.0\overline{73}$ to eight decimal places.

solution First we write the number in expanded form, circle the digit in the eighth decimal place, and mark the next digit with an arrow.

$$718.0737373\,\textcircled{7}\,\overset{\downarrow}{3}73$$

Since the arrow-marked digit is less than 5, we do not change the circled digit when we round off. Thus, the answer is

718.07373737

15.C fractions to decimals

To write a fraction as a decimal, we perform the indicated division.

example 15.C.1 Write as decimals: (a) $\frac{1}{8}$ (b) $\frac{1}{30}$

solution (a) We divide 1 by 8.

$$\begin{array}{r} .125 \\ 8\overline{)1.000} \\ \underline{8} \\ 20 \\ \underline{16} \\ 40 \\ \underline{40} \end{array}$$

(b) We divide 1 by 30.

$$\begin{array}{r} .0333 \\ 30\overline{)1.000} \\ \underline{90} \\ 100 \\ \underline{90} \\ 100 \\ \underline{90} \\ 10 \end{array}$$

On the left the number terminated. On the right the 3s repeated. Thus our answers are

(a) **.125** (b) **$.0\overline{3}$**

example 15.C.2 Write as decimals: (a) $\frac{2}{5}$ (b) $\frac{5}{7}$

solution We will perform the indicated divisions.

(a)
$$\begin{array}{r} \mathbf{.4} \\ 5\overline{)2.0} \\ \underline{2\,0} \end{array}$$

(b)
$$\begin{array}{r} \mathbf{.7142857} \\ 7\overline{)5.0} \\ \underline{4\,9} \\ 10 \\ \underline{7} \\ 30 \\ \underline{28} \\ 20 \\ \underline{14} \\ 60 \\ \underline{56} \\ 40 \\ \underline{35} \\ 50 \\ \underline{49} \end{array}$$

The answer to (a) is .4 and the answer to (b) is a 6 digit repeater, but we decide we want a shorter answer so we round off to two digits.

(a) **.4** (b) **.71**

problem set 15

1. Their approach was inexorable, and they numbered fourteen million, seven thousand, nine hundred twenty. In the second wave, there were only nine hundred thousand, sixty-seven. How many were there in all?

2. Harriet ran 14 times as far after she got her second wind. If she ran three thousand and seven hundred eighty-seven millionths feet on her first wind, how far did she run in all?

3. Ward 4 reported 8 times as many votes as Ward 7 reported. Ward 6 reported 12 times as many votes as Ward 7 reported. If Ward 7 reported nine thousand forty-three votes, how many votes did the three wards report in all?

4. The ladybugs grouped themselves in bunches of 20. If there were nine thousand forty-two ladybugs, how many bunches did they form?

5. Which of the following numbers are divisible (a) by 3, and (b) by 2: 212, 2133, 312, 610, 630?

6. Find the sum of the prime numbers that are between 30 and 42.

7. Find the perimeter of this figure. Dimensions are in feet.

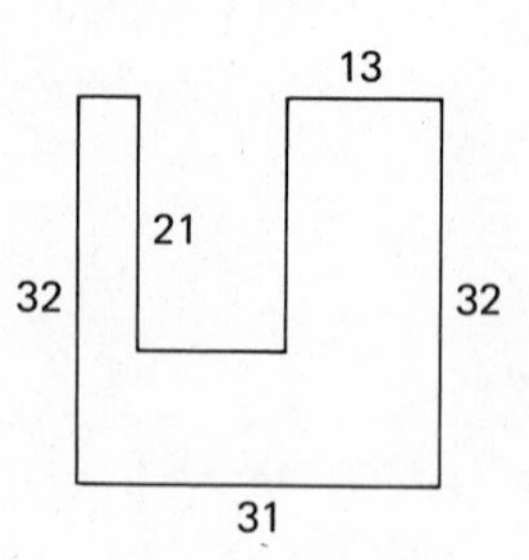

Write the following fractions as decimals. Round off to two decimal places.

*8. $\frac{5}{7}$ *9. $\frac{1}{30}$ 10. $\frac{3}{11}$

11. Reduce the following fractions to lowest terms:

(a) $\frac{70}{60}$ (b) $\frac{16}{24}$ (c) $\frac{24}{36}$

Simplify mentally:

12. 169,211 ÷ 10,000 13. 123.61311 × 100

Multiply:

14. 89.21 × 62.1 15. 2.16 × 32.8

16. Write each of the following numbers as a product of primes: (a) 3600 (b) 450 (c) 4500

Subtract. Add to check.

17. 1131.13 − 131.98 18. 192.68 − 6.321

19. 118.62 − 117.99 20. 1234.78 − 123.87

Divide and round off the answers to two places:

21. $\frac{629.1}{2.31}$ 22. 117.2 ÷ .012

23. 7.81 ÷ 3.11 24. $\frac{2310}{13}$

*25. Round off $42.\overline{617}$ to the nearest millionth.

*26. Round off $718.0\overline{73}$ to eight decimal places.

27. Use words to write the number 1876211.32

Add:

28. 1921.6 + 1872.7 + 1321.3 + 62.1

29. 613.1 + 7214.6 + 11.2 + 3.1

30. .1792 + .1492 + .1777 + 31.6215

LESSON 16 Decimals to fractions

16.A decimals to fractions

Both terminating and repeating decimal numbers can be written as fractions of whole numbers. The method of writing repeaters as fractions is rather complicated and will be taught in a later course. We will discuss terminating decimals now.

Remember from Lesson 14 that a particular number can be multiplied and divided by another number without changing the value of the original number. This is because a number divided by itself equals 1.

$$\frac{47}{47} = 1 \qquad \frac{5}{5} = 1 \qquad \frac{100}{100} = 1 \qquad \frac{1000}{1000} = 1$$

We use this fact and the rule for multiplying by powers of 10 to write terminating decimal numbers as fractions. We multiply and divide by the power of 10 necessary to make the top number a whole number.

example 16.A.1 Write .041 as a fraction.

solution If we multiply .041 by 1000, we will make it a whole number.

$$.041 \times 1000 = 41$$

But we must also divide by 1000 so we do not change the value of .041.

$$.041 \times \frac{1000}{1000} = \mathbf{\frac{41}{1000}}$$

example 16.A.2 Write 43.21657 as a fraction.

solution This time the decimal point must be moved five places so we multiply and divide by 100,000, which has five 0s.

$$43.21657 \times \frac{100{,}000}{100{,}000} = \mathbf{\frac{4{,}321{,}657}{100{,}000}}$$

problem set 16

1. Only fourteen could crawl into a single space. If three thousand eight hundred eight had to be sheltered, how many spaces were necessary?

2. Gene could muster up a total of fourteen and seven hundred forty-two ten-thousandths. Mary could muster up seven times that much. How much could the two of them muster up all together?

3. When the shakedown was finished, Roberto found that he had shaken ninety-one thousand, forty-two. Raoul was chagrined because this exceeded his total by twelve thousand, fifteen. What was Raoul's total?

4. Each box would hold 142 apples. If the crew filled 432 boxes one shift and had 5 apples left over, how many apples were there in all?

5. Which of the following are divisible (a) by 5, (b) by 10: 650, 625, 15, 20, 30?

6. What are the prime numbers that are greater than 40 but less than 64?

7. Find the perimeter of this figure. The dimensions are in yards.

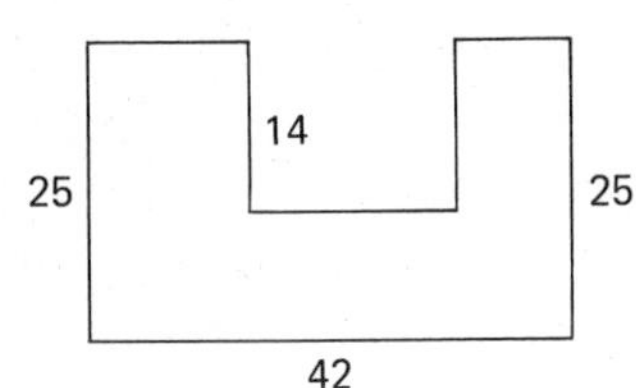

Write the following fractions as decimals. Round off to two decimal places.

8. $\frac{11}{16}$ **9.** $\frac{1}{17}$ **10.** $\frac{4}{7}$

Reduce the following fractions to lowest terms.

11. $\frac{36}{42}$ **12.** $\frac{72}{120}$ **13.** $\frac{81}{243}$

Multiply:

14. 13.61×71.3 **15.** 61.3×11.2

16. Simplify mentally: $12{,}389.32 \div 100$

Subtract. Add to check.

17. $169{,}211.36 - 1892.98$ **18.** $181{,}131.62 - 1.9876$

Divide and round off the answer to two places.

19. $\frac{613.1}{3.17}$ **20.** $\frac{123.8}{9.98}$ **21.** $181.3 \div 1.2$

Write as a product of primes:

22. 288 **23.** 1080 **24.** 10,800

25. Round off 87,621.32178939 to the nearest millionth.

26. Round off $437.00\overline{621}$ to the nearest hundred-millionth.

27. Use words to write the number 172.312.

Add:

28. $1361.31 + 21.14 + 112.17 + 1.18$

29. $12{,}348.36 + 19{,}998.76$

30. $.36121 + 1.1 + 2.312$

LESSON 17 Rectangular area

17.A area

The left-hand diagram represents a table top that is 4 feet long and 3 feet wide. On the right we show how this top can be divided into 12 squares. All the sides of the squares are 1 foot (1 ft) long. We say that the area of each square is 1 square foot (sq. ft.) or 1 ft^2.

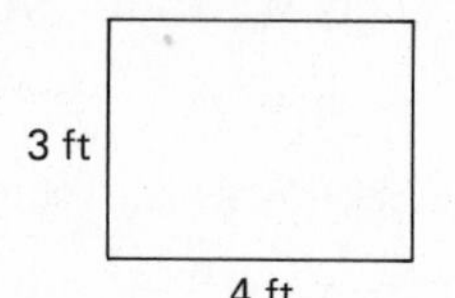

3 ft
4 ft
4 squares
4 squares
4 squares

There are four squares in each row and 3 rows so there are 12 squares.

$$4 \text{ squares} \times 3 = 12 \text{ squares}$$

Thus, we say that the total area is 12 square feet or 12 ft^2. If there were 6 rows of 4 squares each, then there would be a total of 24 squares.

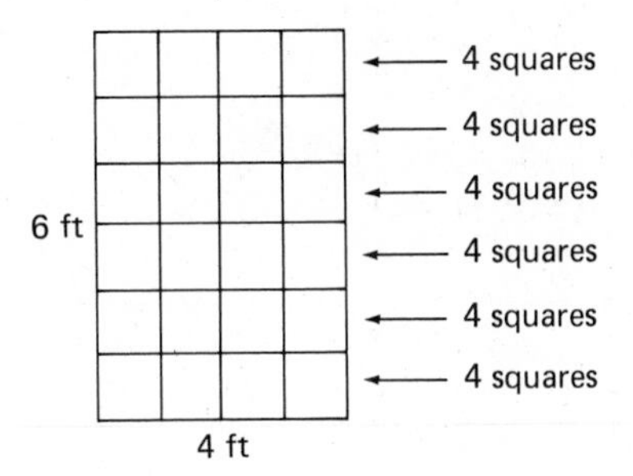

$$4 \text{ squares} \times 6 = 24 \text{ squares}$$

The length of the first row tells us the number of squares in this row, and the width tells us the number of rows. Thus, the number of squares is the length times the width

$$\text{Number of squares} = \text{length} \times \text{width}$$

Each square could be covered by a floor tile that is 1 foot on a side. It is helpful to think "floor tiles" when we hear the word area. Floor tiles can be touched and felt and understood, while area is more abstract.

example 17.A.1 How many 1-inch square floor tiles would it take to cover this figure?

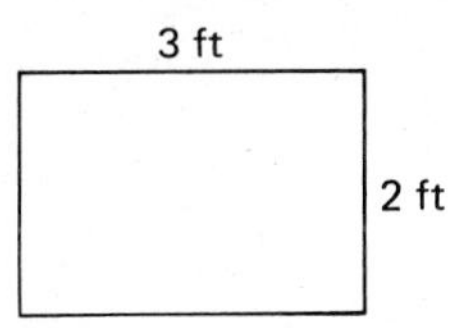

solution The dimensions are in feet so we first change them to inches (in).

$$3 \text{ ft} = 3 \times 12 \text{ in} = 36 \text{ in} \qquad 2 \text{ ft} = 2 \times 12 \text{ in} = 24 \text{ in}$$

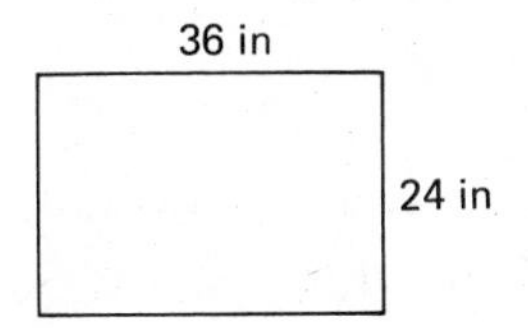

Each row will have 36 tiles, and there will be 24 rows.

$$24 \times 36 \text{ tiles} = \textbf{864 tiles}$$

example 17.A.2 Find the area of this figure in square centimeters (cm^2). Dimensions are in centimeters (cm).

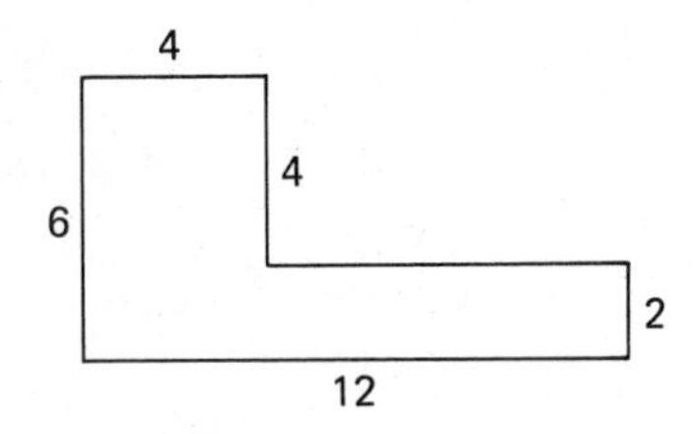

solution We will divide the given area into rectangles and find the areas of the rectangles. Two ways are shown.

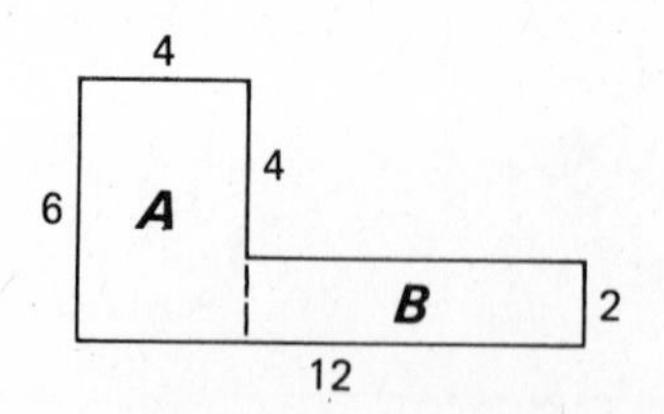

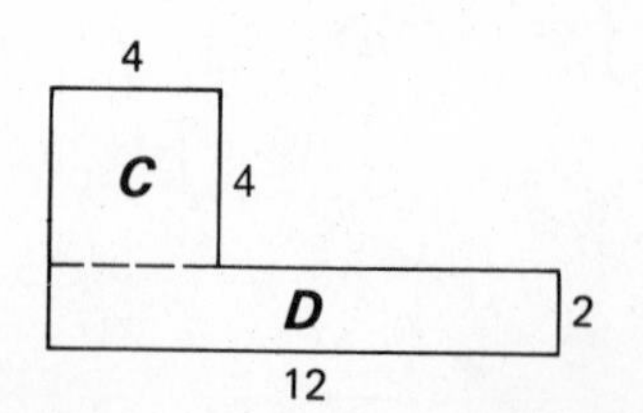

$$\begin{aligned} A &= 6 \times 4 = 24 \text{ cm}^2 \\ B &= 2 \times 8 = 16 \text{ cm}^2 \\ \hline \text{Total} &= 40 \text{ cm}^2 \end{aligned}$$

$$\begin{aligned} C &= 4 \times 4 = 16 \text{ cm}^2 \\ D &= 2 \times 12 = 24 \text{ cm}^2 \\ \hline \text{Total} &= 40 \text{ cm}^2 \end{aligned}$$

The area of the figure is **40 cm²**. Thus 40 tiles, each 1-centimeter square, would be required to cover the figure.

problem set 17

1. The third avatar of Vishnu was a boar. Vishnu is said to have 11 avatars in all. If he used one avatar a day how many times would he appear as a boar in 10,802 days?

2. There were four thousand eight hundred forty-two ants in each anthill. If there were three hundred thirty anthills, how many ants were there in all?

3. In the next valley the anthills each contained one hundred eight thousand fifteen ants. How many more ants lived in one of these anthills than lived in an anthill from Problem 2?

4. Which of the following numbers are divisible (a) by 3, (b) by 5: 135, 1050, 335, 4145, 1010?

5. Find the sum of the first 13 prime numbers.

6. Find the perimeter of this figure. Dimensions are in meters.

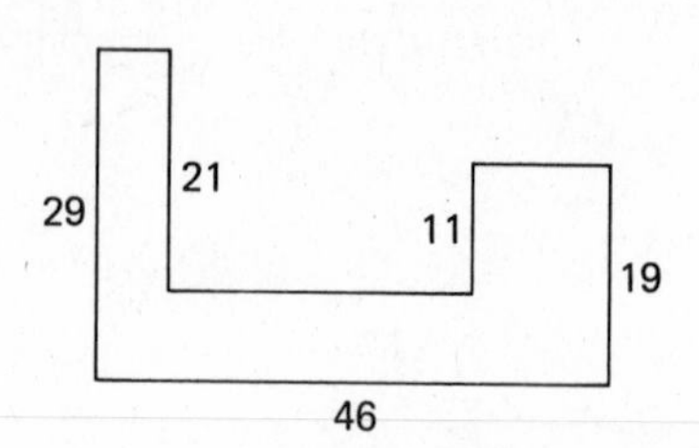

*7. Find the area of this figure. Dimensions are in centimeters.

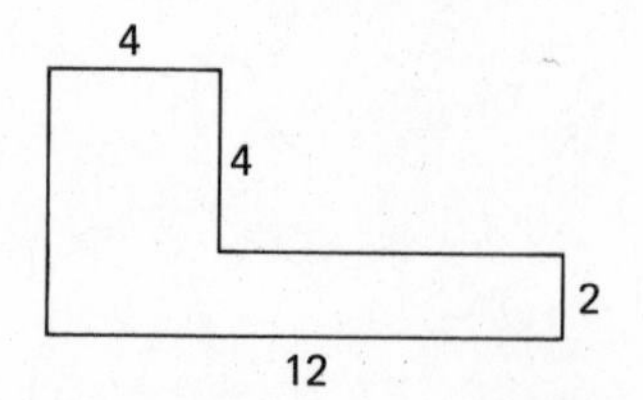

8. How many 1-inch square floor tiles would cover this figure? Dimensions are in inches.

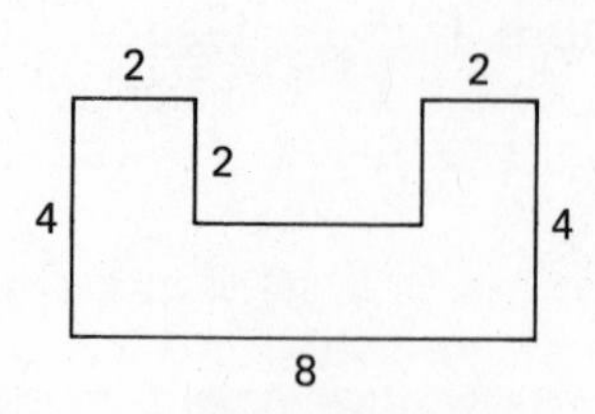

Write the following fractions as decimals. Round off to two decimal places.

9. $\frac{3}{17}$ 10. $\frac{7}{11}$ 11. $\frac{12}{13}$

Write the following numbers as a product of primes.

12. 540 **13.** 30

14. 210 **15.** 1260

Reduce the following fractions to lowest terms:

16. $\frac{21}{49}$ **17.** $\frac{36}{48}$ **18.** $\frac{30}{210}$

Multiply:

19. 17.31×31.11 **20.** 62.13×7.81

21. Simplify mentally: 625.3876×100

Subtract. Add to check.

22. $185.3617 - 17.3169$ **23.** $179.3622 - 111.8979$

Divide and round off to two places.

24. $\frac{231.2}{3.89}$ **25.** $190.1 \div 2.1$

26. Round off 3.141592654 to the nearest ten-millionth.

27. Round off $2.0\overline{70}$ to seven decimal places.

28. Use words to write the number 62987621.1.

Add:

29. $1381.62 + 1.82 + 32.61 + 1.99$ **30.** $3.14159 + 2.621 + 82.71$

LESSON 18 Products of primes in cancellation

18.A products of primes in cancellation

If we look at the expression

$$\frac{2 \times 3 \times 7 \times 5}{2 \times 3 \times 7 \times 7}$$

We see that both the top and the bottom contain $2 \times 3 \times 7$, which equals 42.

$$\frac{2 \times 3 \times 7 \times 5}{2 \times 3 \times 7 \times 7} = \frac{42 \times 5}{42 \times 7}$$

Now 42 over 42 equals 1, so the original expression can be written in lowest terms as $\frac{5}{7}$. We can use this thought to reduce some fractions to lowest terms. We write both the top and the bottom as products of prime factors and cancel the factors that appear both above and below.

example 18.A.1 Use products of primes to reduce $\frac{360}{900}$ to lowest terms.

solution We begin by writing both numbers as products of primes.

$$\frac{360}{900} = \frac{2 \times 2 \times 2 \times 3 \times 3 \times 5}{2 \times 2 \times 3 \times 3 \times 5 \times 5}$$

We note the combinations of factors that equal 1.

$$\frac{2 \times 2 \times 2 \times 3 \times 3 \times 5}{2 \times 2 \times 3 \times 3 \times 5 \times 5}$$

Now $\frac{2}{2}$ and $\frac{2}{2}$ and $\frac{3}{3}$ and $\frac{3}{3}$ and $\frac{5}{5}$ all have a value of 1. All that is left is a 2 on top and a 5 on the bottom. Thus, this fraction reduces to

$$\frac{2}{5}$$

example 18.A.2 Use products of primes to reduce $\frac{42}{78}$ to lowest terms.

solution We write both numbers as products of primes and cancel the common factors.

$$\frac{42}{78} = \frac{\cancel{2} \times \cancel{3} \times 7}{\cancel{2} \times \cancel{3} \times 13} = \frac{7}{13}$$

problem set 18

1. Paul picked 5482 peaches. Peter picked four times as many as Paul picked. Wilbur picked twice as many as Peter picked. How many peaches did they pick altogether?
2. Nineteen million and forty-seven hundred-thousandths was Ahab's first guess. His second guess was seventeen thousand, five and two hundred thirty-seven ten-thousandths greater than his first guess. What was his second guess?
3. Even with a pry bar Henry could not force more than sixty-three into each container. If he had seven hundred thousand, thirty-three containers, how many did he need to fill all the containers?
4. When he came to work the next day, he was delighted because four hundred sixteen thousand, nine hundred forty-three containers had been delivered. How many things (Problem 3) would fill them all?
5. What prime numbers are less than 50 but are greater than 25?
6. Find the perimeter of this figure. Dimensions are in centimeters. Be careful—the problem requires some thinking.

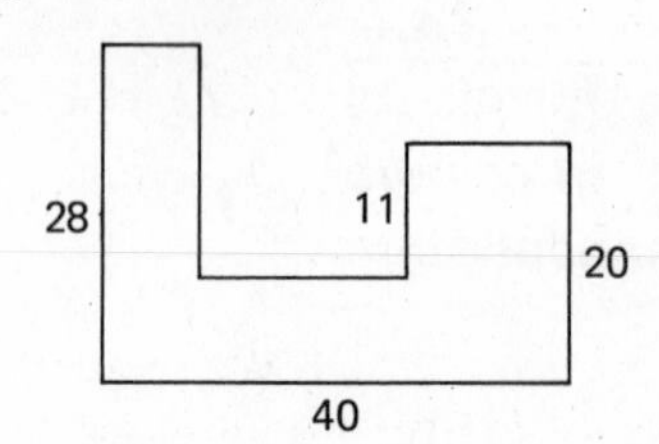

Find the areas of the following figures. Dimensions are in centimeters.

7.

5
30
14
35

8.

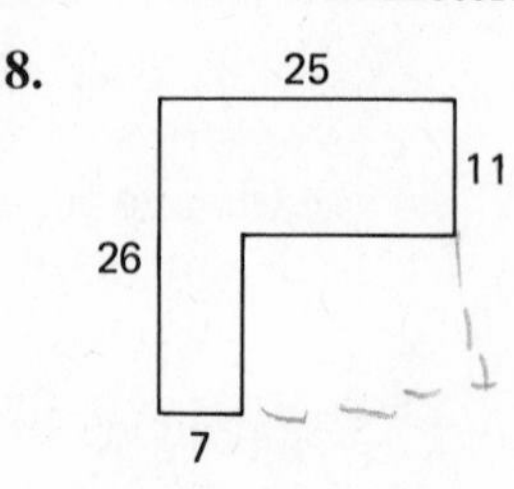

Write both numbers as products of prime factors and simplify.

*9. $\frac{360}{900}$

*10. $\frac{42}{78}$

11. $\frac{210}{294}$ 12. $\frac{738}{882}$

Write the following fractions as decimals. Round off to two decimal places.

13. $\frac{16}{17}$ 14. $\frac{6}{7}$ 15. $\frac{11}{13}$

Write the following numbers as the product of primes.

16. 5040 17. 7560

Multiply:

18. 62.13×11.31 19. 78.11×113

20. Simplify mentally: $621.1378 \div 10{,}000$

Subtract. Add to check.

21. $183.1782 - 1.8999$ 22. $6211.31 - 689.13$

Divide and round off to two places.

23. $\frac{6.89}{2.17}$ 24. $\frac{62.15}{2.1}$

25. Round off 9.8696044 to five decimal places.

26. Use words to write the number 1123689113.

Add:

27. $6251.13 + 231.61 + 81.62 + 998.11$

28. $6.91 + 8.23 + .1135 + 113$

29. $123{,}456 + 654{,}321$

LESSON 19 Multiplying fractions • Dividing fractions

19.A symbols for multiplication

Thus far, we have used a cross to designate multiplication. To write 4 times 3, we have written

$$4 \times 3$$

In algebra, the letter x is used often instead of using a particular number. The symbol for the letter x is similar to the cross $\times$ used to indicate multiplication. Therefore, in algebra we sometimes do not use the cross and designate multiplication in other ways.† Sometimes we use a dot to designate multiplication, and sometimes we use parentheses. Sometimes we use both a dot and parentheses. All the following indicate that 4 and 3 are to be multiplied.

$$4 \cdot 3 \qquad 4(3) \qquad (4)3 \qquad (4)(3) \qquad (4) \cdot (3)$$

None of these forms is necessarily a preferred form since all of them designate the same thing.

† But we will still use the cross when it is convenient.

19.B multiplication of fractions

To multiply fractions we multiply the tops to get the new top and multiply the bottoms to get the new bottom.

example 19.B.1 Simplify: $\frac{4}{9} \cdot \frac{5}{7}$

solution We get the new top by multiplying 4 and 5 and the new bottom by multiplying 9 and 7.

$$\frac{4}{9} \cdot \frac{5}{7} = \mathbf{\frac{20}{63}}$$

If the tops and bottoms contain equal factors, it is often helpful to simplify before multiplying, as shown in the next problem.

example 19.B.2 Simplify: $\frac{4}{9} \cdot \frac{18}{30}$

solution We cancel before we multiply.

$$\frac{\overset{2}{\cancel{4}}}{\cancel{9}} \cdot \frac{\overset{2}{\cancel{18}}}{\underset{15}{\cancel{30}}} = \mathbf{\frac{4}{15}}$$

19.C division of fractions

We divide fractions by inverting the divisor and multiplying. Both of these expressions indicate that $\frac{5}{7}$ is to be divided by $\frac{4}{5}$.

$$\text{(a)} \quad \frac{\frac{5}{7}}{\frac{4}{5}} \qquad \text{(b)} \quad \frac{5}{7} \div \frac{4}{5}$$

Thus, in both expressions, the divisor is $\frac{4}{5}$. We simplify both expressions in the same way—by inverting the divisor and multiplying.

$$\frac{5}{7} \cdot \frac{5}{4} = \frac{25}{28}$$

example 19.C.1 Simplify: $\frac{21}{6} \div \frac{27}{4}$

solution First we invert the divisor. Then we cancel and multiply.

$$\frac{21}{6} \cdot \frac{4}{27} = \frac{\overset{7}{\cancel{21}}}{\underset{3}{\cancel{6}}} \cdot \frac{\overset{2}{\cancel{4}}}{\underset{9}{\cancel{27}}} = \mathbf{\frac{14}{27}}$$

problem set 19

1. The prize was for the closest guess. The exact answer was 14,016.2163. Charles guessed thirteen thousand, forty-one and six ten-thousandths. Mary guessed fourteen thousand, nine hundred ninety-one and two hundred thirty thousandths. By how much did each of them miss the exact answer? Whose guess was closer?

2. They crawled in until 51,416 had arrived. Then 4 times this number crawled in on the next day. How many of them crawled in in all?

3. Five thousand, seven hundred fifteen stood and roared their approval. Nineteen thousand seven remained seated and were discontented. How many more were discontented than approved?

4. The fire department had ruled that occupancy by more than 315 in a room was unlawful. If 14,321 attended, how many rooms would it take to hold them lawfully?

5. What is the smallest prime number that is at least 14 greater than 21?

Simplify. Cancel if possible before multiplying.

*6. $\frac{4}{9} \cdot \frac{5}{7}$

*7. $\frac{4}{9} \cdot \frac{18}{30}$

*8. $\frac{5}{7} \div \frac{4}{5}$

*9. $\frac{21}{6} \div \frac{27}{4}$

10. Find the perimeter of this figure. Dimensions are in meters.

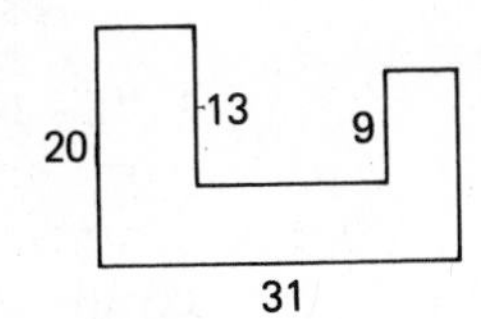

Find the areas of the following figures. Dimensions are in yards.

11.

20
40
10
50

12.

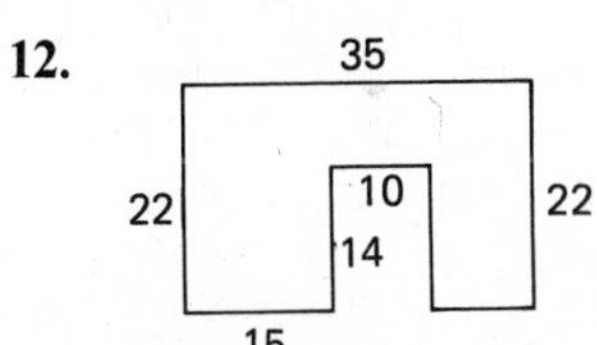

Write both numbers as products of prime factors and simplify.

13. $\frac{360}{540}$

14. $\frac{196}{360}$

15. $\frac{256}{720}$

Write the following fractions as decimals. Round off to two decimal places.

16. $\frac{17}{23}$

17. $\frac{5}{17}$

Write each of the following numbers as a product of prime numbers.

18. 4200

19. 10,080

Multiply:

20. 11.23 × 62.14

21. 1.28 × 62.1

Subtract. Add to check.

22. 1921.761 − 18.32

23. 612.81 − 19.362

Divide and round off to two places.

24. $\frac{7.32}{.12}$

25. $\frac{61.31}{3.1}$

26. Round off $28.30\overline{57}$ to the nearest ten-millionth.

27. Use words to write the number 9699690.

Add:

28. 31.613 + 1.812 + 61.324 + 111.621

29. 118.32 + 116.15 + 311.13

30. Simplify mentally: 62,158.3211 × 100

LESSON 20 Multiples

20.A multiples

Multiples, factors, and **products** are words that go together. We know that 3 times 2 equals 6.

$$3 \cdot 2 = 6$$

In this example the factors are 3 and 2 and the product is 6. Because we can multiply 3 by 2 and get 6, we say that 6 is a multiple of 3 and that 6 is also a multiple of 2. The composite numbers 6, 9, 12, 15, and 18 are also multiples of 3 because they can be composed by multiplying 3 by itself or by another whole number.

$$3 \cdot 2 = 6$$
$$3 \cdot 3 = 9$$
$$3 \cdot 4 = 12$$
$$3 \cdot 5 = 15$$
$$3 \cdot 6 = 18$$

Similarly, the numbers 4, 6, 8, 10, 12 and 14 are all multiples of 2 because they can be composed by multiplying 2 by itself or by another whole number as shown here.

$$2 \cdot 2 = 4$$
$$2 \cdot 3 = 6$$
$$2 \cdot 4 = 8$$
$$2 \cdot 5 = 10$$
$$2 \cdot 6 = 12$$
$$2 \cdot 7 = 14$$

The words multiple, factor, and product are not difficult, but they are new and thus are different. Problems that use these words will help us remember what they mean.

example 20.A.1 The number 8 is a multiple of which numbers?

solution We can compose 8 by multiplying 2 and 4 or by multiplying 1 and 8.

$$2 \cdot 4 = 8 \qquad 8 \cdot 1 = 8$$

So 8 is a multiple of **1**, **2**, **4**, and **8**. Note that we say that 8 is a multiple of itself and also of 1.

example 20.A.2 Find all multiples of 9 that are less than 50.

solution A multiple of 9 is the product of 9 and a whole number greater than zero.

$$9 \cdot 1 = 9$$
$$9 \cdot 2 = 18$$
$$9 \cdot 3 = 27$$
$$9 \cdot 4 = 36$$
$$9 \cdot 5 = 45$$

Thus the multiples of 9 that are less than 50 are **9, 18, 27, 36, and 45.**

problem set 20

1. Seven thousand, forty-two wore red hats. Ninety-three thousand, nine hundred seventy-five wore blue hats. If one hundred thirty-seven thousand, eight hundred forty-two attended, how many did not wear either red hats or blue hats?

2. When the first box was opened, the chapeaux inside totaled one thousand, nine hundred three. If 47,300 students were expected to attend, how many boxes of chapeaux would be required so that every student could have one?

3. When the first frost came, nineteen thousand, two shriveled and died. If five times this number survived the frost, how many had there been before the frost came?

4. For 8 minutes Ralph could do 17 every minute. Then he got tired and could only do 14 a minute for the next 6 minutes. How many did he do in all?

***5.** Find all multiples of 9 that are less than 50.

Simplify. Cancel before multiplying if possible.

6. $\frac{42}{15} \cdot \frac{3}{7}$

7. $\frac{5}{18} \cdot \frac{90}{120}$

8. $\frac{5}{8} \div \frac{7}{4}$

9. $\frac{7}{18} \div \frac{14}{9}$

10. Find the perimeter of this figure. Dimensions are in feet.

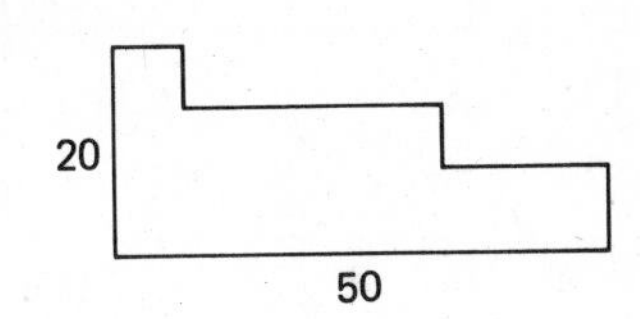

Find the areas of the following figures. Dimensions are in meters.

11.

20
20
12
7

12.

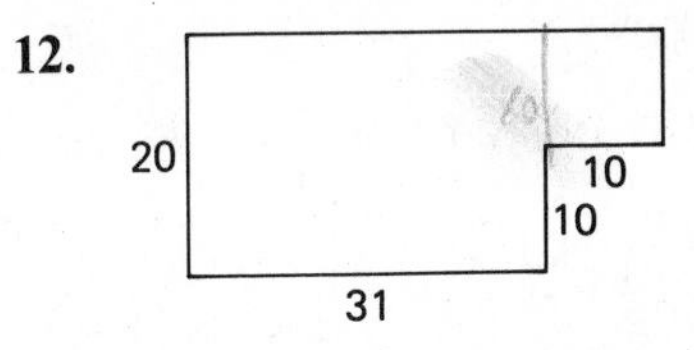

Write both numbers as products of prime factors and simplify.

13. $\frac{540}{720}$

14. $\frac{144}{360}$

15. $\frac{432}{720}$

Write the following fractions as decimals. Round off to two decimal places.

16. $\frac{21}{23}$

17. $\frac{6}{13}$

Write the following numbers as the products of primes.

18. 1200

19. 12,600

Multiply:

20. 11.92×12.32

21. 1792×61.2

Subtract. Add to check.

22. $62.0325 - 7.9813$

23. $17.3252 - 8.9211$

Divide and round off to two places.

24. $\frac{7.91}{3.14}$

25. $\frac{9.86}{3.14}$

26. Round off 194,591,014.62 to the nearest million.

27. Use words to write the number 111,546,435.

28. Simplify mentally: $8361.2361 \div 100$

Add:

29. $16.17 + 181.325 + .00131$

30. $14.159 + 2.71828 + 2236.21$

LESSON 21 Average

21.A average

The word **average** is difficult to define. In a way it means "middle" or the result that would be obtained if everything were leveled out. In mathematics we say that the average of a group of numbers is the sum of the numbers divided by the number of numbers.

$$\text{Average} = \frac{\text{sum of the numbers}}{\text{number of numbers}}$$

Using this definition, we can find the average of the five numbers 7, 18, 29, 143, and 285 by adding the numbers and then dividing by 5.

$$\text{Average} = \frac{7 + 18 + 29 + 143 + 285}{5} = \frac{482}{5} = 96.4$$

example 21.A.1 Find the average of 1765, 93, 742, and 21,050.

solution We add the four numbers and divide the sum by 4.

$$\text{Average} = \frac{1765 + 93 + 742 + 21{,}050}{4} = \frac{23{,}650}{4} = \mathbf{5912.5}$$

example 21.A.2 Find the average of 8, 17, 14, 10, 18, 6, 13, and 12.

solution We add the numbers and divide by 8.

$$\text{Average} = \frac{8 + 17 + 14 + 10 + 18 + 6 + 13 + 12}{8} = \frac{98}{8} = \mathbf{12.25}$$

problem set 21

1. Jed guessed ten billion, four hundred seventy-five thousand, fifteen and nine hundred twenty-three ten-thousandths. Ned guessed eight billion, four hundred seventy-three million, forty-two and seventy-five thousandths. How much greater was Jed's guess than Ned's guess?

2. On the first two days the counts were 4012.06 and .00418. On the next day the counts were 732.05 and 9.016. What was the total of the counts for all three days?

3. Only 243 could squeeze through at one time. If 416,202 came, how many times would it take all of them to squeeze through?

4. Which of the following numbers are divisible (a) by 3, (b) by 5: 11,682; 10,517; 2193; 4200?

5. List the prime numbers that are greater than 30 and are less then 50.

*6. Find the average of 1765, 93, 742 and 21,050.

*7. Find the average of 8, 17, 14, 10, 18, 6, 13, and 12.

Simplify. Cancel before multiplying if possible.

8. $\frac{36}{28} \cdot \frac{7}{6}$

9. $\frac{21}{32} \cdot \frac{4}{7}$

10. $\frac{5}{7} \div \frac{15}{21}$

11. $\frac{9}{25} \div \frac{27}{15}$

12. Find the perimeter of the following figure. Dimensions are in meters.

13. Find the area of the following figure. Dimensions are in centimeters.

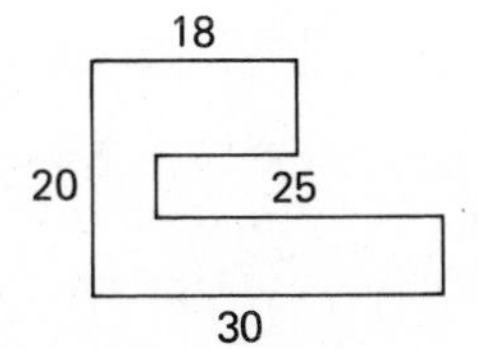

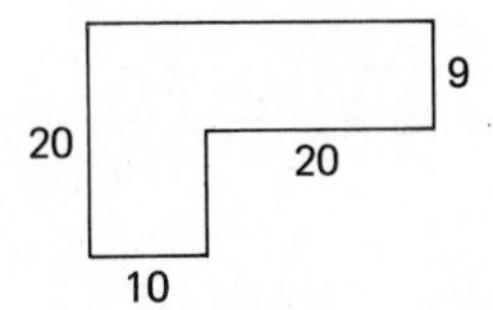

Write both numbers as products of prime factors and simplify.

14. $\frac{270}{360}$

15. $\frac{35}{42}$

16. $\frac{120}{450}$

Write the following fractions as decimals. Round off to two decimal places.

17. $\frac{11}{17}$

18. $\frac{5}{13}$

Write the following numbers as products of primes:

19. 2520

20. 8400

Multiply:

21. 1.83×61.3

22. 66.12×1.7

Subtract. Add to check.

23. $123.789 - 1.899$

24. $12.876 - 1.7829$

Divide and round off to two places:

25. $\frac{6.25}{.87}$

26. $\frac{71.3}{3.7}$

27. Round off $34.7\overline{18}$ to the nearest ten-thousandth.

28. Use words to write the number 2718.2818.

29. Simplify mentally: $9.8762 \div 1000$

30. Add: $7.36 + 99.825 + 1070.0327$

LESSON 22 Multiple fractional factors

22.A repeated multiplication

There is no change in the procedure when a problem has three or more fractions to be multiplied. First we cancel common factors and then we multiply.

example 22.A.1 Simplify: $\frac{56}{27} \cdot \frac{14}{21} \cdot \frac{15}{22}$

solution We begin by writing each number as a product of prime factors. Then we cancel common factors and multiply as the last step.

$$\frac{7 \cdot 2 \cdot 2 \cdot 2}{3 \cdot 3 \cdot \cancel{3}} \cdot \frac{\cancel{2} \cdot \cancel{7}}{3 \cdot \cancel{7}} \cdot \frac{\cancel{3} \cdot 5}{\cancel{2} \cdot 11} = \mathbf{\frac{280}{297}}$$

example 22.A.2 Simplify: $\frac{26}{30} \cdot \frac{21}{39} \div \frac{4}{15}$

solution First we must invert $\frac{4}{15}$ and change the division symbol to a dot for multiplication.

$$\frac{26}{30} \cdot \frac{21}{39} \cdot \frac{15}{4}$$

Now we simplify and then multiply.

$$\frac{\cancel{2} \cdot \cancel{13}}{2 \cdot \cancel{3} \cdot \cancel{5}} \cdot \frac{\cancel{3} \cdot 7}{\cancel{3} \cdot \cancel{13}} \cdot \frac{\cancel{3} \cdot \cancel{5}}{\cancel{2} \cdot 2} = \frac{7}{4}$$

problem set 22

1. Ninety million, four thousand and sixty-two millionths is a big number. By how much is it greater than eighty-six million, four hundred twenty-seven thousand and fourteen ten-thousandths?

2. In the first wave there were 742,000. The second and third waves contained 96,016 and 1,001,892, respectively. How many were there in all?

3. Jim tried as hard as he could, but he could only do 1421 in an hour. If there were 18,473 in all, how long would it take Jim to do them all?

4. Harry cornered 146 in the backyard. Jennet cornered five times that number in the side yard. How many did they corner all together?

5. Find the average of 1862, 1430, and 276.

6. (a) What are the prime numbers between 80 and 90? (b) List the multiples of 3 that are greater than 80 and less than 90.

Simplify. Cancel where possible as the first step.

*7. $\frac{56}{27} \cdot \frac{14}{21} \cdot \frac{15}{22}$

*8. $\frac{26}{30} \cdot \frac{21}{39} \div \frac{4}{15}$

9. $\frac{36}{24} \cdot \frac{8}{6} \div \frac{3}{6}$

10. Find the perimeter of the following figure. Dimensions are in inches.

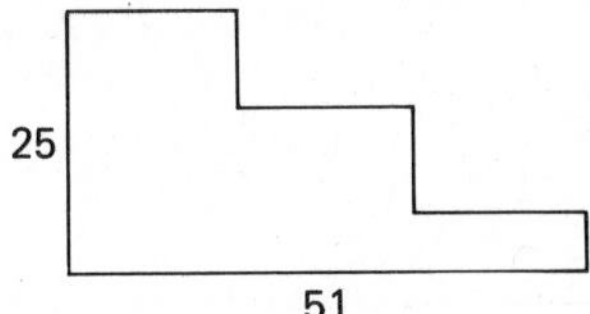

11. Find the area of the following figure. Dimensions are in feet.

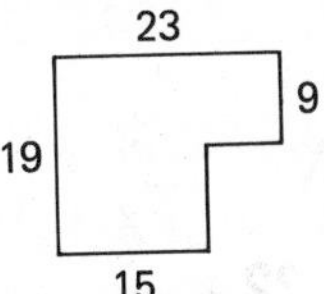

Write both numbers as products of prime factors and simplify.

12. $\frac{540}{1440}$

13. $\frac{128}{360}$

14. $\frac{720}{840}$

15. $\frac{240}{256}$

Write the following fractions as decimals. Round off to two decimal places.

16. $\frac{13}{17}$

17. $\frac{3}{11}$

18. $\frac{3}{7}$

Write the following numbers as the products of primes.

19. 14,000

20. 5040

Multiply:

21. 1.91×31.7

22. $581 \times .163$

Subtract. Add to check.

23. $6854.32 - 1.871$

24. $725.113 - 26.91$

Divide and round off to two places:

25. $\frac{11.7}{3.1}$

26. $\frac{1132.1}{.7}$

27. Round off .301029996 to the nearest hundredth.

28. Use words to write the number 3678922117.32.

29. Simplify mentally: 123.7136×1000

30. Add: $11.762 + .8171 + 1162.18$

LESSON 23 English units · Unit multipliers

23.A English units

When we attach words to numbers such as

4 feet 13 inches 27 yards 42 miles

we call each combination a **denominate number,** and we call the words **units.** We say that the units of the denominate number

4 feet

are feet. In the denominate number

32 miles per hour

we say that the units are miles per hour. The basic units of length in the English system are the inch, foot, yard, and mile. They are related by the following equivalences.

$$12 \text{ inches} = 1 \text{ foot} \qquad 3 \text{ feet} = 1 \text{ yard} \qquad 5280 \text{ feet} = 1 \text{ mile}$$

23.B unit multipliers

If we begin with the number 5

$$5$$

and multiply and divide by 7, we end up with 5 again.

$$\frac{5 \cdot 7}{7} = \frac{35}{7} = 5$$

We remember that one explanation is that multiplication and division are inverse operations because they undo one another. Another explanation is that 7 over 7 has a value of 1 and multiplying 5 by 1 equals 5.

$$5\left(\frac{7}{7}\right) = 5(1) = 5$$

We say that something that has a value of 1 has a value of unity, for unity is another word for 1. The expression

$$\frac{1 \text{ foot}}{12 \text{ inches}}$$

has a value of unity because 12 inches is the same length as 1 foot. We will call expressions such as this one **unit multipliers.** We can use this multiplier to change inches to feet.

example 23.B.1 Use a unit multiplier to convert 72 inches to feet.

solution We will multiply by 1 foot over 12 inches and cancel the units.

$$72 \cancel{\text{inches}} \times \frac{1 \text{ foot}}{12 \cancel{\text{inches}}} = \frac{72}{12} \text{ feet} = \mathbf{6 \text{ feet}}$$

We note that we cancel units just as if they were numbers.

example 23.B.2 Use a unit multiplier to convert 430 feet to yards (yd).

solution The two unit multipliers that we can use are

$$\text{(a)} \quad \frac{3 \text{ ft}}{1 \text{ yd}} \qquad \text{and} \qquad \text{(b)} \quad \frac{1 \text{ yd}}{3 \text{ ft}}$$

Beginners sometimes don't know which one to use. We decide to use (a).

$$430 \text{ ft} \times \frac{3 \text{ ft}}{1 \text{ yd}} = 430 \cdot 3 \frac{(\text{ft})(\text{ft})}{\text{yd}}$$

This is not incorrect but is not what we want. Let's try (b).

$$430 \cancel{\text{ft}} \times \frac{1 \text{ yd}}{3 \cancel{\text{ft}}} = \mathbf{\frac{430}{3} \text{ yd}}$$

Notice that this time we could cancel the unit "ft" above with the "ft" below. We could write $\frac{430}{3}$ as a mixed number, but we decide to leave it as it is.

There are six unit multipliers that we can get from the three relationships

$$12 \text{ in} = 1 \text{ ft} \qquad 3 \text{ ft} = 1 \text{ yd} \qquad 5280 \text{ ft} = 1 \text{ mi}$$

These multipliers are:

$$\frac{12 \text{ in}}{1 \text{ ft}} \text{ and } \frac{1 \text{ ft}}{12 \text{ in}} \qquad \frac{3 \text{ ft}}{1 \text{ yd}} \text{ and } \frac{1 \text{ yd}}{3 \text{ ft}} \qquad \frac{5280 \text{ ft}}{1 \text{ mi}} \text{ and } \frac{1 \text{ mi}}{5280 \text{ ft}}$$

The use of these multipliers will prevent multiplying when we should divide and dividing when we should multiply. When the units don't cancel, we know we have tried the wrong unit multiplier.

problem set 23

1. Hortense ran nineteen thousand, seven and four hundred twenty-three thousandths centimeters in the allotted time. Then in the same time Jim ran eight thousand, forty-two and seven hundred sixty-five ten-thousandths centimeters. What was the sum of the distances they ran?

2. At first there were only 2,046,021 microbes. Then Muzowbe added three times this many. Then how many microbes were there in all?

3. The first factory could produce 243 cars in an allotted period. If 19,197 cars were required, how many periods had to be allotted?

4. From this list Kunta selected the numbers that were divisible by 3: 40,225; 93,663; 71,205; 20,163. What was the sum of the numbers he selected?

5. (a) List the prime numbers between 60 and 70. (b) List the multiples of 9 between 60 and 70.

6. Find the average of 14, 183, 75, 911, and 42.

Use unit multipliers to convert.

*7. 72 inches to feet

*8. 430 feet to yards

9. 42 feet to inches

Simplify. Cancel if possible as the first step.

10. $\frac{42}{12} \cdot \frac{36}{14} \div \frac{18}{2}$

11. $\frac{12}{16} \cdot \frac{4}{3} \div \frac{5}{6}$

12. $\frac{16}{24} \cdot \frac{12}{4} \cdot \frac{1}{2}$

13. Find the perimeter of the following figure. Dimensions are in yards.

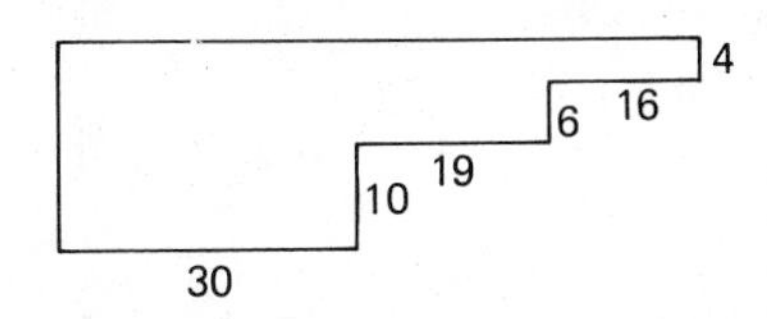

14. Find the area of the following figure. Dimensions are in meters.

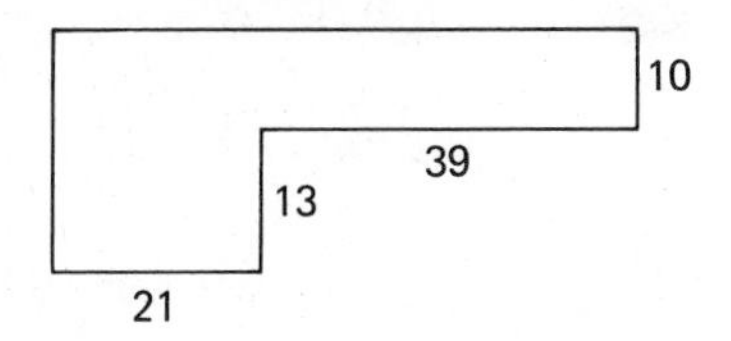

Write both numbers as products of prime factors and simplify.

15. $\frac{180}{256}$

16. $\frac{240}{600}$

17. $\frac{540}{2160}$

Write the following fractions as decimals. Round off to two decimal places.

18. $\frac{13}{19}$

19. $\frac{17}{19}$

Write the following numbers as products of primes.

20. 2880 **21.** 6750

Multiply:

22. $71.2 \times .173$ **23.** 61.7×11.2

Subtract. Add to check.

24. $1812.721 - 729.871$ **25.** $61{,}131.2 - 71.823$

Divide and round off to two places.

26. $\dfrac{713.7}{2.5}$ **27.** $\dfrac{7181.3}{.3}$

28. Round off 728,651.32 to the nearest thousand.

29. Use words to write the number 67211361.72.

30. Add: $71.161 + .8121 + 177.62$

LESSON 24 *Metric length conversions*

24.A metric unit conversions

The advantage to the metric system is its simplicity. All you have to do to convert from one metric unit of length to another metric unit of length is to move the decimal point. The basic unit of length in the metric system is the meter. A meter is just a little longer than a yard. If we divide a meter into 100 equal parts, we call each of the parts a centimeter. One thousand meters is one kilometer, which is about six-tenths of a mile.

100 centimeters = 1 meter (m) 1000 meters = 1 kilometer (km)

The four unit multipliers that we get from these two equivalences are:

$$\frac{100 \text{ cm}}{1 \text{ m}} \text{ and } \frac{1 \text{ m}}{100 \text{ cm}} \qquad \frac{1000 \text{ m}}{1 \text{ km}} \text{ and } \frac{1 \text{ km}}{1000 \text{ m}}$$

There are other units of length such as the millimeter, the decimeter, and the micrometer. These units are special-purpose units and can be researched when their use is required. We will concentrate on the three basic metric units of length.

example 24.A.1 Convert 528 centimeters to meters.

solution We have our choice of using

$$\text{(a)}\ \frac{100 \text{ cm}}{1 \text{ m}} \quad \text{or} \quad \text{(b)}\ \frac{1 \text{ m}}{100 \text{ cm}}$$

We will use (b) because "cm" is on the bottom and will cancel with the "cm" on the top.

$$528 \cancel{\text{cm}} \times \frac{1 \text{ m}}{100 \cancel{\text{cm}}} = \frac{528}{100} \text{ m} = \mathbf{5.28 \text{ m}}$$

Note that 528 cm equals 5.28 m. All we had to do was to move the decimal point!

example 24.A.2 Convert 486 kilometers to meters.

solution We will use the unit multiplier that has kilometers on the bottom.

$$486\ \cancel{\text{km}} \times \frac{1000\text{ m}}{1\ \cancel{\text{km}}} = \mathbf{486{,}000\text{ m}}$$

Again the digits are the same. The only difference is in the position of the decimal point.

problem set 24

1. At the turn of the century fourteen thousand, seven things were on the proscribed list. Thirty years later only seven thousand, nine hundred forty-two remained on this list. How many had been removed?

2. Mary measured it first. She got one hundred forty-seven and nine hundred twenty-three millionths inches. Then Jim tried. He got one hundred forty-six and three thousand, one hundred forty-two ten-thousandths inches. By how much was Mary's measurement greater?

3. Each slot could hold only 47 balls. If there were 1982 balls, how many slots were necessary to hold them all?

4. Which of the following numbers are divisible by (a) 2, (b) 10: 40,613; 90,528; 71,423; 4020?

5. (a) List the prime numbers between 20 and 30. (b) List the multiples of 4 between 20 and 30.

6. Find the average of 98, 142, 76, 81, and 6.

Use unit multipliers to convert.

***7.** 528 centimeters to meters

***8.** 486 kilometers to meters

9. 204 inches to feet

10. 47 yards to feet

Simplify. Cancel where possible as the first step.

11. $\frac{16}{24} \cdot \frac{12}{4} \cdot \frac{1}{3}$

12. $\frac{18}{24} \cdot \frac{8}{9} \cdot \frac{3}{2}$

13. $\frac{14}{21} \cdot \frac{7}{2} \div \frac{14}{6}$

14. Find the perimeter of the following figure. Dimensions are in feet.

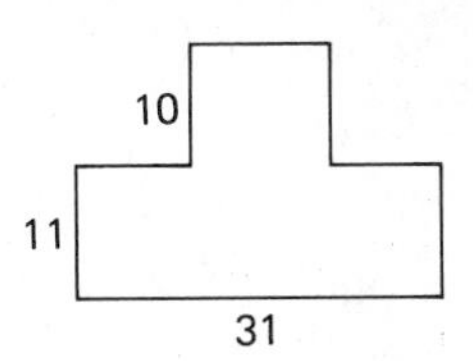

15. Find the area of the following figure. Dimensions are in inches.

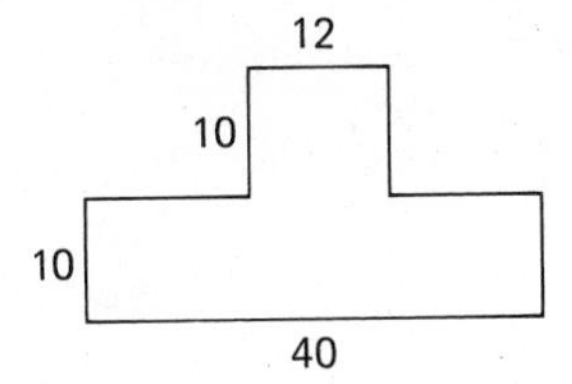

Write both numbers as products of prime factors and simplify.

16. $\frac{24}{72}$

17. $\frac{18}{24}$

18. $\frac{81}{135}$

Write the following fractions as decimals. Round off to two decimal places.

19. $\frac{3}{14}$

20. $\frac{4}{17}$

Write the following numbers as the products of primes:

21. 480

22. 2400

Multiply:

23. 61.7×121

24. $712(1.23)$

Subtract. Add to check.

25. 712.61 − 699.781 **26.** 1211.7 − 3.78

Divide and round off to two places.

27. $\frac{211}{3.1}$ **28.** $\frac{7.162}{.7}$

29. Round off 3,817,321.127 to the nearest hundred.

30. Use words to write the number 1317621.13.

LESSON 25 *Area as a difference*

25.A

area as a difference We have been finding the areas of figures by adding rectangular areas. Sometimes we can find total area by taking the difference of rectangular areas.

example 25.A.1 Use the difference of areas to find the area of this figure. The dimensions are in meters.

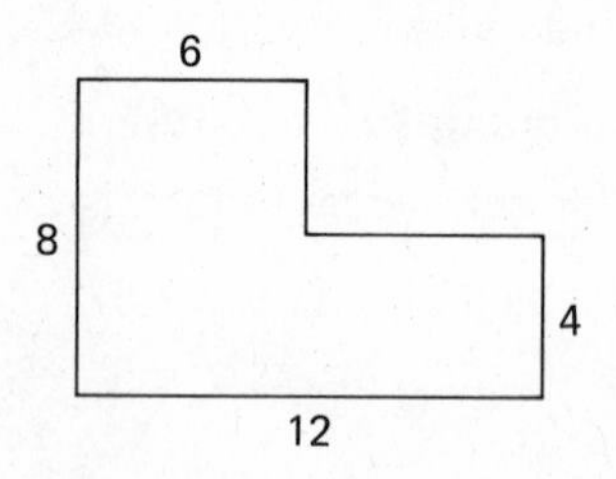

solution We add dashed lines to complete the rectangle.

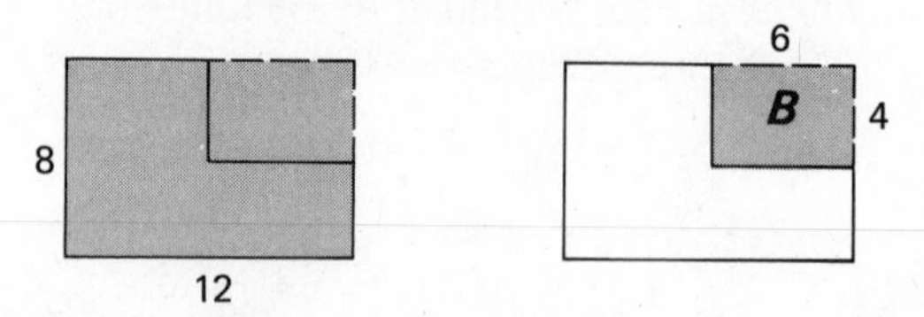

Then we can find the area of the whole rectangle shown on the left and subtract from it the area of the small rectangle B on the right.

$$8 \text{ m} \times 12 \text{ m} = 96 \text{ m}^2 \qquad 6 \text{ m} \times 4 \text{ m} = 24 \text{ m}^2$$

Thus the area of the figure is

$$96 \text{ m}^2 - 24 \text{ m}^2 = \mathbf{72\ m^2}$$

example 25.A.2 Use the difference of areas to find the area of this figure. The dimensions are in feet.

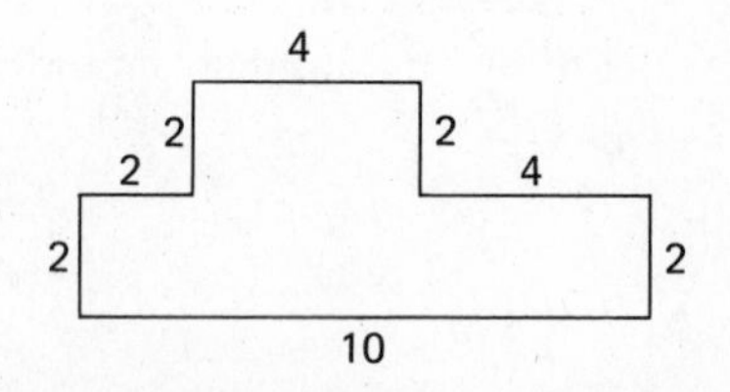

solution We use dashed lines to complete the big rectangle. Then we find the area of the big rectangle and from this subtract the areas of the two small rectangles we have labeled A and B.

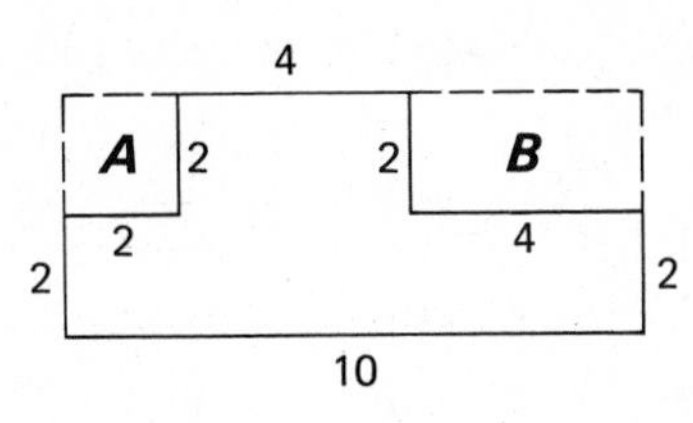

$$\begin{aligned} \text{Big rectangle:} \quad & 4 \text{ ft} \times 10 \text{ ft} = 40 \text{ ft}^2 \\ A: \quad & 2 \text{ ft} \times 2 \text{ ft} = 4 \text{ ft}^2 \\ B: \quad & 2 \text{ ft} \times 4 \text{ ft} = 8 \text{ ft}^2 \end{aligned}$$

$$\text{Area} = 40 \text{ ft}^2 - 4 \text{ ft}^2 - 8 \text{ ft}^2 = \mathbf{28 \text{ ft}^2}$$

problem set 25

1. In the first month ten thousand, forty-two students used the library. In the second month 4 times this number used the library. In the third month, only four thousand, seventy-five used the library. How many students used the library in all?
2. The flowers proliferated until they numbered 17,842. Then the first frost killed 9085. How many flowers were still alive?
3. The first night 482 came. Five times this number came on the second night. On the third night only 392 came. How many came in all?
4. Which of the following numbers are divisible by 3: 315,234; 906,185; 21,387; 4072?
5. (a) List the prime numbers between 40 and 50. (b) List the multiples of 3 between 40 and 50.
6. Find the average of 32, 75, 143, and 91.

Use unit multipliers to convert:

7. 46.31 meters to centimeters
8. 48 miles to feet
9. 416 meters to kilometers

Simplify. Cancel where possible as the first step.

10. $\frac{42}{24} \cdot \frac{6}{7} \cdot \frac{2}{3}$
11. $\frac{72}{48} \cdot \frac{8}{9} \cdot \frac{3}{4}$
12. $\frac{2}{3} \cdot \frac{27}{64} \div \frac{3}{16}$

13. Find the perimeter of the following figure. Dimensions are in feet.

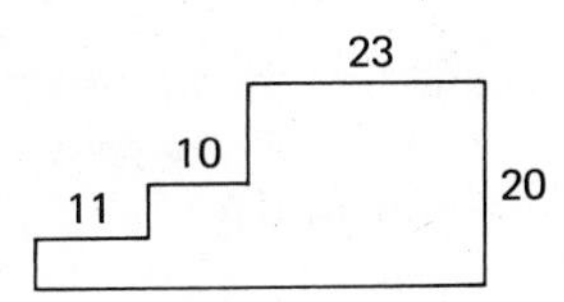

14. Find the area of the following figure in square meters.

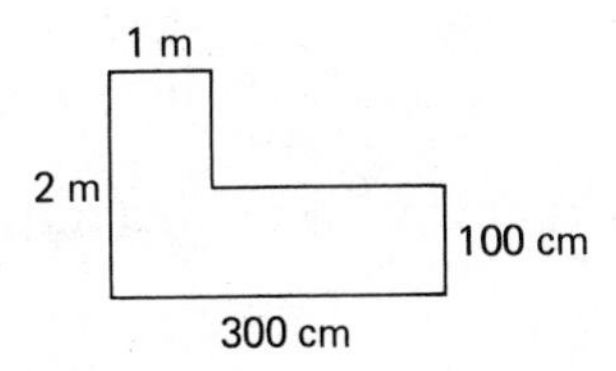

Write the following fractions as products of prime factors and simplify.

15. $\frac{39}{72}$
16. $\frac{210}{420}$
17. $\frac{125}{360}$

Write the following fractions as decimals. Round off to two decimal places.

18. $\frac{3}{19}$ **19.** $\frac{8}{23}$

Write the following numbers as products of primes:

20. 288 **21.** 630

Multiply:

22. 1123.1 × 3.1 **23.** .187 × 92.8

Subtract. Add to check.

24. 718.87 − 9.876 **25.** 31.67 − 13.8

Divide and round off to two places:

26. $\frac{625.8}{1.7}$ **27.** $\frac{1361.4}{.4}$

28. Round off $18{,}976{,}511.31\overline{052}$ to the nearest hundred thousand.

29. Use words to write the number 111321654.7.

30. Add: 713.85 + .3162 + 3.14

LESSON 26 Mode, median, and mean · Average in word problems

26.A mode, median, and mean

In a statistics course, there are three words that will be used to describe the distribution of a group of numbers. In this list of numbers

2, 2, 2, 2, 2, 7, 8, 8, 9, 10, and 784

the number 2 appears more than any other number. In statistics we will say that the number that appears the most often is the **mode** of a group of numbers. In this list the number 7 is the middle number. In statistics we will call the middle number the **median** of a group of numbers. If we add the numbers and divide this sum by the number of numbers, we will find what will be called in statistics the **mean** of a group of numbers.

$$\text{Mean} = \frac{2 + 2 + 2 + 2 + 2 + 7 + 8 + 8 + 9 + 10 + 784}{11} = 76$$

In statistics the mode, the median, and the mean are considered to be three different kinds of averages. In this book we will use the word **average** instead of the word **mean** and will not consider the mode or the median.

26.B average in word problems

Many word problems require that we find the average of a group of numbers.

example 26.B.1 When the boys measured the distance three times, they got three different answers. If the measurements were 4.017 meters, 4.212 meters, and 3.996 meters, what was the average of their measurements?

solution We add the numbers and then divide the sum by 3.

$$\begin{array}{r} 4.017 \\ 4.212 \\ \underline{3.996} \\ 12.225 \end{array} \qquad \frac{12.225}{3} = \mathbf{4.075}$$

example 26.B.2 When Athena sprang from the head of Zeus in full armor, two other gods guessed the weight of her panoply. One guess was 41.802 kilograms and the other guess was 37.408 kilograms. What was the average of the two guesses?

solution We add the two numbers and divide the sum by 2.

$$\begin{array}{r} 41.802 \\ \underline{37.408} \\ 79.210 \end{array} \qquad \frac{79.210}{2} = \mathbf{39.605\ kg}$$

problem set 26

*1. The boys measured the distance three times and they got three different answers. If the measurements were 4.017 meters, 4.212 meters, and 3.996 meters, what was the average of the measurements?

*2. When Athena sprang from the head of Zeus in full armor, two other gods guessed the weight of her panoply. One guess was 41.802 kilograms, and the other guess was 37.408 kilograms. What was the average of the two guesses?

3. Find the average of the first eight prime numbers.

4. Which of the following numbers are divisible by 3? 41,625; 9081; 20,733; 10,662. What is the average of the numbers that are divisible by 3?

5. There were 325 pellets in every shotgun shell. If 495,625 pellets came in the shipment, how many shells could be reloaded?

6. The first toss was 60,152.035 units. The second toss was 82,006.012 units. By how many units was the second toss greater?

Use unit multipliers to convert:

7. 136.15 centimeters to meters

8. 12 miles to feet

9. 1899 meters to kilometers

Simplify.

10. $\frac{21}{36} \cdot \frac{12}{14} \cdot \frac{2}{3}$

11. $\frac{36}{42} \cdot \frac{21}{9} \cdot \frac{1}{2}$

12. $\frac{28}{36} \cdot \frac{8}{21} \div \frac{16}{27}$

Find the areas of the following figures. Units are in meters.

13.

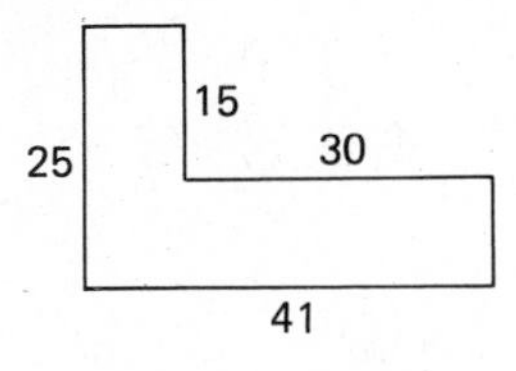

14.

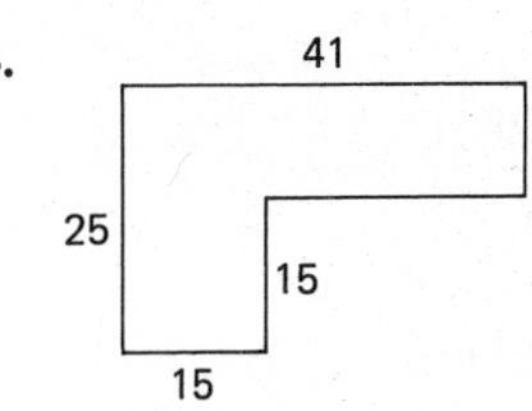

Write the following fractions as products of prime factors and simplify.

15. $\frac{24}{28}$ **16.** $\frac{240}{300}$ **17.** $\frac{144}{256}$

Write the following fractions as decimals. Round off to two decimal places.

18. $\frac{7}{13}$ **19.** $\frac{11}{17}$

Write the following numbers as the products of primes.

20. 6300 **21.** 1050

Multiply:

22. 132.1×6.1 **23.** $17.21 \times .11$

Subtract. Add to check.

24. $1361.78 - 31.921$ **25.** $789.891 - 77.892$

Divide and round off to two places.

26. $\frac{7621.1}{2.1}$ **27.** $\frac{3.623}{.02}$

28. Round off 781,136.36293 to three decimal places.

29. Use words to write the number 6211357.5.

30. Add: $613.7211 + .0178 + 17.369$

LESSON 27 *Areas of triangles*

27.A areas of triangles

The **altitude,** or **height,** of a triangle is the perpendicular distance from the base of the triangle to the opposite corner. Any one of the three sides can be designated as the base.

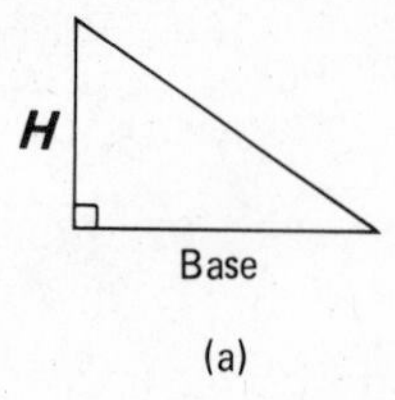

(a)

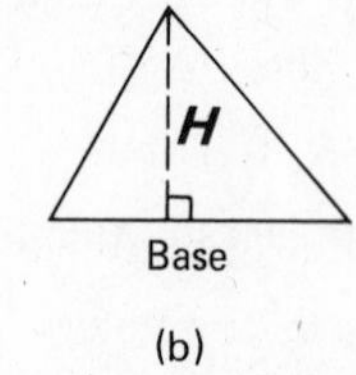

(b)

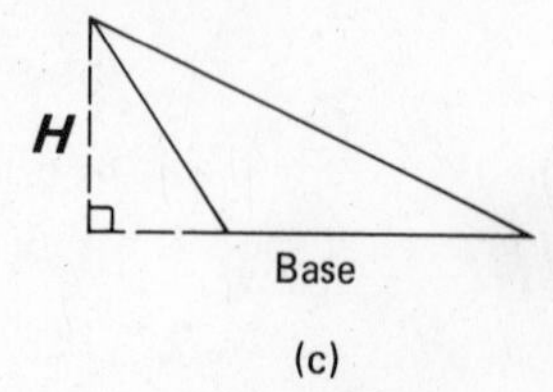

(c)

The altitude can (a) be one of the sides of the triangle, (b) fall inside the triangle, or (c) fall outside of the triangle. When the altitude falls outside of the triangle, we have to extend the base so that the altitude can be drawn.

The formula for the area of any triangle is the product of the base and altitude divided by 2.

$$\text{Area} = \frac{\text{base} \times \text{height}}{2}$$

example 27.A.1 Find the areas of these triangles. Dimensions are in inches.

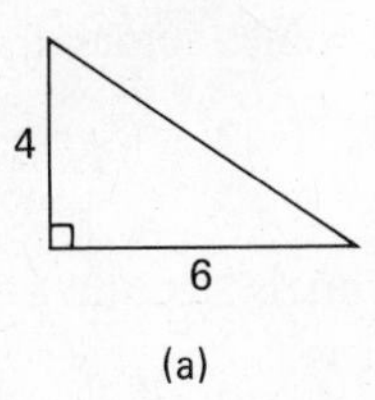

(a)

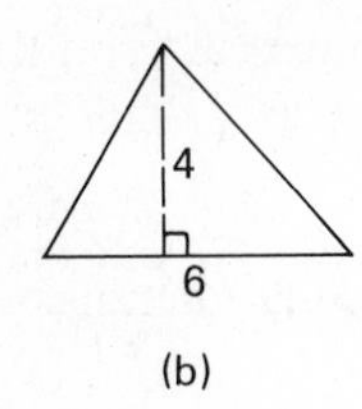

(b)

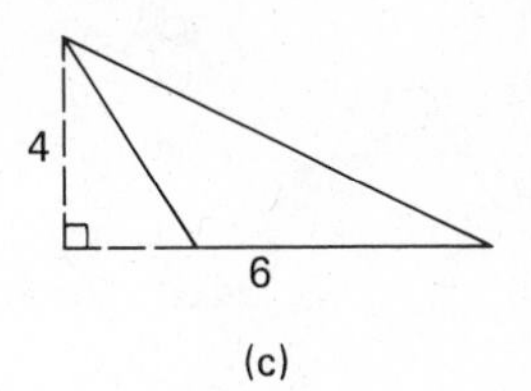

(c)

solution The base of each triangle is 6 in and the height of each triangle is 4 in. Thus, all the triangles have the same area.

$$\text{Area} = \frac{\text{base} \times \text{height}}{2} = \frac{6 \text{ in} \times 4 \text{ in}}{2} = \mathbf{12 \text{ in}^2}$$

problem set 27

1. The first shift produced 14,782 units. The second shift produced 18,962 units. The third shift produced only 2085 units. By how many units did the production of the second shift exceed the sum of the production of the first and third shifts?

2. Seven thousand, four hundred eight hundred-thousandths was Mary's guess. Tom guessed six thousand, five hundred eighty-two ten-thousandths. Whose guess was larger and by how much?

3. Nine hundred eighty-four was increased by one hundred forty-seven thousand, seventeen. Then this sum was doubled. What was the final result?

4. Find the average of the prime numbers that are greater than 20 but less than 42.

5. The first number was 743. The second number was 486. The third number was twice the sum of the first two. What was the average of the numbers?

6. Which of the following numbers is divisible by 5: 41,320; 90,183; 76,142; 9405; 50,172?

Use unit multipliers to convert:

7. 81 yards to feet

8. 81 feet to yards

9. 4899 meters to centimeters

Simplify.

10. $\frac{20}{24} \cdot \frac{6}{15} \cdot \frac{2}{3}$

11. $\frac{16}{20} \cdot \frac{5}{8} \div \frac{1}{4}$

12. $\frac{18}{20} \cdot \frac{5}{3} \div \frac{6}{9}$

13. Find the perimeter of this figure. Dimensions are in feet.

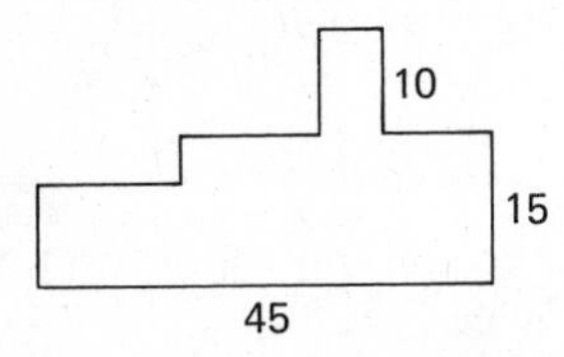

Find the areas of the following figures. Dimensions are in feet.

14.

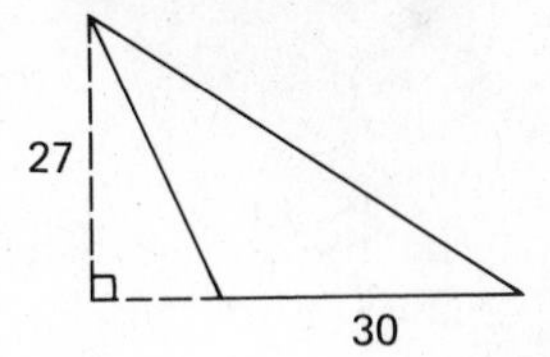

15.

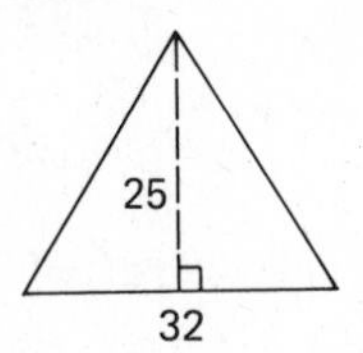

Write both numbers as products of prime factors and simplify.

16. $\dfrac{72}{120}$ **17.** $\dfrac{280}{360}$

Write the following fractions as decimals. Round off to two decimal places.

18. $\dfrac{7}{17}$ **19.** $\dfrac{7}{13}$ **20.** $\dfrac{6}{11}$

Write the following numbers as the product of primes.

21. 324 **22.** 2916

Multiply:

23. 612.1×2.8 **24.** 625×1.93

Subtract. Add to check.

25. $18{,}521.62 - 3.8972$ **26.** $362.772 - 1.999$

Divide and round off to two places.

27. $\dfrac{611.32}{.04}$ **28.** $\dfrac{.0062}{.07}$

29. Round off $781.10\overline{563}$ to the nearest hundred-thousandth.

30. Use words to write the number 7811.7821.

LESSON 28 Improper fractions and mixed numbers

28.A improper fractions and mixed numbers

If the numerator of the fraction is less than the denominator of the fraction, the fraction represents a number less than 1 and is called a **proper fraction.** The fraction $\frac{2}{3}$ has a numerator that is less than the denominator and is therefore a proper fraction.

$\dfrac{2}{3}$ designates a number less than 1, as we see in the figure

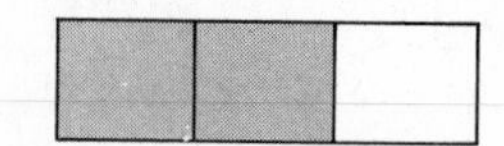

If the numerator of a fraction is greater than the denominator of the fraction, the fraction represents a number greater than 1 and the fraction is called an **improper fraction.** Thus the number

$$\frac{10}{3}$$

can be called an improper fraction. We see that the denominator is 3 so the basic part is $\frac{1}{3}$ of a whole.

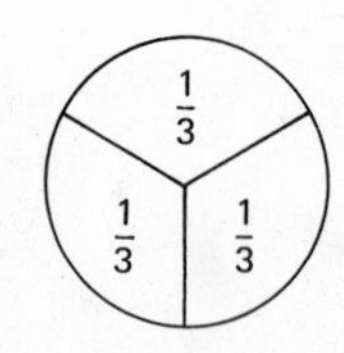

Each part of this circle is $\frac{1}{3}$ of the whole, and the fraction $\frac{10}{3}$ tells us that we are considering 10 of these parts, each of which is $\frac{1}{3}$ of a whole.

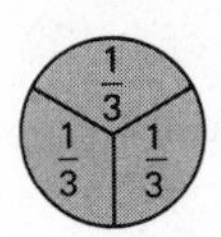

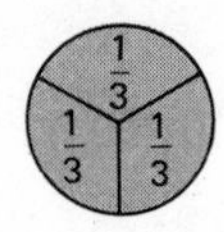

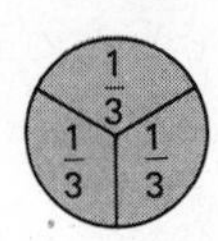

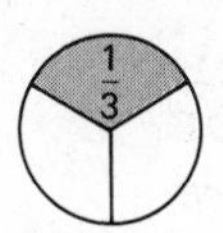

Each of the first three circles has three shaded parts equal to $\frac{1}{3}$ of a whole. The last circle has one shaded part for a total of 10 shaded parts. From this we can see that the number represented by $\frac{10}{3}$ is the same number as the number represented by $3\frac{1}{3}$.

$$\frac{10}{3} \quad \text{equals} \quad 3\frac{1}{3}$$

The number $3\frac{1}{3}$ has two parts. One part is the number 3, which is a whole number. The other part is the fraction $\frac{1}{3}$. We call a number that has **both a whole part and a fractional part a mixed number.** We note that $3\frac{1}{3}$ really means

$$\text{three plus one-third} \qquad \text{or} \qquad 3 + \frac{1}{3}$$

Because we use mixed numbers so often, we omit the plus sign to make the number easier to write. But we must remember that

$$3\frac{1}{3} \quad \text{means} \quad 3 + \frac{1}{3}$$

28.B improper fraction to mixed number

To change an improper fraction to a mixed number, we divide the denominator into the numerator. The whole part of the answer is the whole part of the mixed number. The remainder will be the numerator of the fraction of the mixed number.

To convert the improper fraction $\frac{10}{3}$ to a mixed number, we divide 10 by 3.

$$3\overline{)10} = 3, \text{R}1 \qquad \begin{array}{r} 3 \\ 3\,\overline{)\,10} \\ 9 \\ \hline 1 \end{array}$$

This tells us that $\frac{10}{3}$ represents 3 wholes, with a remainder of $\frac{1}{3}$. So

$$\frac{10}{3} = 3\frac{1}{3}$$

example 28.B.1 Convert the following improper fractions to mixed numbers or to whole numbers.

(a) $\frac{14}{5}$ (b) $\frac{21}{8}$

solution In each case we divide the denominator into the numerator to find the number of wholes represented by the fraction. The remainder will be the numerator of the fraction of the mixed number.

(a) $\frac{14}{5} \longrightarrow \begin{array}{r} 2 \\ 5\,\overline{)\,14} \\ 10 \\ \hline 4 \end{array}$ so $\frac{14}{5} = 2, \text{R}4$ Thus $\frac{14}{5}$ can be written as $\mathbf{2\frac{4}{5}}$.

(b) $\frac{21}{8} \longrightarrow \begin{array}{r} 2 \\ 8\,\overline{)\,21} \\ 16 \\ \hline 5 \end{array}$ so $\frac{21}{8} = 2, \text{R}5$ Thus $\frac{21}{8}$ can be written as $\mathbf{2\frac{5}{8}}$.

28.C mixed number to improper fraction

In the preceding paragraph we found that $\frac{10}{3}$ and $3\frac{1}{3}$ were equal numbers,

$$\frac{10}{3} = 3\frac{1}{3}$$

and we found that the mixed number that is equal to $\frac{10}{3}$ can be found by dividing 10 by 3.

$$3\overline{)10} = 3 \text{ R } 1 \qquad \text{so } \frac{10}{3} \text{ equals } 3\frac{1}{3}$$

To go the other way and find the improper fraction represented by the mixed number $3\frac{1}{3}$, we will use our picture of $3\frac{1}{3}$ again.

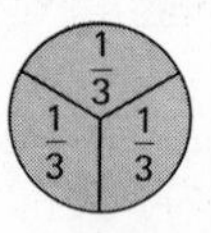
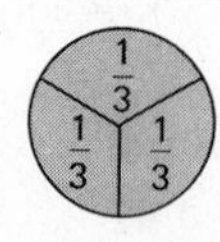
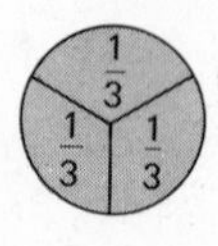
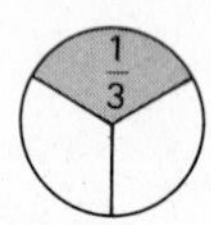

We see here that the number $3\frac{1}{3}$ is composed of 10 parts each with a size of $\frac{1}{3}$. Nine of these one-thirds come from the whole number 3 and one of them comes from the fraction $\frac{1}{3}$. We can say, therefore, that the mixed number $3\frac{1}{3}$ is the same as the improper fraction $\frac{10}{3}$.

For another example, we will write $2\frac{1}{8}$ as an improper fraction. To do this we must find out how many $\frac{1}{8}$'s there are in 2 and then add one more $\frac{1}{8}$.

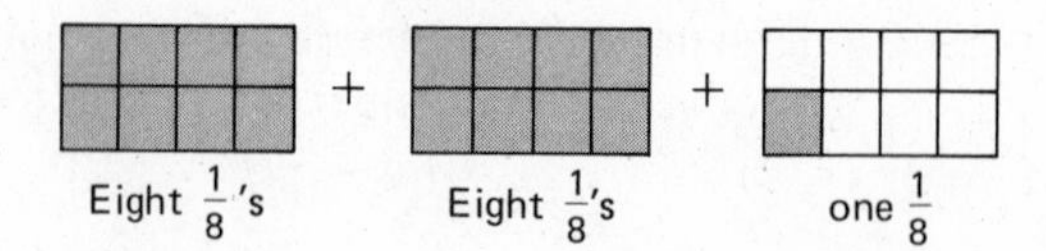

Eight $\frac{1}{8}$'s plus eight $\frac{1}{8}$'s plus one $\frac{1}{8}$ equals seventeen $\frac{1}{8}$'s or $\frac{17}{8}$

Thus, the procedure for finding the improper fraction represented by a mixed number is to (a) find the number of fractional parts in the whole number and then (b) add to this the number of fractional parts designated by the fraction. We can do this in two steps by

(a) Multiplying the denominator of the fraction by the whole number, and then
(b) Adding to this product the numerator of the fraction and recording the sum over the denominator of the fraction.

Thus to write $5\frac{3}{11}$ as an improper fraction, we write

$$\frac{5 \times 11 + 3}{11} = \frac{58}{11}$$

Some people find it helpful to remember the procedure by thinking of it as a circular process.

$$5\frac{3}{11} = 5 \begin{matrix} + \\ \times \end{matrix} \frac{3}{11} = \frac{11 \times 5 + 3}{11} = \frac{55 + 3}{11} = \frac{58}{11}$$

We begin at the bottom with 11, multiply by 5, then add 3, and record over 11.

example 28.C.1 Write $7\frac{4}{9}$ as an improper fraction.

solution We will think of it as a circular process.

$$7\frac{4}{9} = \frac{9 \times 7 + 4}{9} = \frac{63 + 4}{9} = \frac{\mathbf{67}}{\mathbf{9}}$$

We multiply 9 by 7 to get 63 and then add 4 to get the numerator of 67; then we record this over the denominator of 9.

problem set 28

1. Four hundred seventeen ten-thousandths is how much greater than four hundred seventeen hundred-thousandths?

2. The first flock contained 5283 birds. The second flock contained 5 times as many birds. The third flock had twice as many birds as did the second flock. How many birds were there in all?

3. Seven times 4,820,718 is a very large number. Write this number in words.

4. Find the average of the prime numbers that are greater than 10 and less than 25.

Convert these fractions to mixed numbers.

*5. $\frac{14}{5}$ *6. $\frac{21}{8}$

*7. Draw a diagram that shows why $3\frac{1}{3}$ equals $\frac{10}{3}$.

Convert the following mixed numbers to improper fractions.

*8. $5\frac{3}{11}$ *9. $7\frac{4}{9}$ 10. $5\frac{2}{8}$

Use unit multipliers to convert:

11. 192.72 centimeters to meters 12. 17 miles to feet

Simplify.

13. $\frac{28}{36} \cdot \frac{24}{21} \cdot \frac{3}{16}$ 14. $\frac{18}{24} \cdot \frac{36}{28} \cdot \frac{14}{27}$ 15. $\frac{16}{18} \cdot \frac{16}{12} \div \frac{8}{9}$

16. Find the area of the following figure. (Total area = area of rectangle + area of triangle). Dimensions are in feet.

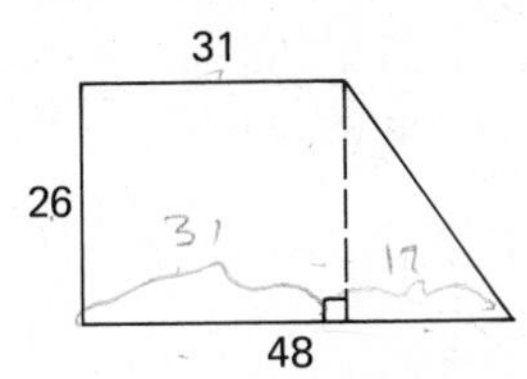

17. Express the perimeter in centimeters.

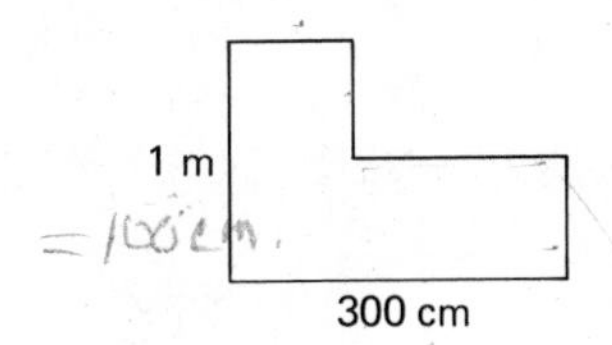

Write both numbers as products of prime factors and simplify.

18. $\frac{36}{42}$ 19. $\frac{280}{320}$ 20. $\frac{125}{175}$

Write the following fractions as decimals. Round off to two decimal places.

21. $\frac{16}{23}$ 22. $\frac{7}{19}$

23. Write 1440 as a product of prime factors.

24. List the multiples of 3 that are greater than 15 and are less than 30.

25. Multiply: 61.72×1.89

26. Subtract. Add to check: $172.325 - 61.89$

27. Divide and round off to two places: $\dfrac{7811.3}{.03}$

28. Round off 6,789,211.82 to the nearest million.

29. Use words to write 78256113.7.

30. Simplify mentally: $62{,}562.13 \div 100$

LESSON 29 Multiplying fractions and whole numbers · Fractional part of a number

29.A multiplying fractions and whole numbers

When we multiply two fractions, we multiply the tops to get the new top. We multiply the bottoms to get the new bottom.

$$\frac{1}{4} \cdot \frac{3}{5} = \frac{1 \cdot 3}{4 \cdot 5} = \frac{3}{20}$$

We remember that every whole number can be written as a fraction by writing a 1 below the number. Thus both of these

$$\frac{4}{1} \qquad 4$$

are ways to write the number 4. To multiply 4 by $\frac{2}{9}$, we can either write the 1 under the 4 as we do on the left or omit the 1 as we do on the right.

$$\frac{2}{9} \cdot \frac{4}{1} = \frac{8}{9} \qquad \text{or} \qquad \frac{2}{9} \cdot 4 = \frac{8}{9}$$

29.B top, bottom, and middle

A fraction has a top. A fraction has a bottom. A fraction does not have a middle. Some people are confused by the notation used to indicate the multiplication of a fraction and a whole number such as

$$4 \cdot \frac{3}{5}$$

because they cannot remember whether to multiply the 4 by 3 or by 5. The way it is written, it appears that the 4 is somehow in the middle. Of course,

$$4 \cdot \frac{3}{5} \qquad \text{means} \qquad \frac{4}{1} \cdot \frac{3}{5}$$

For this reason, students often find it helpful to write the 1 below the whole number as we have done here.

29.C fractional part of a number

When we multiply a whole number by a fraction, the answer is a part of the whole number. In (a) we show the number 15 and see that if we divide it into 5 equal parts (b) each part is 3. Each of these parts is $\frac{1}{5}$ of 15.

15

(a)

3	3	3	3	3

(b)

3	3	3	3	3

(c)

3	3	3	3	3

(d)

$$\frac{1}{5} \times 15 = 3 \qquad \frac{2}{5} \times 15 = 6 \qquad \frac{4}{5} \times 15 = 12$$

In (c) we show $\frac{2}{5} \times 15$ means two of the parts, and the answer is 6. In (d) we show that $\frac{4}{5} \times 15$ means four of the parts, and the answer is 12.

We use the word **of** to designate multiplication. Thus,

$$\frac{2}{5} \text{ of } 15 \quad \text{means} \quad \frac{2}{5} \times 15$$

$$\frac{4}{5} \text{ of } 15 \quad \text{means} \quad \frac{4}{5} \times 15$$

example 29.C.1 Find three-sevenths of 42.

solution The word **of** means to multiply. So we multiply 42 by $\frac{3}{7}$.

$$\frac{3}{7} \times 42 = \frac{126}{7} = \mathbf{18}$$

example 29.C.2 Find five-eighths of 40.

solution The word **of** means to multiply. So we multiply 40 by $\frac{5}{8}$.

$$\frac{5}{8} \times 40 = \frac{200}{8} = \mathbf{25}$$

problem set 29

1. The first try flew nineteen thousand and seventy-five millionths feet. The second try flew twenty-one thousand and one thousand, three millionths feet. What was the sum of the two tries?
2. First came fourteen thousand nine hundred eighty-two. Then 10 times this number came. How many came in all?
3. What is the average of the prime numbers greater than 30 but less than 45?
4. Eight hundred forty units could be put in one truck. If 19,420 units had to be transported, how many trucks were required?
5. In the preceding problem, if only 18 trucks could be located, how many units had to be left on the loading dock?

*6. Find three-sevenths of 42.

*7. Find five-eighths of 40.

Convert these fractions to mixed numbers:

8. $\frac{15}{7}$

9. $\frac{21}{5}$

10. Draw a diagram that shows why $2\frac{1}{4}$ equals $\frac{9}{4}$.

Convert the following mixed numbers to improper fractions:

11. $7\frac{3}{8}$ **12.** $6\frac{2}{3}$ **13.** $5\frac{7}{11}$

Use unit multipliers to convert:

14. 199.62 meters to centimeters **15.** 18 miles to feet

Simplify. Cancel, if possible, as the first step.

16. $\frac{42}{48} \cdot \frac{32}{14} \cdot \frac{3}{4}$ **17.** $\frac{18}{20} \cdot \frac{16}{8} \div \frac{27}{2}$

18. Find the area of this figure in square meters.

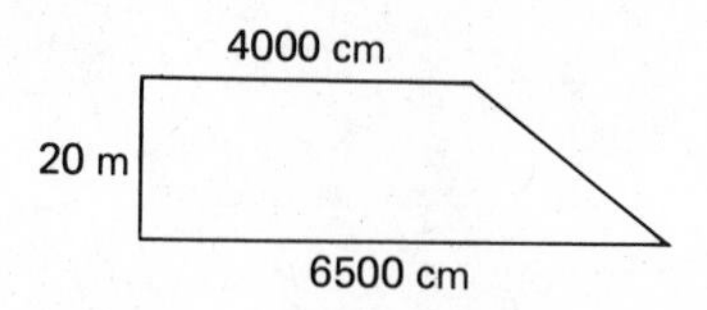

Write both numbers as products of prime factors and simplify.

19. $\frac{180}{270}$ **20.** $\frac{360}{420}$

Write the following fractions as decimals. Round off to two decimal places.

21. $\frac{16}{23}$ **22.** $\frac{8}{13}$

Write the following numbers as the product of primes:

23. 2100 **24.** 5250

Multiply:

25. $31.25 \times .0012$ **26.** $17.01 \times .12$

27. Subtract. Add to check: $61.892 - 9.299$

28. Divide and round off to two places: $\frac{6218.3}{.007}$

29. Round off 2831.8211261 to the hundred-thousandths' place.

30. Use words to write 13.82567.

LESSON 30 Graphs

30.A graphs

Graphs are used to present numerical information in picture form. The two most common forms of graphs are bar graphs and broken-line graphs. The two graphs shown here present the same information.

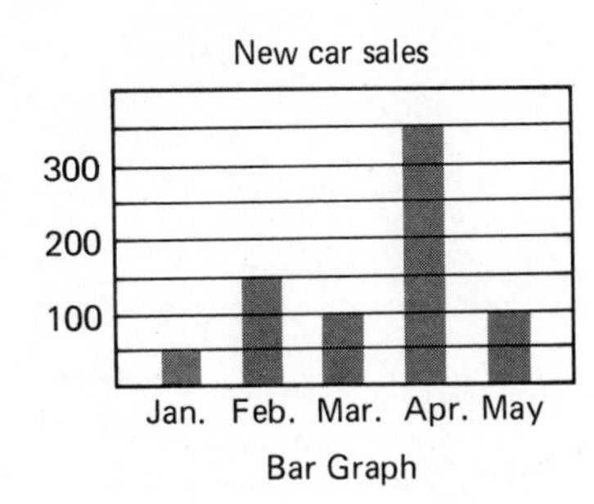

Bar Graph

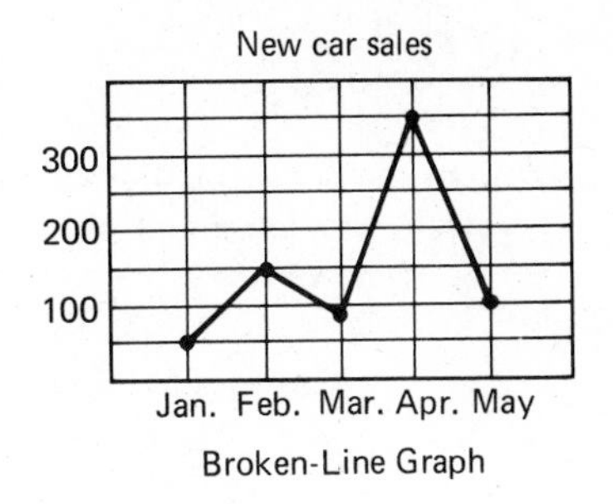

Broken-Line Graph

Both graphs show sales of new cars each month from January through May. Some people believe that the bar graph is easier to interpret while others prefer broken-line graphs. Note that exact information cannot be obtained from either kind of graph.

When we draw graphs, the quantities plotted vertically and horizontally should be clearly labeled. Also, care should be taken when choosing the scale. Whenever possible, the major divisions on the scales should be divisible by 2 or by 10. Scales in which the major divisions are numbers such as 3 or 7 are often difficult to read.

example 30.A.1 In the graphs above, how many more cars were sold in April than were sold in January?

solution We cannot read the graphs exactly, but it appears that 350 cars were sold in April while only 50 were sold in January.

$$350 - 50 = \mathbf{300} \text{ more sold in April}$$

example 30.A.2 The highest temperatures recorded during the week were as follows: Monday, 86°; Tuesday, 80°; Wednesday, 75°; Thursday, 82°; Friday, 88°; Saturday, 74°; and Sunday, 84°. Make a bar graph and a broken-line graph that present this information.

solution It is not necessary to show the full vertical scale. We will use a vertical scale that goes between 70° and 100°.

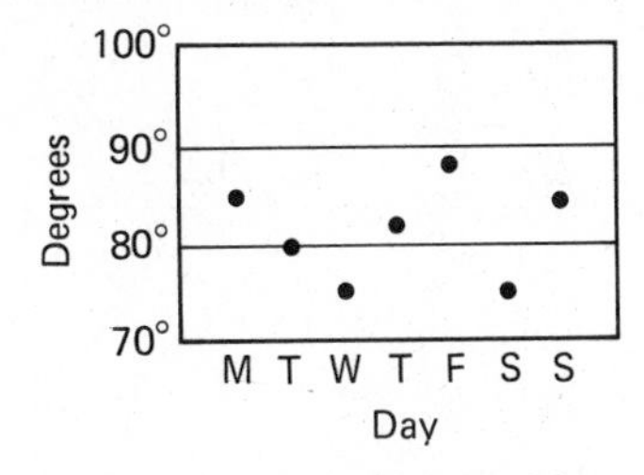

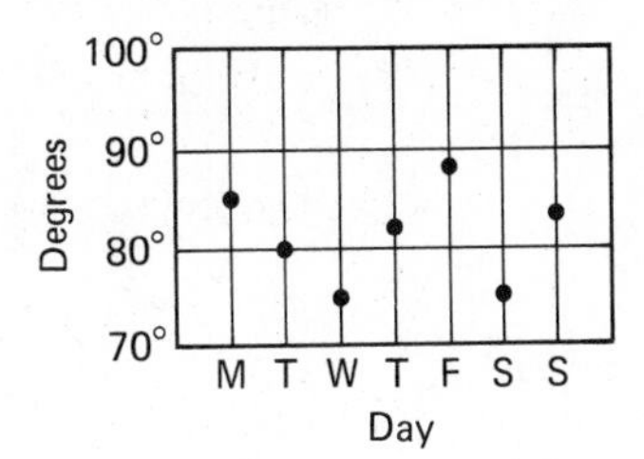

In the left figure we draw bars of proper heights to make a bar graph, and in the right figure, we connect the points with straight lines to make a broken-line graph.

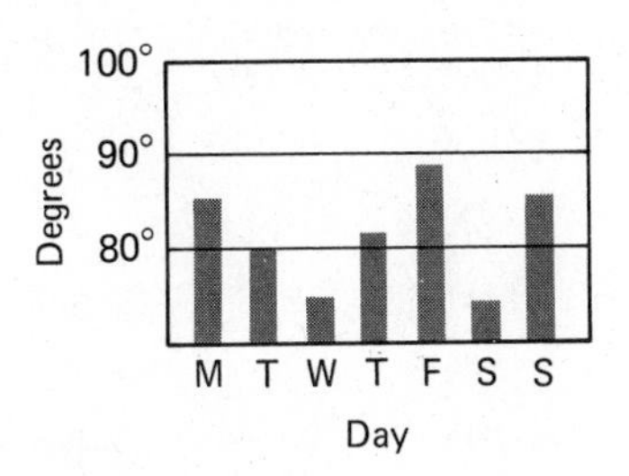

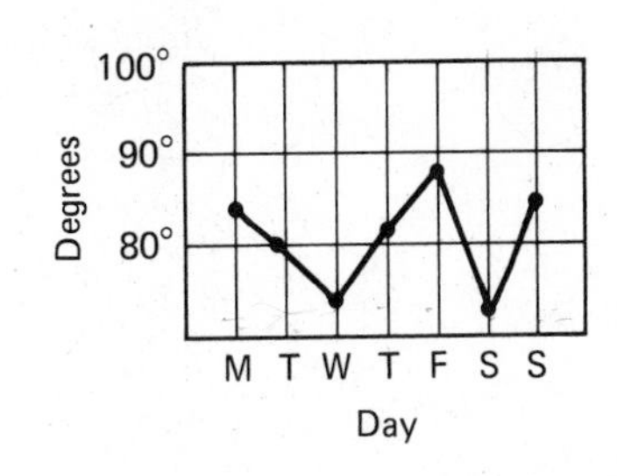

30.B graphs for problem sets

The five graphs shown here will be referred to in problems in future problem sets. They are included at this point so that they may be also used for classroom discussion with this lesson as is necessary.

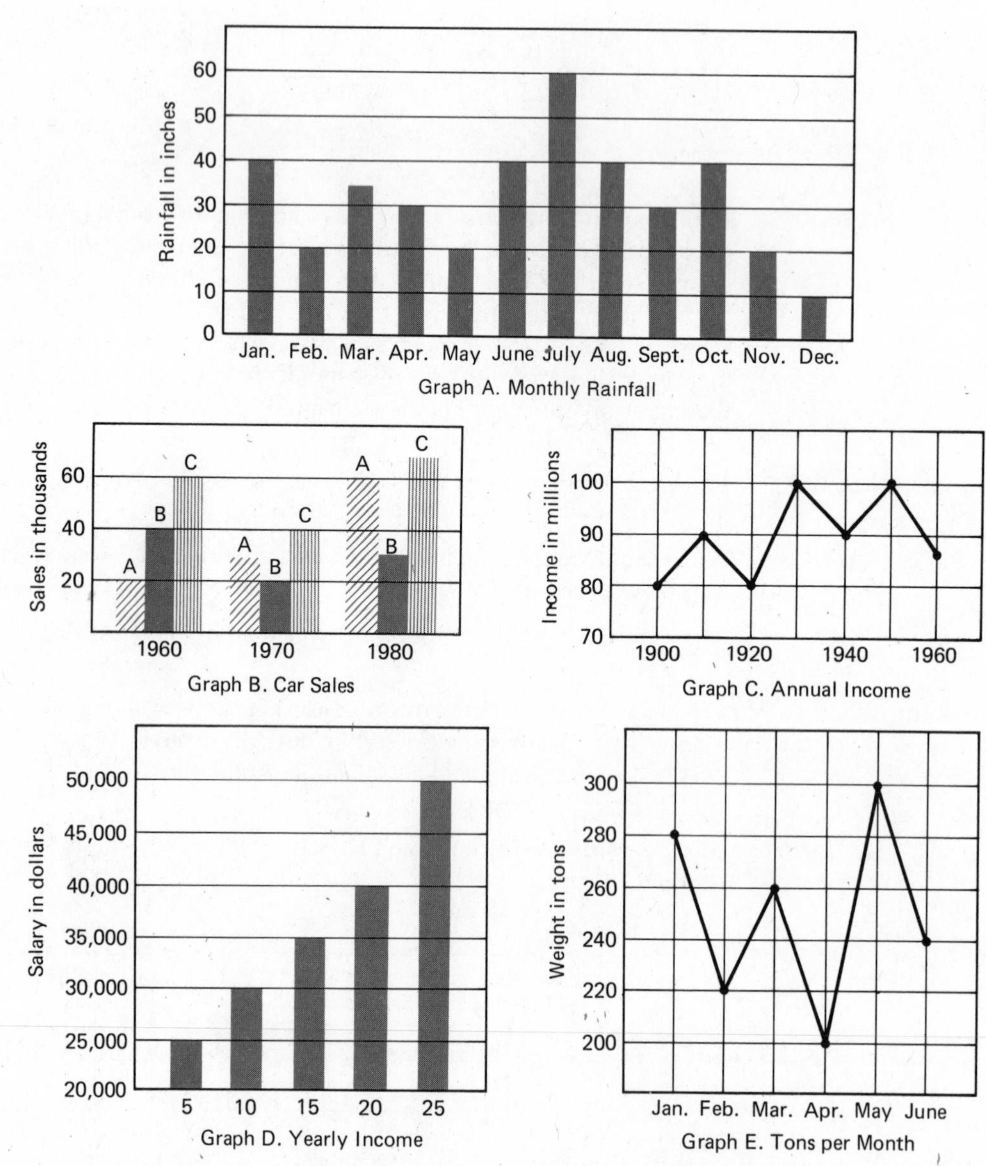

Graph A. Monthly Rainfall

Graph B. Car Sales

Graph C. Annual Income

Graph D. Yearly Income

Graph E. Tons per Month

problem set 30

*1. In the graphs at the beginning of this lesson, how many more cars were sold in April than were sold in January?

*2. The highest temperatures recorded during the week were as follows: Monday, 86°; Tuesday, 80°; Wednesday, 75°; Thursday, 82°; Friday, 88°; Saturday, 74°; and Sunday, 84°. Make a bar graph and a broken-line graph that present this information.

3. In Graph A, by how many inches did the rainfall in January exceed the rainfall in December?

4. When Hanibal opened the first box, he found that it contained 53 spear points. If 5062 spears needed new points, how many boxes did he have to open?

5. The first batch contained 5142 reds and 7821 greens. The second batch had twice as many reds and 3 times as many greens. How many total reds and greens were there in both batches?

6. Find three-eighths of 42.

7. Find six-sevenths of 49.

Convert these fractions to mixed numbers:

8. $\frac{17}{8}$ **9.** $\frac{22}{5}$

10. Draw a diagram that shows that $3\frac{1}{5}$ equals $\frac{16}{5}$.

Convert the following mixed numbers to improper fractions:

11. $7\frac{2}{3}$ **12.** $7\frac{6}{7}$ **13.** $4\frac{5}{8}$

Use unit multipliers to convert:

14. 721 yards to feet **15.** 19,262 centimeters to meters

Simplify. Cancel, if possible, as the first step.

16. $\frac{28}{32} \cdot \frac{24}{21} \cdot \frac{3}{4}$ **17.** $\frac{16}{18} \cdot \frac{20}{24} \div \frac{10}{9}$

18. Express the area of this figure in square centimeters.

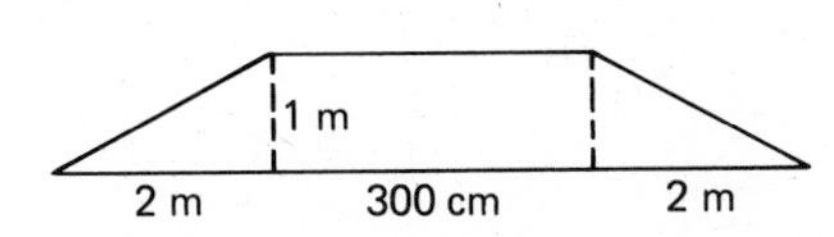

19. Express the perimeter of this figure in meters.

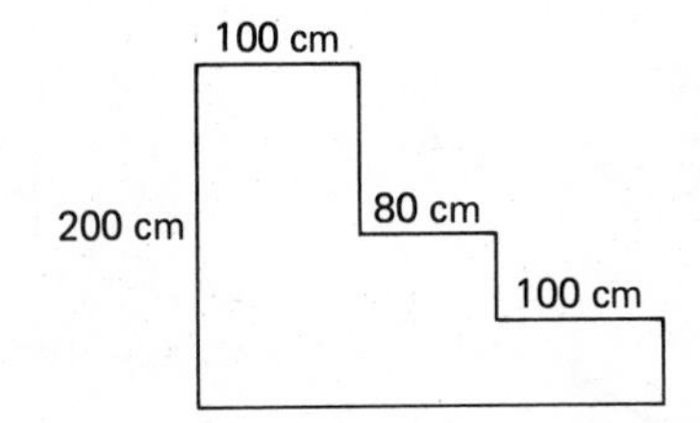

Write both numbers as products of prime factors and simplify.

20. $\frac{180}{200}$ **21.** $\frac{256}{720}$

Write the following fractions as decimals. Round off to two decimal places.

22. $\frac{16}{21}$ **23.** $\frac{9}{17}$

Write the following numbers as the product of primes:

24. 2450 **25.** 1650

Multiply:

26. $612.7 \times .0017$ **27.** $1821.1 \times .025$

28. Subtract. Add to check: $71.328 - 19.6214$

29. Divide and round off to two places: $\frac{689.2}{.005}$

30. Round off $.02\overline{735}$ to the nearest ten-millionth.

31. Use words to write 68211351.

LESSON 31 Least common multiple

31.A least common multiple

The smallest whole number into which several other whole numbers will divide evenly is called the **least common multiple** (LCM) of the several whole numbers. The least common multiple of the numbers

2, 3, and 5

is 30 because 30 is the smallest number into which the three numbers will divide and have a remainder of zero. We can find the LCM of some numbers by making mental calculations, but it is nice to have a special procedure to use if some of the numbers are large numbers. To demonstrate this procedure, we will find the LCM of

18, 81, and 500

The procedure is to

1. Write each number as a product of prime factors.

$$18 = 2 \cdot 3 \cdot 3 \qquad 81 = 3 \cdot 3 \cdot 3 \cdot 3 \qquad 500 = 2 \cdot 2 \cdot 5 \cdot 5 \cdot 5$$

2. Compute the LCM by using every factor of the given numbers as a factor of the LCM. Use each factor the greatest number of times it is a factor in any of the numbers.

The number 2 is a factor of both 18 and 500. It is a factor twice in 500 so it will appear twice in the LCM.

$$2 \cdot 2$$

The number 3 is a factor of both 18 and 81. It appears four times in 81 so it will appear four times in the LCM.

$$2 \cdot 2 \cdot 3 \cdot 3 \cdot 3 \cdot 3$$

The number 5 is the other factor. It appears three times in 500 so it will appear three times in the LCM.

$$\text{LCM} = 2 \cdot 2 \cdot 3 \cdot 3 \cdot 3 \cdot 3 \cdot 5 \cdot 5 \cdot 5 = \mathbf{40{,}500}$$

example 31.A.1 Find the LCM of 24, 55, and 80.

solution First we write the numbers as products of prime factors.

$$24 = 2 \cdot 2 \cdot 2 \cdot 3 \qquad 55 = 5 \cdot 11 \qquad 80 = 2 \cdot 2 \cdot 2 \cdot 2 \cdot 5$$

To compute the LCM, we use each factor the greatest number of times it is a factor of any of the numbers.

$$\text{LCM} = 3 \cdot 5 \cdot 11 \cdot 2 \cdot 2 \cdot 2 \cdot 2 = \mathbf{2640}$$

example 31.A.2 Here we show three numbers expressed as products of prime factors. (a) What are the numbers? (b) What is the LCM of the numbers?

$$(1)\ 2 \cdot 2 \cdot 3 \cdot 7 \cdot 7 \qquad (2)\ 2 \cdot 2 \cdot 2 \cdot 2 \cdot 5 \cdot 7 \qquad (3)\ 5 \cdot 5 \cdot 5 \cdot 5 \cdot 2$$

solution (a) We multiply to find the numbers.

$$2 \cdot 2 \cdot 3 \cdot 7 \cdot 7 = 588 \qquad 2 \cdot 2 \cdot 2 \cdot 2 \cdot 5 \cdot 7 = 560 \qquad 5 \cdot 5 \cdot 5 \cdot 5 \cdot 2 = 1250$$

(b) $\text{LCM} = 2 \cdot 2 \cdot 2 \cdot 2 \cdot 3 \cdot 5 \cdot 5 \cdot 5 \cdot 5 \cdot 7 \cdot 7 = \mathbf{1{,}470{,}000}$

Thus, we find that the smallest number that is evenly divisible by 588, 560, and 1250 is 1,470,000.

problem set 31

1. Mildred bought twenty show tickets for \$2.50 each. If she gave the agent a \$100 bill, how much change did she receive?

2. In Graph B (Lesson 30), how many more type C cars were sold in 1980 than were sold in 1960?

3. Try as he might Waltersheid found that he could do only 175 in an hour. If he had to do 2975 in all, how many hours would it take him?

4. The first batch of 40 cost 20 cents each. The second batch contained 60 and they cost 50 cents each. What was the average cost per item?

5. Each mustafa contained 14 raguts. If Ahab wanted to build 104 mustafas, how many raguts did he need?

6. What number is $\frac{5}{6}$ of 960?

7. What number is $\frac{11}{5}$ of 55?

Convert to mixed numbers:

8. $\frac{41}{3}$

9. $\frac{52}{16}$

10. Draw a diagram that shows why $4\frac{1}{3}$ equals $\frac{13}{3}$.

Find the least common multiple of:

*11. 18, 81, and 500

*12. 24, 55, and 80

13. 40, 85, and 100

14. Write $15\frac{2}{3}$ as an improper fraction.

15. Use unit multipliers to convert 14,780,000 centimeters to meters.

Simplify:

16. $\frac{15}{18} \cdot \frac{6}{30} \div \frac{3}{2}$

17. $\frac{160}{180} \cdot \frac{10}{12} \div \frac{20}{18}$

18. Find the area of this figure in square centimeters. Dimensions are in meters.

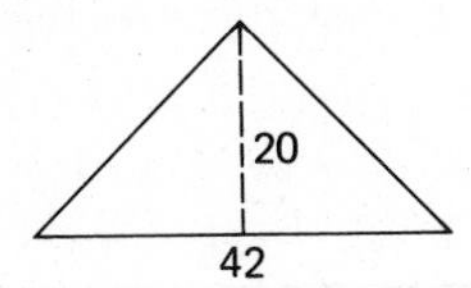

19. Find the perimeter of this figure in centimeters. Dimensions are in meters.

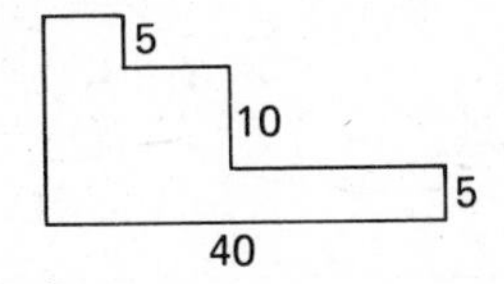

20. Write 460 as a product of prime factors.

21. Simplify $\frac{210}{240}$. Begin by writing both numbers as products of prime factors.

Write as decimal numerals. Round off to two decimal places.

22. $\frac{42}{17}$

23. $\frac{19}{23}$

Multiply:

24. $4.016 \times .0027$

25. $30.60 \times .0409$

26. 3.0162×4.008

27. Subtract. Add to check: $71.428 - 14.649$

28. Divide and round off to two places: $\frac{413.02}{.0061}$

29. Round off 415.62852 to the nearest thousandth.

30. Use words to write 42000.00162.

LESSON 32 *Adding fractions*

32.A adding fractions

The denominator (bottom) of a fraction tells us into how many parts the whole is divided. The numerator (top) of a fraction indicates how many of these parts we are considering. Since the fractions $\frac{1}{5}$ and $\frac{3}{5}$ both have 5 in the denominator, we know that each of the wholes has been divided into 5 parts. In one of the fractions we are considering 1 of the parts, and in the other we are considering 3 of the parts.

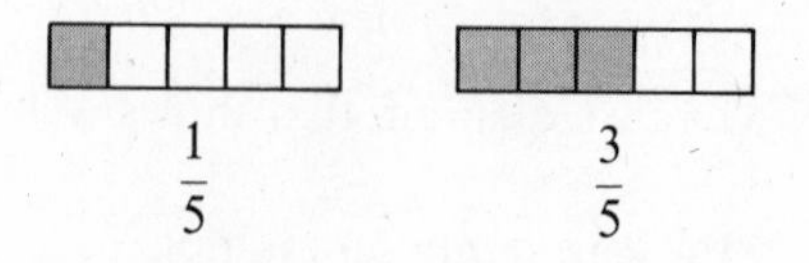

$$\frac{1}{5} \qquad \frac{3}{5}$$

If we add the fractions $\frac{1}{5}$ and $\frac{3}{5}$, we get $\frac{4}{5}$,

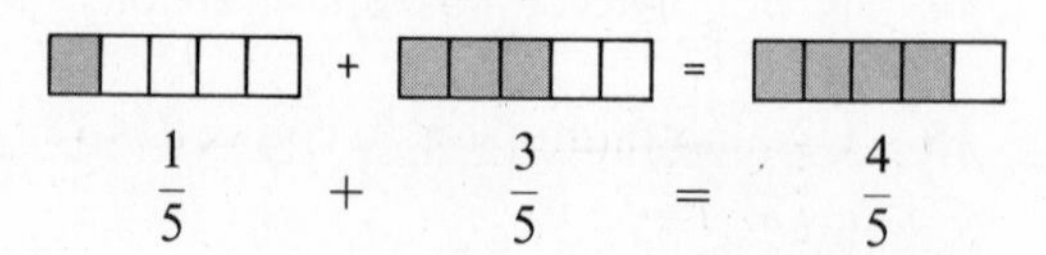

$$\frac{1}{5} + \frac{3}{5} = \frac{4}{5}$$

for we have one $\frac{1}{5}$ from the first and three $\frac{1}{5}$'s from the second for a total of four $\frac{1}{5}$'s. This gives us a picture of why we add the numerators when we add fractions whose denominators are equal.

> FRACTIONS WITH EQUAL DENOMINATORS
>
> To add or subtract fractions **whose denominators are equal,** the numerators are added or subtracted as indicated by the + and − signs, and the results are recorded over a single denominator.

32.B unequal denominators

When we wish to add fractions whose denominators are not equal

$$\frac{1}{4} + \frac{1}{2}$$

we must first transform one or both fractions as necessary so that the fractions will have equal denominators. If we multiply the top and bottom of the second fraction by 2

$$\frac{1}{4} + \frac{1(2)}{2(2)} = \frac{1}{4} + \frac{2}{4}$$

we can change it to $\frac{2}{4}$. Now both bottoms are equal, and we can add.

$$\frac{1}{4} + \frac{2}{4} = \frac{3}{4}$$

We will call the rule we used to do this the bottom-and-top rule.

> BOTTOM-AND-TOP RULE
>
> Both the bottom and top of a fraction can be multiplied by the same number (except zero) without changing the value of the fraction.

In later courses, this rule is called the fundamental theorem of fractions. We will call it the bottom-and-top rule because this name is descriptive and is easy to remember. We will use three steps to add fractions with unequal denominators.

1. Write the LCM as the new bottoms.
2. Determine the new tops.
3. Add the fractions.

example 32.B.1 Add $\frac{1}{4} + \frac{5}{9} + \frac{2}{3}$.

solution As the first step, we write the new denominators as the LCM of the old denominators.

$$\frac{}{36} + \frac{}{36} + \frac{}{36}$$

In the left fraction, we have multiplied the old bottom by 9 to get 36 so we also multiply the old top by 9.

$$\frac{9}{36} + \frac{}{36} + \frac{}{36}$$

In the center fraction, we have multiplied the old bottom by 4 to get 36 so we must also multiply the old top by 4.

$$\frac{9}{36} + \frac{20}{36} + \frac{}{36}$$

In the right fraction, we have multiplied the old bottom by 12 to get 36 so we must also multiply the old top by 12. Then we add.

$$\frac{9}{36} + \frac{20}{36} + \frac{24}{36} = \frac{53}{36} = \mathbf{1\frac{17}{36}}$$

example 32.B.2 Simplify: $\frac{5}{11} - \frac{1}{5}$

solution We use the same three-step procedure. As the last step we will subtract.

$$\text{(a)} \quad \frac{}{55} - \frac{}{55}$$

$$\text{(b)} \quad \frac{25}{55} - \frac{}{55}$$

$$\text{(c)} \quad \frac{25}{55} - \frac{11}{55} = \mathbf{\frac{14}{55}}$$

problem set 32

1. Franklin did 42 a minute for the first hour. In the next 30 minutes his rate dropped to 30 a minute and in the next 50 minutes he could only do 20 a minute. How many did he do in all?

2. The attendance at the first six home games was 10,400; 8000; 14,600; 7000; 12,000; 15,700. Make a broken-line graph that presents this information.

3. In the first cache Hazel uncovered 1481 dull ones. The next cache held 1300 shiny ones, and the last cache had 300 that were both shiny and dull. What was the average number per cache?

4. The time for the first increment was seven hundred thousand and fourteen hundred forty hundred-thousandths' seconds. The time for the second increment was only six hundred forty-two thousand, fourteen and seventy-five ten-thousandths' seconds. By how much was the time for the first increment greater?

5. Four thousand, fifty-seven screaming football fans could be seated in a single section of the stadium. If the stadium had 17 sections, how many screaming football fans would it hold?

6. What number is $\frac{4}{13}$ of 39?

7. What number is $\frac{11}{5}$ of 500?

Convert to mixed numbers:

8. $\frac{93}{13}$

9. $\frac{41}{7}$

10. Draw a diagram that shows why $3\frac{4}{5}$ equals $\frac{19}{5}$.

Find the least common multiple of:

11. 100, 40, 90

12. 200, 120, 180

13. 12, 20, 72

Simplify:

*14. $\frac{1}{4} + \frac{5}{9} + \frac{2}{3}$

*15. $\frac{5}{11} - \frac{1}{5}$

16. $\frac{3}{5} + \frac{4}{7} - \frac{1}{3}$

17. $\frac{4}{5} \cdot \frac{20}{40} \div \frac{5}{10}$

18. $\frac{2}{8} \cdot \frac{16}{24} \div \frac{15}{8}$

19. $.0023 \times 1.047$

20. $3.016 \div .042$

21. $483.124 - 1.632$

22. $\frac{4}{7} - \frac{1}{5}$

23. Find the area of this figure. Dimensions are in feet.

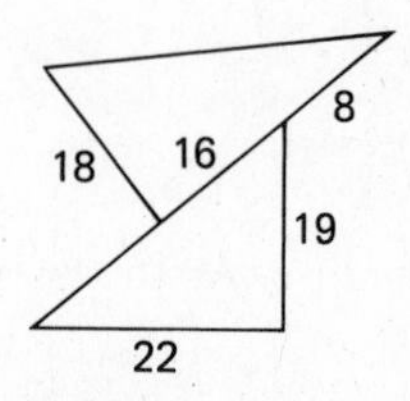

24. Find the perimeter of this figure. Dimensions are in centimeters.

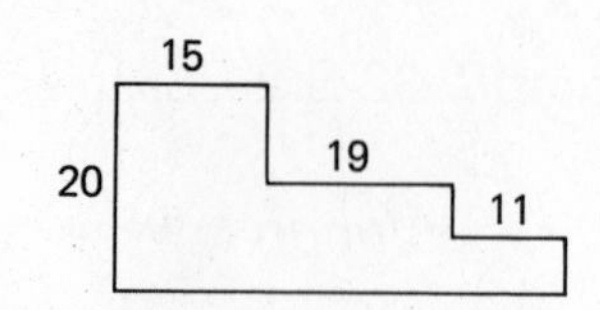

25. Simplify. Begin by writing both numbers as products of prime factors. $\frac{260}{305}$

26. (a) List the prime numbers between 50 and 60. (b) List the multiples of 4 between 50 and 60.

Write as decimals. Round off.

27. $\frac{52}{63}$

28. $\frac{14}{19}$

29. Round off 42,062,918,004.01372 to the nearest thousand.

30. Use words to write 42062918004.001372.

LESSON 33 Order of operations

33.A order of operations

There seem to be two ways that the following expression can be simplified.

$$4 + 3 \times 5$$

It looks as though we could add first and then multiply, or multiply first and then add.

(a) Add first:	(b) Multiply first:
$4 + 3 \times 5$	$4 + 3 \times 5$
7×5	$4 + 15$
35	**19**

Thus, there seem to be two possible answers to this problem. Since two answers to the same problem would cause considerable confusion, mathematicians have had to agree on one way or the other. They have agreed to always

Multiply before adding or subtracting

Thus, the answer to the problem above is 19 and is not 35.

example 33.A.1 Simplify $41 - 3 \cdot 5 + 4 \cdot 6$.

solution Two multiplications are indicated. We do these first and then move from left to right, performing the additions and subtractions in the order we encounter them.

$41 - 15 + 24$	multiplied
$26 + 24$	subtracted 15 from 41
50	added

example 33.A.2 Simplify: $4 + 5 \cdot 3 - 2 \cdot 4 + 6(3)$

solution First we do the multiplications and get

$$4 + 15 - 8 + 18$$

We finish by doing the additions and subtractions from left to right.

$19 - 8 + 18$	added 4 and 15
$11 + 18$	subtracted 8 from 19
29	added 11 and 18

problem set 33

1. For every 37 tickets she sold, Jezebel was given a free ticket for one of the children. If 175 children wanted to go, how many tickets did Jezebel have to sell?

2. In Graph E (Lesson 30), by how many pounds was the April production less than the January production?

3. Ralph had taken five tests. His scores were 85, 73, 92, 66, and 94. What was his average score for these tests?

4. Nineteen million, four hundred eighty-four thousand and seventy-five thousandths is a big number. Twenty-two million, thirteen and nine hundred eighty-four ten-thousandths is even bigger. What is the difference of these numbers?

5. Reds cost \$5 each so Don bought 7 reds. Blues were \$3.40 each, and he bought 9 of these. Greens were only \$1.30 each so he bought 20 of these. How much money did Don spend?

6. What number is $\frac{5}{16}$ of 128?

7. What number is $\frac{14}{3}$ of 30?

Write as mixed numbers:

8. $\frac{93}{12}$

9. $\frac{40}{7}$

Find the least common multiple of:

10. 24, 36, and 40

11. 8, 108, and 180

Simplify:

12. $\frac{3}{7} + \frac{2}{5} - \frac{3}{10}$

13. $\frac{5}{8} + \frac{3}{5} - \frac{1}{4}$

*14. $41 - 3 \cdot 5 + 4 \cdot 6$

*15. $4+5 \cdot 3-2 \cdot 4+6(3)$

16. $3 + 2 \cdot 5 - 3 \cdot 2$

17. $\frac{4}{5} \cdot \frac{25}{20} \div \frac{5}{10}$

18. $\frac{25}{36} \cdot \frac{12}{5} \div \frac{5}{6}$

19. $.016 \times .0023$

20. $\frac{400.7}{.0016}$

21. $\frac{7}{11} - \frac{1}{3}$

22. $41.003 - 22.404$

23. Find the area of this figure. Dimensions are in feet.

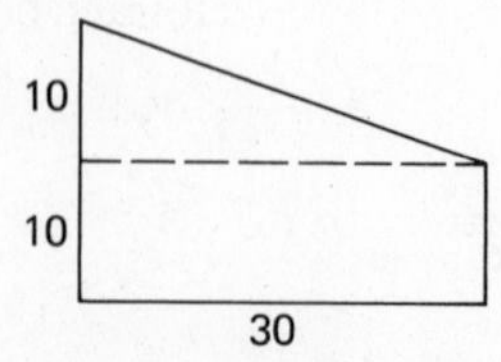

24. Find the perimeter of this figure. Dimensions are in yards.

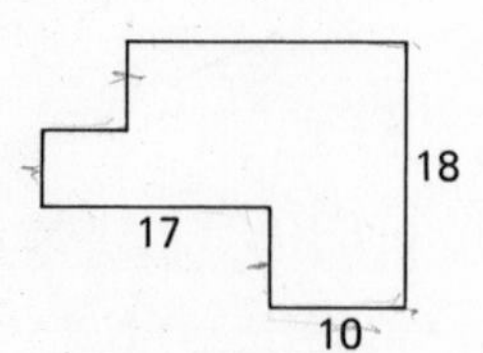

25. Write 4160 as a product of prime factors.

Write as decimals. Round off.

26. $\frac{13}{5}$

27. $\frac{7}{18}$

28. $\frac{17}{4}$

29. Round off 99,535,620 to the nearest ten thousand.

30. Use words to write .0021587.

LESSON 34 Variables and evaluation

34.A variables and evaluation

In algebra we often use letters as **variables** to stand for or to take the places of numbers. The letters themselves have no value. The value of the expression

$$x + 4$$

depends on the number we use for x. If we replace x with 11, the value of the expression is 15.

$$11 + 4 = 15$$

When two letters are written together such as

$$xy$$

the notation means that x and y are to be multiplied.

example 34.A.1 Evaluate $xy + x$ if $x = 2$ and $y = 4$.

solution The word **evaluate** means "find the value of." We replace x with 2 and y with 4.

$$xy + x = 2 \cdot 4 + 2$$

We remember to multiply before we add. Thus, we get

$$8 + 2 = \mathbf{10}$$

example 34.A.2 Evaluate $xmy - xy$ if $x = 2$, $m = 5$, and $y = 4$.

solution We replace x with 2, replace m with 5, and replace y with 4.

$$xmy - xy = 2 \cdot 5 \cdot 4 - 2 \cdot 4$$

We multiply before we subtract and get

$$40 - 8 = \mathbf{32}$$

example 34.A.3 Evaluate $mx + 4m$ if $x = \frac{2}{3}$ and $m = \frac{9}{11}$.

solution First we replace m and x with $\frac{9}{11}$ and $\frac{2}{3}$.

$$mx + 4m = \frac{9}{11} \cdot \frac{2}{3} + 4 \cdot \frac{9}{11}$$

Now we multiply and then add.

$$\frac{6}{11} + \frac{36}{11} = \frac{42}{11} = \mathbf{3\frac{9}{11}}$$

problem set 34

1. Silicon chips were packed 420 to the box. If 130,420 chips were to be shipped, how many boxes would be needed?

2. In the catalog, apple trees were \$15.95 each and Sam bought three. He bought seven nectarine trees at \$17.75 each and four peach trees for \$11.95 each. How much money did he spend for fruit trees?

3. The monthly car sales for the first 6 months of 1982 were 340, 410, 310, 440, 500, 550. Make a broken-line graph that presents this information.

4. The weight of the first four loads was 14,000 pounds, 12,000 pounds, 18,200 pounds, and 16,280 pounds. What was the average weight of these loads?

5. The repairman could fix 32 broken sets each hour. If 544 sets were broken, how long would it take him to repair them?

6. What number is $\frac{4}{9}$ of 72?

7. What number is $\frac{5}{17}$ of 136?

Write as mixed numbers:

8. $\frac{37}{3}$

9. $\frac{421}{5}$

Find the least common multiple of:

10. 12, 210, and 600

11. 60, 84, and 120

Simplify:

12. $\frac{3}{4} + \frac{5}{8} + \frac{2}{3} - \frac{1}{6}$

13. $\frac{5}{8} + \frac{1}{16} + \frac{1}{2} - \frac{1}{4}$

14. $3 + 2 \cdot 6 - 4(3)$

15. $5(2) - 3 \cdot 2 + 4(3)$

16. $\frac{16}{25} \times \frac{15}{8} \div \frac{3}{2}$

17. $\frac{30.02}{.0021}$

18. $\frac{4}{3} - \frac{7}{10}$

19. $1742.05 - 11.06$

20. $\frac{4}{5} \div \frac{2}{7}$

21. $.0016 \times 4000.085$

Evaluate:

*22. $xy + x$ if $x = 2$ and $y = 4$

*23. $xym - xy$ if $x = 2$, $m = 5$, and $y = 4$

24. $mx + 4m$ if $x = \frac{2}{3}$, $m = 5$

25. Find the area of this figure. Dimensions are in feet.

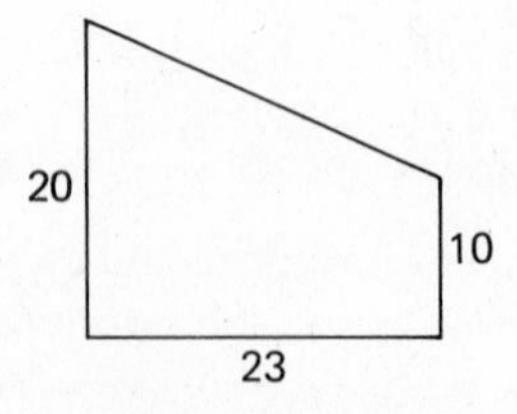

26. Find the perimeter of this figure. Dimensions are in meters.

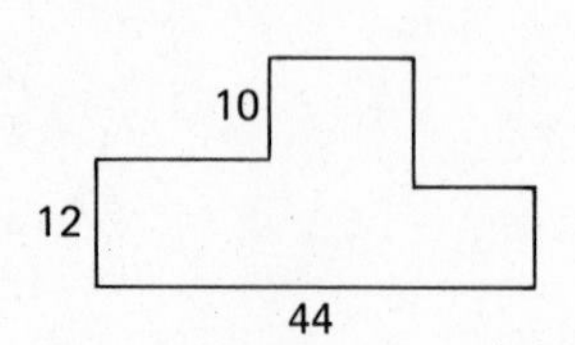

Write as decimals. Round off.

27. $\frac{3}{5}$

28. $\frac{1.3}{5}$

29. Round off 41,062,315 to the nearest ten thousand.

30. Use words to write 41002.00164.

LESSON 35 Multiple unit multipliers

35.A multiple unit multipliers

Unit multipliers are used to help prevent mistakes. We can use unit multipliers almost without thinking and be sure the answer we get is correct. We will often use an extra step in unit conversions if the extra step makes the process more automatic. To convert from inches to yards, we could remember that 1 yard equals 36 inches. This conversion would require only one unit multiplier but would require a little extra thought. A more automatic approach would be to go from inches to feet and from feet to yards.

example 35.A.1 Convert 360 inches to yards.

solution We will use one unit multiplier to convert from inches to feet. Then we use another to convert from feet to yards.

$$360\,\cancel{\text{in}} \cdot \frac{1\,\cancel{\text{ft}}}{12\,\cancel{\text{in}}} \cdot \frac{1\text{ yd}}{3\,\cancel{\text{ft}}} = \frac{360}{36}\text{ yd} = \mathbf{10\ yd}$$

example 35.A.2 Convert 1.4 kilometers to centimeters.

solution We will use one multiplier to convert to meters and another to convert to centimeters.

$$1.4\,\cancel{\text{km}} \cdot \frac{1000\,\cancel{\text{m}}}{1\,\cancel{\text{km}}} \cdot \frac{100\text{ cm}}{1\,\cancel{\text{m}}} = \mathbf{140{,}000\ cm}$$

We might have remembered that 1 kilometer equals 100,000 centimeters but attempting shortcuts like this often leads to errors.

example 35.A.3 Convert 24 miles to inches.

solution We will go from miles to feet to inches.

$$24\,\cancel{\text{mi}} \cdot \frac{5280\,\cancel{\text{ft}}}{1\,\cancel{\text{mi}}} \cdot \frac{12\text{ in}}{1\,\cancel{\text{ft}}} = 24(5280)(12)\text{ in} = \mathbf{1{,}520{,}640\ in}$$

problem set 35

1. The first measurement was four hundred seventeen ten-thousandths. The second measurement was forty-five thousandths. Which measurement was larger and by how much?

2. The two hundred tatterdemalions had an average of 45 cents each. The ragamuffins numbered 450, and they had an average of 55 cents each. What was the total value of the money of both groups?

3. In Graph D (Lesson 30), how much more did a 25-year employee make than a 10-year employee?

4. The legal limit for the number of cars in a parade was 450 cars. If 9000 drivers wanted to be in a parade, how many parades were required to accommodate all of them?

5. The fat ones weighed 417 lb, 832 lb, and 619 lb. The leans weighed only 148 lb, 212 lb, and 184 lb. What was the average weight of the six fats and leans?

6. What number is $\frac{4}{7}$ of 28?

7. What number is $\frac{5}{12}$ of 48?

Write as a mixed number:

8. $\dfrac{214}{5}$

9. $\dfrac{47}{2}$

10. Find the least common multiple of 50, 60, and 72.

Simplify:

11. $\dfrac{3}{4} + \dfrac{5}{8} + \dfrac{3}{16}$

12. $\dfrac{13}{15} - \dfrac{1}{5}$

13. $3 \cdot 12 - 4(2) + 3(5)$

14. $2 \cdot 5(2) - 3(5) + 2 - 5$

15. $4.014 \times .027$

16. $\dfrac{3.062}{.07}$

17. $513.002 - .0009$

18. $\dfrac{4}{6} \times \dfrac{9}{14} \div \dfrac{2}{5}$

19. $\dfrac{3}{7} \times \dfrac{21}{6} \div \dfrac{2}{4}$

Evaluate:

20. $xy - y$ if $x = 5$ and $y = 4$

21. $m - xy$ if $m = 10$, $x = 2$, and $y = 3$

22. $xym - m$ if $x = \dfrac{1}{2}$, $y = 4$ and $m = 2$

Use unit multipliers to convert:

***23.** 360 inches to yards

***24.** 1.4 kilometers to centimeters

***25.** 24 miles to inches

26. Express the area of this figure in square centimeters.

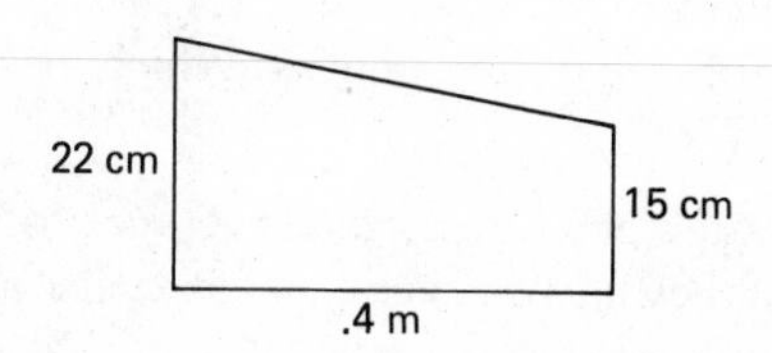

27. Express the perimeter of this figure in meters.

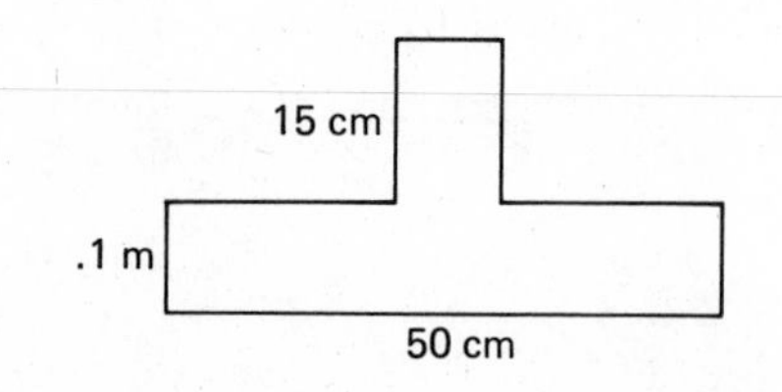

28. Write $\dfrac{7}{8}$ as a decimal number.

29. Round off $5.\overline{5}$ to the nearest hundred-thousandth.

30. Use words to write 71081902.003.

LESSON 36 *Adding mixed numbers · Two-step problems*

36.A adding mixed numbers

There are two ways to add mixed numbers. When the fractions are simple and the whole parts are small, it is convenient to begin by changing the mixed numbers to improper fractions. Then we find a common denominator and add the fractions.

example 36.A.1 Add: $2\frac{1}{4} + 3\frac{1}{8} + 7\frac{1}{2}$

solution First we rewrite each number as an improper fraction.

$$\frac{9}{4} + \frac{25}{8} + \frac{15}{2}$$

Next we change the fractions to equivalent fractions whose denominator is 8 and add.

$$\frac{18}{8} + \frac{25}{8} + \frac{60}{8} = \frac{103}{8} = \mathbf{12\frac{7}{8}}$$

36.B adding like parts

When the whole numbers are large numbers, it is convenient to add the whole numbers and the fractions separately.

example 36.B.1 Add: $528\frac{1}{3} + 7142\frac{3}{4}$

solution We rewrite the problem in a vertical format. Then we rewrite the fractions with a common denominator and add.

$$528\frac{1}{3} = 528\frac{4}{12}$$

$$7142\frac{3}{4} = 7142\frac{9}{12}$$

$$7670\frac{13}{12} = \mathbf{7671\frac{1}{12}} \quad \text{added and simplified}$$

example 36.B.2 Add: $421\frac{3}{5} + 274\frac{1}{20}$

solution Again we will use the vertical format.

$$421\frac{3}{5} = 421\frac{12}{20}$$

$$274\frac{1}{20} = 274\frac{1}{20}$$

$$\mathbf{695\frac{13}{20}} \quad \text{added}$$

36.C two-step problems

Many word problems are two-step problems. One part of the problem is worked, and then the answer for this part is used to work the second part of the problem.

example 36.C.1 Wilbur worked for 40 hours and was paid \$80. Jasper was paid twice as much per hour as Wilbur was paid, and Jasper worked 30 hours. How much was Jasper paid?

solution First we find Wilbur's pay. If he made \$80 and worked 40 hours, he was paid \$2 per hour. Jasper was paid twice as much per hour so he made \$4 an hour.

$$\frac{\$4}{\cancel{hr}} \times 30\ \cancel{hr} = \mathbf{\$120}$$

Jasper was paid \$120 for 30 hours at \$4 per hour.

problem set 36

1. The weights of the linebackers were 215, 305, 265, 214, 196, 221, and 236 lb, respectively. What was their average weight?

2. The value of the crops sold each year was 1940, \$700,000; 1945, \$800,000; 1950, \$1,400,000; 1955, \$2,000,000; 1955, \$3,000,000. Draw a bar graph that presents this information.

3. The girls bought 7 notebooks for \$5.40 each, 200 pencils at 30 cents each, and 40 reams of paper at \$22.50 a ream. How much did they spend in all?

4. If 460 quarts could be packaged in one shift and if 10,120 quarts were needed, how many shifts would be required?

*5. Wilbur worked for 40 hours and was paid \$80. Jasper was paid twice as much per hour as Wilbur, and he worked 30 hours. How much was Jasper paid?

6. What number is $\frac{3}{11}$ of 33?

7. What number is $\frac{4}{5}$ of 200?

Write as mixed numbers:

8. $\frac{21}{4}$

9. $\frac{86}{11}$

10. Find the least common multiple of 40, 50, and 70.

Simplify:

11. $\frac{3}{5} - \frac{1}{15}$

12. $\frac{1}{10} + \frac{3}{5} - \frac{1}{20}$

13. $3 + 5 - 2(4) + 3(5)$

14. $4 + 3(2) + 5 \cdot 4$

15. $2\frac{1}{4} + 3\frac{1}{8}$

*16. $528\frac{1}{3} + 7142\frac{3}{4}$

*17. $421\frac{3}{5} + 274\frac{1}{20}$

18. $51.67 \times .081$

19. $\frac{716.2}{.008}$

20. $713.891 - .8917$

21. $\frac{3}{8} \times \frac{24}{9} \div \frac{3}{7}$

Evaluate:

22. $zy - z$ if $z = 3$ and $y = 4$

23. $xyz + yz$ if $x = \frac{1}{3}, y = 9, z = 2$

Use unit multipliers to convert:

24. 540 inches to yards

25. 187,625.8 centimeters to kilometers

26. Find the area of this figure. Dimensions are in inches.

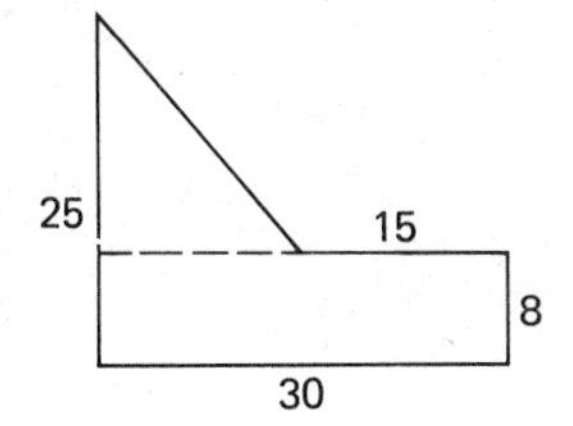

27. Find the perimeter of this figure. Dimensions are in centimeters.

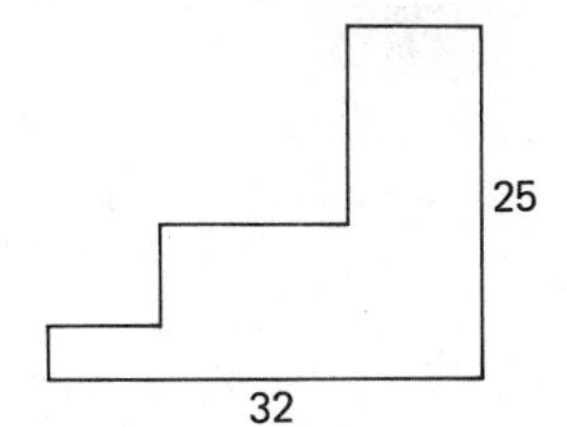

28. (a) List the prime numbers between 20 and 36. (b) List the multiples of 7 between 20 and 36.

29. Round off 658,976,581 to the nearest hundred thousand.

30. Use words to write 62181788.01.

LESSON 37 Mixed number subtraction

37.A mixed number subtraction

When the numbers are small and the fractions are simple, both numbers can be written as improper fractions and then subtracted.

example 37.A.1 Subtract: $4\frac{1}{4} - 1\frac{1}{8}$

solution We first write both numbers as improper fractions.

$$\frac{17}{4} - \frac{9}{8}$$

Next we rewrite both numbers with equal denominators and subtract.

$$\frac{34}{8} - \frac{9}{8} = \frac{25}{8} = \mathbf{3\frac{1}{8}}$$

37.B subtracting like parts

When the numbers are large, it is convenient to subtract the parts individually. A vertical format is often used.

example 37.B.1 Subtract: $416\frac{3}{4} - 21\frac{1}{16}$

solution We use the vertical format and rewrite the fractions with equal denominators. Then we subtract.

$$\begin{array}{rcr} 416\frac{3}{4} & = & 416\frac{12}{16} \\ 21\frac{1}{16} & = & -\ 21\frac{1}{16} \\ \hline & & \mathbf{395\frac{11}{16}} \end{array}$$

Sometimes it is necessary to borrow before we can subtract, as we see in the next example.

example 37.B.2 Subtract: $461\frac{1}{3} - 82\frac{13}{15}$

solution We use the vertical format and equal denominators.

$$\begin{array}{rcr} 461\frac{1}{3} & = & 461\frac{5}{15} \\ -\ 82\frac{13}{15} & = & -\ 82\frac{13}{15} \\ \hline \end{array}$$

We can't subtract $\frac{13}{15}$ from $\frac{5}{15}$. Thus, we borrow 1 unit from 461.

$$461\frac{5}{15} = 460 + \frac{15}{15} + \frac{5}{15} = 460\frac{20}{15}$$

Now we can subtract.

$$\begin{array}{r} 460\frac{20}{15} \\ -\ 82\frac{13}{15} \\ \hline \mathbf{378\frac{7}{15}} \end{array}$$

problem set 37

1. Winifred paid \$180 for the first 10 items. Then the price was increased \$6 per item. How much did she have to pay for the next 20 items?

2. What was the average price Winifred paid for all 30 items?

3. In Graph C (Lesson 30), by how much did the income in 1950 exceed the income in 1910?

4. If 415 came in each unit and if 29,050 were required, how many units had to be obtained?

5. What number is $\frac{5}{16}$ of 32?

6. What number is $\frac{4}{7}$ of 280?

Write as a mixed number:

7. $\frac{31}{5}$

8. $\frac{93}{7}$

9. Find the least common multiple of 8, 36, and 70.

Simplify:

10. $\frac{3}{5} - \frac{2}{10}$ **11.** $\frac{3}{8} + \frac{1}{2} - \frac{1}{4}$ **12.** $4 + 3 \cdot 2 - 5$

13. $3\frac{1}{8} + 2\frac{1}{4} + 5\frac{1}{2}$ **14.** $428\frac{1}{11} + 22\frac{1}{44}$ **15.** $3\frac{2}{5} + 748\frac{2}{10}$

***16.** $4\frac{1}{4} - 1\frac{1}{8}$ ***17.** $416\frac{3}{4} - 21\frac{1}{16}$ ***18.** $461\frac{1}{3} - 82\frac{13}{15}$

19. 71.82×8.01 **20.** $\frac{936.7}{.06}$ **21.** $691.872 - 17.816$

22. $\frac{21}{8} \times \frac{4}{14} \div \frac{9}{2}$

Evaluate:

23. $xy + yz - z$ if $x = 1, y = 7, z = 2$

24. $xyz - xy$ if $x = 2, y = 3, z = 3$

25. Use unit multipliers to convert 10 miles to inches.

26. Find the area of this figure Dimensions are in yards.

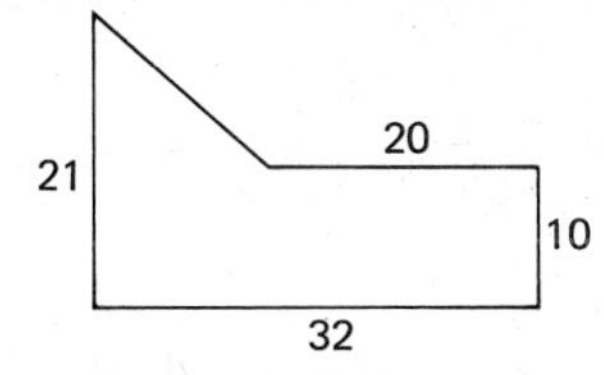

27. Find the perimeter of this figure. Dimensions are in inches.

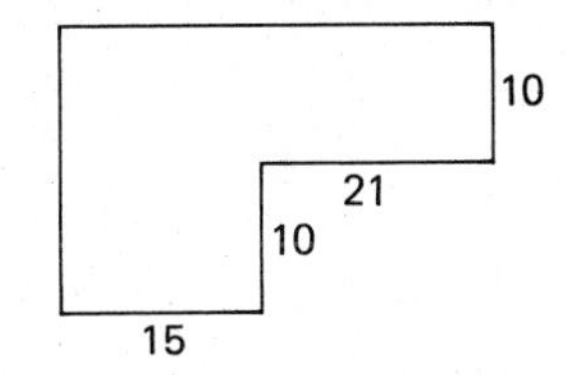

28. Write $\frac{11}{13}$ as a decimal number.

29. Round off $1.05\overline{2}$ to six decimal places.

30. Use words to write 71326811.23.

LESSON 38 Understanding fractional parts

38.A fractional parts

In Lesson 29, we discussed fractional parts of numbers. Since this concept is so important, we will review it here and review the use of diagrams to understand these problems. When we multiply a number by a fraction, we divide the number into equal parts and designate a number of these parts. When we multiply a number by $\frac{1}{6}$ or $\frac{2}{6}$ or $\frac{3}{6}$ or $\frac{4}{6}$ or $\frac{5}{6}$, we divide the number into 6 parts. Then we select 1 or 2 or 3 or 4 or 5 of these parts. Suppose the number is 48.

48

8	8	8
8	8	8

If we divide 48 into 6 equal parts, each part equals 8. Three of these parts equals 8 + 8 + 8. This is the reason that

$$\frac{3}{6} \times 48 = 24$$

48

8	8	8
8	8	8

example 38.A.1 Draw a diagram that depicts the multiplication of 22 by $\frac{3}{11}$.

solution When we multiply by $\frac{3}{11}$, we first divide the number into 11 parts.

22

2	2	2	2	2	2	2	2	2	2	2

Each of the parts equals 2. Three of the parts is 2 + 2 + 2 = 6. This shows why

$$\frac{3}{11} \times 22 = \mathbf{6}$$

example 38.A.2 Draw a diagram that depicts the multiplication of 17 by $\frac{4}{5}$.

solution Since 17 divided by 5 does not come out even, we will write each part as a fraction.

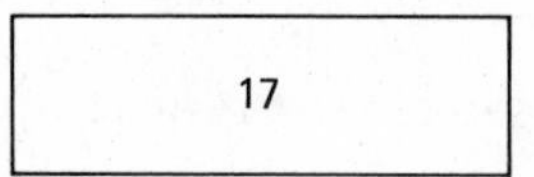

$\frac{17}{5}$	$\frac{17}{5}$	$\frac{17}{5}$	$\frac{17}{5}$	$\frac{17}{5}$

Each of the five parts is $\frac{17}{5}$. Four of the parts is

$$\frac{17}{5} + \frac{17}{5} + \frac{17}{5} + \frac{17}{5} = \frac{68}{5}$$

This shows why

$$\frac{4}{5} \times 17 = \mathbf{\frac{68}{5}}$$

problem set 38

1. Rubella was paid \$28 for working 7 hours. Tommie was a faster worker so she was paid \$3 an hour more than Rubella. How much was Tommie paid for a 40-hour week?

2. The first number was eight million, four hundred thirty-two thousand, four hundred sixteen, and one hundred thirty-seven ten-thousandths. The second number was three hundred seven thousand and two thousand two hundred-millionths. What was the sum of the numbers?

3. The three weights were 4016 tons, 7132 tons, and 9831 tons. What was the average of the three weights?

4. If 942 tries were permitted and John averaged 462 points on each try, how many points could John get?

5. What number is $\frac{5}{17}$ of 34?

6. What number is $\frac{1}{8}$ of 320?

Write as mixed numbers:

7. $\frac{428}{17}$ **8.** $\frac{521}{3}$

9. Find the least common multiple of 50, 47, and 120.

Simplify:

10. $\frac{15}{17} - \frac{2}{34}$ **11.** $\frac{3}{8} + 1\frac{1}{5} - \frac{1}{10}$ **12.** $4 + 13 - 5 \cdot 2 + 7$

13. $2 \cdot 8 - 3 \cdot 1 + 2 \cdot 4$ **14.** $3\frac{1}{5} + 2\frac{1}{8}$ **15.** $376\frac{4}{5} + 142\frac{3}{10}$

16. $42\frac{7}{8} - 15\frac{3}{4}$ **17.** $513\frac{11}{20} - 21\frac{4}{5}$ **18.** $813.2 \times .032$

19. $\frac{13.62}{.05}$ **20.** $1317.2 - 18.861$ **21.** $\frac{14}{16} \cdot \frac{24}{21} \div \frac{2}{3}$

***22.** Draw a diagram that depicts the multiplication of 22 by $\frac{3}{11}$.

23. Draw a diagram that depicts the multiplication of 20 by $\frac{4}{5}$.

Evaluate:

24. $xz - yz$ if $x = 7, y = 1, z = 3$

25. $xyz + yz - y$ if $x = 6, y = 3, z = \frac{1}{3}$

26. Use unit multipliers to convert 144 inches to yards.

27. Express the area of this figure in square centimeters.

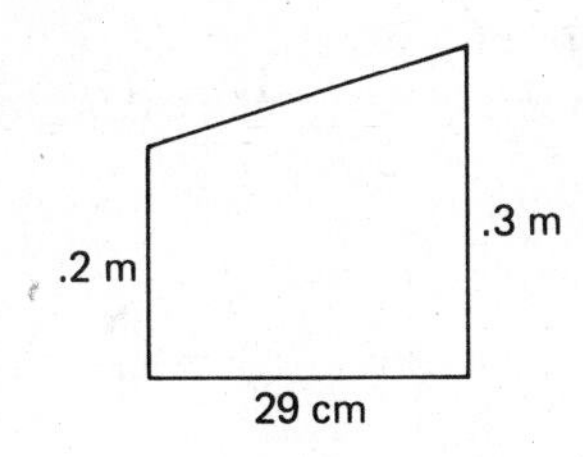

28. Find the perimeter of this figure. Dimensions are in yards.

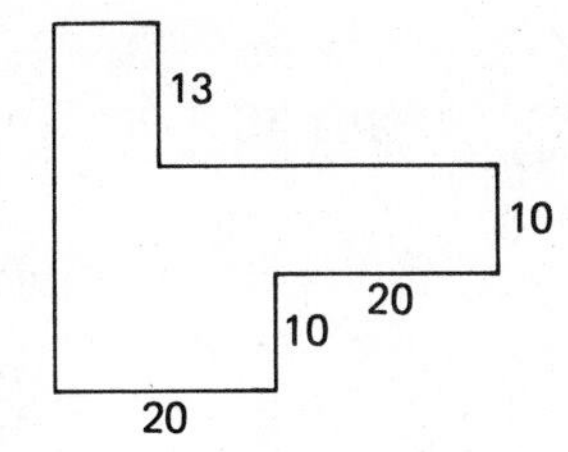

29. Write $\frac{17}{23}$ as a decimal number.

30. Round off 681,972,531 to the nearest hundred thousand.

LESSON 39 Equations; answers and solutions

39.A equations

If we connect two meaningful arrangements of numbers or numbers and letters with an equals sign, we have written an **equation.** Some equations are **true equations,** and some equations are **false equations.**

(a) $4 + 2 = 6$ true (b) $4 + 2 = 3$ false

Equation (a) is a true equation because 4 plus 2 is equal to 6. Equation (b) is a false equation because 4 plus 2 does not equal 3. Some equations that contain variables are called **conditional equations.**

$$\text{(c)}\quad x + 4 = 7 \qquad \text{conditional}$$

This equation is a conditional equation, and it is not a true equation nor a false equation. Its truth or falsity depends on the number we use to replace x. If we replace x with 5, the equation becomes a false equation.

$$5 + 4 = 7 \qquad \text{false}$$

But if we replace x with 3, the equation becomes a true equation.

$$3 + 4 = 7 \qquad \text{true}$$

If replacing x with a number turns a conditional equation into a true equation, we say that the number is the **solution** of the equation. We also say that the number **satisfies the equation.** Many times we can look at a conditional equation and tell by inspection what number will satisfy the equation.

example 39.A.1 Solve $x + 4 = 10$.

solution *Solve* means to find the number that will make this equation a true equation. Since $6 + 4 = 10$, the solution is **6.**

example 39.A.2 Solve $x - 4 = 10$.

solution The equation tells us that 4 subtracted from some number equals 10. Then the number must be 14 because

$$14 - 4 = 10$$

So the solution is **14.**

problem set 39

1. The squirrel could store an average of 48 nuts in a hiding place. If there were 768 nuts to hide, how many hiding places did the squirrel need?
2. The squirrel could store an average of 48 nuts in a hiding place. If there were 768 hiding places, how many nuts could the squirrel store?
3. The average weight of the three classifications of students was: ectomorphs, 110 lb; mesomorphs, 135 lb; endomorphs, 185 lb. Draw a bar graph that gives the same information.
4. The first five numbers drawn were 4165, 320, 7142, 804, and 64. What was the average of the numbers?
5. What number is $\frac{3}{5}$ of 40?
6. What number is $\frac{9}{11}$ of 99?

Write as mixed numbers:

7. $\frac{316}{13}$
8. $\frac{428}{5}$
9. Find the least common multiple of 8, 12, and 72.

Simplify:

10. $\frac{14}{15} - \frac{1}{5}$
11. $4\frac{2}{5} + 3\frac{5}{10}$
12. $3 + 2 + 2 \cdot 5 - 3(2)$

13. $3 - 2(1) + 4(7)$ **14.** $3\frac{2}{7} + 5\frac{20}{21}$ **15.** $420\frac{3}{5} + 262\frac{1}{40}$

16. $43\frac{5}{7} - 12\frac{2}{14}$ **17.** $210\frac{5}{8} - 17\frac{3}{40}$ **18.** $611.2 \times .0061$

19. $\frac{.8136}{.007}$ **20.** $2836.31 - 19.691$ **21.** $\frac{18}{20} \cdot \frac{24}{16} \div \frac{9}{10}$

22. Draw a diagram that depicts the multiplication of 16 by $\frac{5}{8}$.

Solve:

***23.** $x + 4 = 10$ ***24.** $x - 4 = 10$

Evaluate:

25. $x + xy + xyz$ if $x = 1, y = 2, z = 3$

26. $xy + x - y$ if $x = 5, y = 1$

27. Use unit multipliers to convert 625,611 centimeters to kilometers.

28. Find the area of this figure. Dimensions are in feet.

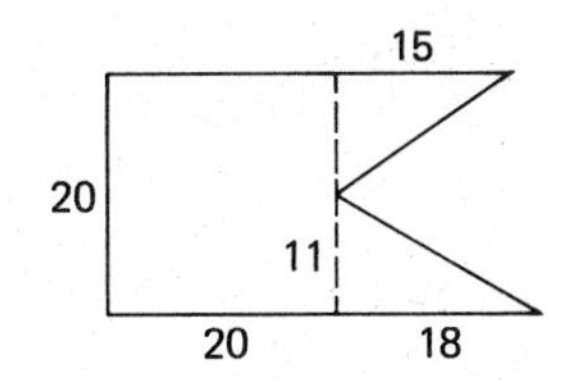

29. Write $\frac{20}{23}$ as a decimal number.

30. Round off 16,998,711 to the nearest thousand.

LESSON 40 Equivalent equations · Addition-subtraction rule for equations

40.A equivalent equations

The equations shown here are different equations.

(a) $x = 6$ (b) $x + 3 = 9$ (c) $x - 3 = 3$

(d) $3x = 18$ (e) $\frac{x}{3} = 2$

Yet the number 6 is a solution to all five equations, for if we replace x with 6, each of the equations becomes a true equation.

(a′) $6 = 6$ (b′) $6 + 3 = 9$ (c′) $6 - 3 = 3$

$6 = 6$ $9 = 9$ $3 = 3$

(d′) $3(6) = 18$ (e′) $\frac{6}{3} = 2$

$18 = 18$ $2 = 2$

Equation (a) was the easiest because it told us that the answer was 6. We didn't even have to think. We say that all five equations are **equivalent equations** because equivalent equations are equations that have the same solutions (answers).

Some equations are very easy to solve. We can mentally calculate the numbers that are solutions to the simple equations (b) through (e) shown here.

(b) $x + 3 = 9$ (c) $x - 3 = 3$ (d) $3x = 18$ (e) $\frac{x}{3} = 2$

However, many equations are not easy to solve mentally. It is helpful to have rules that can be used to solve difficult equations. The first rule is the addition-subtraction rule.

> ADDITION-SUBTRACTION RULE
>
> 1. The same number can be added to both sides of an equation without changing the solution to the equation.
> 2. The same number can be subtracted from both sides of an equation without changing the solution to the equation.

We will begin practicing the use of this rule by using it to solve rather simple equations.

example 40.A.1 Solve $x - 3 = 7$.

solution We can undo the subtraction of 3 by adding 3. So we add 3 to both sides of the equation.

$$x - 3 + 3 = 7 + 3 \quad \text{add 3 to both sides}$$

$$\mathbf{x = 10} \quad \text{simplified}$$

example 40.A.2 Solve $x + 3 = 7$.

solution This time we must undo the addition of 3 by subtracting 3. We subtract 3 from both sides of the equation.

$$x + 3 - 3 = 7 - 3 \quad \text{subtract 3 from both sides}$$

$$\mathbf{x = 4} \quad \text{simplified}$$

example 40.A.3 Solve $x + \frac{3}{4} = \frac{9}{11}$.

solution We can undo adding $\frac{3}{4}$ by subtracting $\frac{3}{4}$. We subtract $\frac{3}{4}$ from both sides of the equation.

$$x + \frac{3}{4} - \frac{3}{4} = \frac{9}{11} - \frac{3}{4} \quad \text{subtract } \frac{3}{4} \text{ from both sides}$$

$$x = \frac{9}{11} - \frac{3}{4} \quad \text{simplified}$$

$$x = \frac{36}{44} - \frac{33}{44} \quad \text{common denominator}$$

$$\mathbf{x = \frac{3}{44}} \quad \text{subtracted}$$

example 40.A.4 Solve $x - \frac{3}{5} = \frac{2}{10}$.

Solution We can undo subtracting $\frac{3}{5}$ by adding $\frac{3}{5}$. We add $\frac{3}{5}$ to both sides of the equation.

$$x - \frac{3}{5} + \frac{3}{5} = \frac{2}{10} + \frac{3}{5} \qquad \text{add } \frac{3}{5} \text{ to both sides}$$

$$x = \frac{2}{10} + \frac{3}{5} \qquad \text{simplified}$$

$$x = \frac{2}{10} + \frac{6}{10} \qquad \text{common denominator}$$

$$x = \frac{8}{10} = \mathbf{\frac{4}{5}} \qquad \text{simplified}$$

problem set 40

1. Sandra counted them as they slid around the corner. The first hour she counted 415 and the second hour she counted 478. If 526 were counted in the third hour, what was the average for the 3 hours?

2. Watonga picked up 420 in the first 6 minutes. Hobart could pick up 8 more a minute than could Watonga. How many could Hobart pick up in 6 minutes?

3. Only 420 could crawl into one space. If there were 12 spaces, how many could crawl in?

4. Nineteen thousand, four hundred forty came to see the spectacle. If 32 could be seated in a bus, how many buses did the police require to haul them all away?

5. What number is $\frac{3}{8}$ of 40?

6. What number is $\frac{4}{5}$ of 80?

Write as mixed numbers:

7. $\frac{41}{3}$

8. $\frac{93}{21}$

9. Find the least common multiple of 27, 28, and 30.

Simplify:

10. $\frac{5}{16} - \frac{1}{8}$

11. $2\frac{1}{5} + 3\frac{1}{3} - \frac{2}{10}$

12. $14 - 2 \cdot 3 + 4(5)$

13. $2(5) - 2 \cdot 2 + 3 \cdot 5 - 2$

14. $7\frac{1}{8} + 3\frac{2}{5}$

15. $674\frac{2}{5} - 13\frac{7}{10}$

16. $2\frac{1}{4} + 3\frac{1}{8} + 4\frac{5}{12}$

17. $461\frac{3}{4} - 65\frac{7}{8}$

18. 117.1×2.01

19. $\frac{171.6}{.006}$

20. $6132.81 - 621.981$

21. $\frac{6}{21} \cdot \frac{24}{3} \div \frac{8}{14}$

22. Use unit multipliers to convert 1 mile to inches.

Solve:

*23. $x - 3 = 7$

*24. $x + \frac{3}{4} = \frac{9}{11}$

*25. $x - \frac{3}{5} = \frac{2}{10}$

26. Draw a diagram that depicts the multiplication of 14 by $\frac{3}{7}$.

27. Evaluate: $xyz + xy + yz - z$ if $x = \frac{1}{3}$, $y = 12$, $z = 2$

28. Find the area of this figure. Dimensions are in inches.

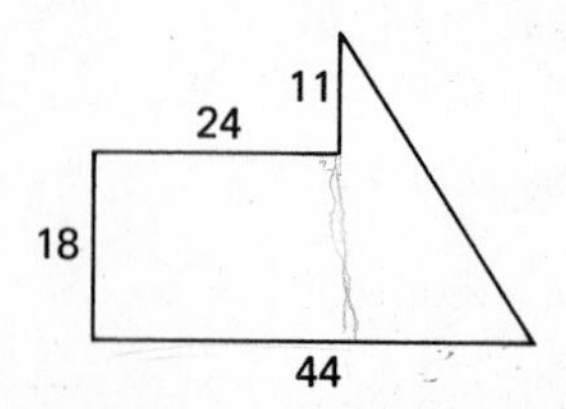

29. (a) List the prime numbers between 37 and 46. (b) List the multiples of 6 between 37 and 46.

30. Round off 8,265,891,131 to the nearest ten thousand.

LESSON 41 Multiplication-division rule

41.A multiplication-division rule

The multiplication-division rule tells us that we can multiply or divide both sides of an equation by any number (except zero) without changing the answer (solution) to the equation.

> MULTIPLICATION-DIVISION RULE
>
> 1. Both sides of an equation can be multiplied by the same number (except zero) without changing the solution to the equation.
> 2. Both sides of an equation can be divided by the same number (except zero) without changing the solution to the equation.

To use either form of this rule, remember that multiplication and division are inverse operations because multiplying by a number undoes dividing by the same number. Also dividing by a number undoes multiplying by the same number.

example 41.A.1 Solve $3x = 12$.

solution In the equation, the letter x is multiplied by 3. We can undo multiplication by 3 by dividing by 3. We divide both sides by 3.

$$\frac{3x}{3} = \frac{12}{3} \quad \text{divide both sides by 3}$$

$$\mathbf{x = 4} \quad \text{simplified}$$

example 41.A.2 Solve $\frac{x}{3} = 12$.

solution In this equation, the letter x is divided by 3. We can undo division by 3 by multiplying by 3. We multiply both sides by 3.

$$3 \cdot \frac{x}{3} = 3 \cdot 12 \quad \text{multiply both sides by 3}$$

$$\mathbf{x = 36} \quad \text{simplified}$$

problem set 41

1. The first shift lasted 4 hours and produced 800 units per hour. The second shift was a 6-hour shift that produced 600 units per hour. The third shift was a 4-hour shift whose production was 400 units per hour. What was the average number of units per hour for the entire time?

2. Ronald paid $600 and got 30 hanging plants. How much would he have had to pay for 70 hanging plants?

3. If 47 could be crammed into each compartment and if 2820 were waiting patiently in line, how many compartments would it take for all of them?

4. In Graph A (Lesson 30), what was the average rainfall for the first 5 months of the year?

5. What number is $\frac{3}{7}$ of 21?

6. What number is $\frac{6}{13}$ of 39?

Write as a mixed number:

7. $\frac{82}{5}$

8. $\frac{121}{15}$

9. Find the least common multiple of 35, 40, and 120.

Solve:

10. $x + \frac{3}{4} = \frac{7}{8}$

11. $x - \frac{1}{2} = \frac{5}{6}$

*12. $3x = 12$

*13. $\frac{x}{3} = 12$

14. $5x = 20$

15. $\frac{x}{7} = 5$

Simplify:

16. $\frac{7}{15} - \frac{1}{5}$

17. $3 \cdot 8 - 2 \cdot 6 + 1 \cdot 7$

18. $36\frac{3}{4} - 21\frac{7}{8}$

19. $\frac{171.6}{.6}$

20. $112.4 \times .071$

21. $6781.8 - 179.89$

22. $\frac{16}{18} \cdot \frac{24}{36} \div \frac{8}{9}$

23. Draw a diagram that depicts the multiplication of 24 by $\frac{5}{6}$.

Evaluate:

24. $x + zx - y$ if $x = 10, y = 3, z = 2$

25. $xy + xz + yz$ if $x = 2, y = 4, z = 6$

26. Use unit multipliers to convert 5280 feet to yards.

27. Find the area of this figure. Dimensions are meters.

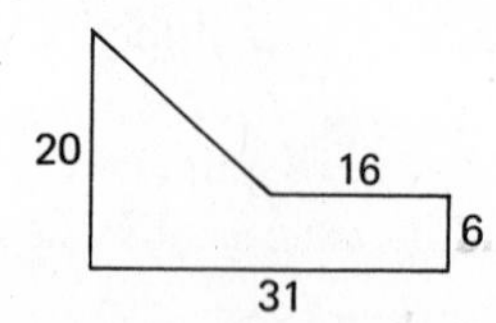

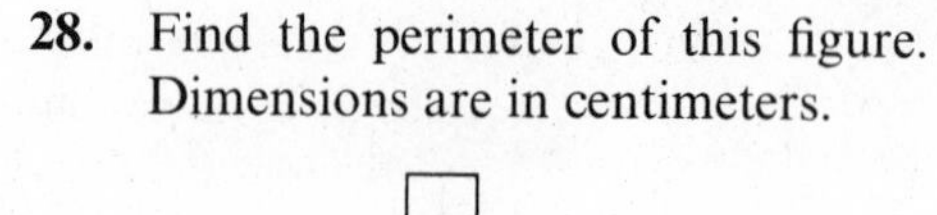

28. Find the perimeter of this figure. Dimensions are in centimeters.

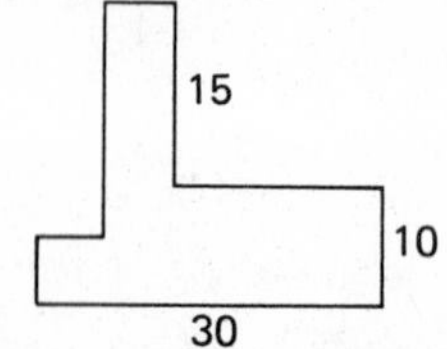

29. Write $\frac{19}{24}$ as a decimal number.

30. Round off $4.0\overline{325}$ to the nearest millionth.

LESSON 42 *Reciprocals · Equations with fractional coefficients*

42.A reciprocals

When we write a fraction upside down, we say that we have written the reciprocal of the fraction. Thus,

$$\frac{2}{5} \text{ is the reciprocal of } \frac{5}{2}$$

$$\frac{5}{2} \text{ is the reciprocal of } \frac{2}{5}$$

$$\frac{3}{7} \text{ is the reciprocal of } \frac{7}{3}$$

$$\frac{7}{3} \text{ is the reciprocal of } \frac{3}{7}$$

Since whole numbers can be written with a denominator of 1, all nonzero whole numbers have a reciprocal.

$$\frac{1}{4} \text{ is the reciprocal of } 4$$

$$4 \text{ is the reciprocal of } \frac{1}{4}$$

$$\frac{1}{16} \text{ is the reciprocal of } 16$$

$$16 \text{ is the reciprocal of } \frac{1}{16}$$

The number 0 can be written with a denominator of 1

$$\frac{0}{1}$$

and this still means zero. When we turn it upside down we get

$$\frac{1}{0}$$

We say that this expression has no meaning because we cannot divide by 0. **For this reason we say that the number 0 does not have a reciprocal. The number 0 is the only number that does not have a reciprocal.**

42.B products of reciprocals

The product of any number and its reciprocal is the number 1. To demonstrate, we will multiply $\frac{3}{7}$ by its reciprocal, which is $\frac{7}{3}$.

$$\frac{3}{7} \cdot \frac{7}{3} = \frac{21}{21} = 1$$

This fact is of great importance in the solution of equations, as we will see.

42.C equations with fractional coefficients

When we look at the equation

$$\frac{4}{5}x = 3$$

we see that x is multiplied by $\frac{4}{5}$. We know that multiplication by $\frac{4}{5}$ can be undone by dividing by $\frac{4}{5}$. Thus, we will divide both sides by $\frac{4}{5}$.

$$\frac{\frac{4}{5}x}{\frac{4}{5}} = \frac{3}{\frac{4}{5}} \qquad \text{divided both sides by } \frac{4}{5}$$

$$x = 3 \cdot \frac{5}{4} \qquad \text{simplified}$$

$$x = \mathbf{\frac{15}{4}} \qquad \text{multiplied}$$

Now we will show another way to work this type of problem.

example 42.C.1 Solve $\frac{2}{3}x = 3$.

solution The coefficient of x is $\frac{2}{3}$. If we multiply $\frac{2}{3}x$ by $\frac{3}{2}$, we can make the coefficient 1. Thus, we multiply both sides of the equation by $\frac{3}{2}$.

$$\frac{3}{2} \cdot \frac{2}{3}x = 3 \cdot \frac{3}{2} \qquad \text{multiplied both sides by } \frac{3}{2}$$

$$x = \mathbf{\frac{9}{2}}$$

Check:

$$\frac{2}{3}\left(\frac{9}{2}\right) = 3 \longrightarrow 3 = 3 \qquad \text{check}$$

example 42.C.2 Solve $\frac{3}{7}x = \frac{2}{9}$.

solution We solve by multiplying both sides by $\frac{7}{3}$, which is the reciprocal of $\frac{3}{7}$.

$$\frac{7}{3} \cdot \frac{3}{7}x = \frac{2}{9} \cdot \frac{7}{3} \qquad \text{multiplied both sides by } \frac{7}{3}$$

$$x = \mathbf{\frac{14}{27}} \qquad \text{simplified}$$

Check:

$$\frac{3}{7}\left(\frac{14}{27}\right) = \frac{2}{9} \longrightarrow \frac{2}{9} = \frac{2}{9} \qquad \text{check}$$

problem set 42

1. Nineteen thousand, one hundred forty-two millionths was the first guess. The second guess was eight thousand, five hundred forty-one hundred-thousandths. Which guess was greater and by how much?

2. They were happy because 80 of them could be seated comfortably in each vehicle. If there were 1420 vehicles in the parking lot, how many could be seated comfortably?

3. The next day there were 6970 in line. How many vehicles would it take to seat them all comfortably?

4. The first go contained 40. The second go contained twice that many. The third go contained four times the number the second go contained. By how many did the third go exceed the sum of the first go and the second go?

5. What number is $\frac{11}{7}$ of 70?

6. What number is $\frac{13}{5}$ of 20?

7. Write $12\frac{5}{7}$ as an improper fraction.

8. Write $\frac{271}{15}$ as a mixed number.

9. Find the least common multiple of 12, 22, and 40.

Solve:

*10. $\frac{2}{3}x = 3$

*11. $\frac{3}{7}x = \frac{2}{9}$

12. $\frac{x}{5} = 60$

13. $5x = 60$

14. $x + \frac{1}{8} = \frac{1}{2}$

15. $x - \frac{1}{4} = \frac{1}{2}$

Simplify:

16. $61\frac{11}{15} - 15\frac{3}{5}$

17. $\frac{5}{8} + 2\frac{1}{4} - \frac{1}{10}$

18. $6 \cdot 8 - 4 \cdot 2 + 6$

19. $611.3 \times .016$

20. $2362.8 - 189.87$

21. $\frac{612.8}{.07}$

22. $\frac{12}{16} \cdot \frac{12}{21} \div \frac{6}{14}$

23. Draw a diagram that depicts the multiplication of 16 by $\frac{7}{8}$.

Evaluate:

24. $xy + yz - z$ if $x = \frac{1}{5}$, $y = 20$, $z = 3$

25. $xyz - x$ if $x = 6$, $y = 16$, $z = \frac{1}{12}$

26. Use unit multipliers to convert 628 kilometers to centimeters.

27. Find the area of this figure. Dimensions are in feet.

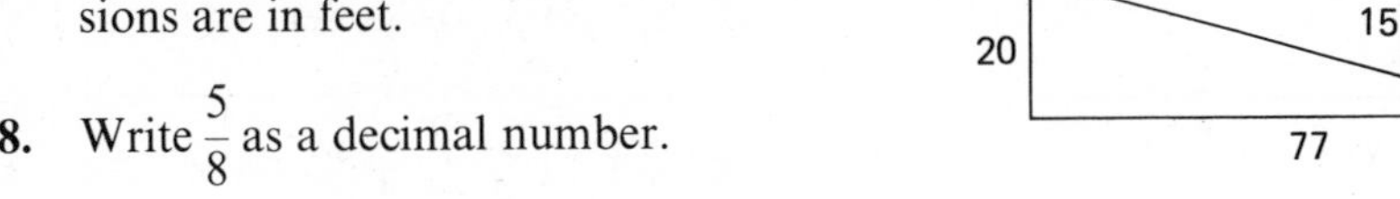

28. Write $\frac{5}{8}$ as a decimal number.

29. Round off 189762581.32 to the nearest million.

30. Use words to write 6213625326.12.

LESSON 43 Symbols of inclusion · Conversion of units of area

43.A symbols of inclusion

We use parentheses, brackets, and braces in algebra to help us make what we write as clear as possible.

( ) parentheses [ ] brackets { } braces

These symbols do not change the values of expressions. For instance

$$4 + 5 \quad \text{and} \quad (4 + 5)$$

both have a value of 9. Parentheses, brackets and braces have the same meaning and any of them may be used when convenient. To show how using symbols of inclusion can clarify expressions we will regard the expression

$$4 + 3 \cdot 5$$

The simplification of this expression is 19 because we remember that we always multiply before we add. But if we use parentheses, we don't have to remember to multiply first because the notation

$$4 + (3 \cdot 5)$$

clearly indicates what we are to do. We can also use symbols of inclusion to indicate nonstandard order of operations. The notation

$$(4 + 3)5$$

clearly indicates that we are to multiply 5 by what is inside the parentheses. Thus, the simplification of this expression is 35.

To simplify expressions that contain symbols of inclusion, we always simplify within the symbols of inclusion first. Then we finish the simplification, remembering to multiply before we add or subtract.

example 43.A.1 Simplify: $4 + (3 \cdot 5)2 + 2(15 - 3)$

solution We begin by simplifying within the two sets of parentheses.

$$4 + (15) \cdot 2 + 2(12)$$

Now we perform all the multiplications and get

$$4 + 30 + 24$$

We finish by adding these numbers and get

$$\mathbf{58}$$

example 43.A.2 Simplify: $8(20 - 3) - (17 - 11 + 3)2 + 5$

solution We begin by simplifying inside the two sets of parentheses.

$$8(17) - (9)2 + 5$$

Now we multiply

$$136 - 18 + 5$$

and finish by adding and subtracting from left to right.

$118 + 5$ subtracted

$\mathbf{123}$ added

43.B conversion of units of area

Two unit multipliers are required to convert units of area, as shown in the next two examples.

example 43.B.1 Convert 4 square feet to square inches.

solution First we write 4 square feet as

$$4 \text{ ft}^2$$

Now since "ft²" means "ft" times "ft," we can write this as

$$4 \text{ ft} \cdot \text{ft}$$

Now one unit multiplier is required to change each "ft" to "in."

$$4 \cancel{\text{ft}} \cdot \cancel{\text{ft}} \times \frac{12 \text{ in}}{1 \cancel{\text{ft}}} \times \frac{12 \text{ in}}{1 \cancel{\text{ft}}} = 4(12)(12) \text{ in}^2 = \mathbf{576 \text{ in}^2}$$

example 43.B.2 Convert 4 square meters to square centimeters.

solution We will use two unit multipliers.

$$4 \cancel{\text{m}^2} \times \frac{100 \text{ cm}}{\cancel{\text{m}}} \times \frac{100 \text{ cm}}{\cancel{\text{m}}} = \mathbf{40{,}000 \text{ cm}^2}$$

problem set 43

1. Pavo bought 20 bunches at \$40 a bunch and 80 bunches at \$30 a bunch. What was the average price of one bunch?

2. If 42 boats could hold 420 of them, how many of them would 80 boats hold?

3. The freshmen raised \$100; the sophomores raised \$140; the juniors raised \$160; and the seniors raised only \$80. Make a bar graph that presents this information.

4. Red ones were only \$4 each so Sam bought 20. Green ones were \$2 each and he bought 60. Yellow ones were \$1 each and he bought 400. How much more did he pay for the yellows than he paid for all of the reds and greens?

5. What number is $\frac{3}{16}$ of 320?

6. What number is $\frac{5}{3}$ of 120?

7. Write $42\frac{3}{8}$ as an improper fraction.

8. Find the least common multiple of 42, 50, and 75.

Solve:

9. $\frac{4}{5}x = 88$

10. $5x = 100$

11. $\frac{x}{5} = 100$

12. $x + \frac{3}{8} = \frac{15}{16}$

13. $x - \frac{1}{4} = \frac{7}{8}$

Simplify:

*14. $4 + (3 \cdot 5)2 + 2(15 - 3)$

*15. $8(20 - 3) - (17 - 11 + 3)2 + 5$

16. $\frac{5}{8} + 1\frac{1}{5} - \frac{1}{10}$

17. $61\frac{7}{18} - 31\frac{4}{9}$

18. $311.61 - 69.893$

19. $23.81 \times .0012$

20. $\frac{611.8}{.09}$

21. $\frac{16}{18} \cdot \frac{36}{24} \div \frac{8}{7}$

22. Draw a diagram that depicts the multiplication of 32 by $\frac{3}{8}$.

Evaluate:

23. $xz + yx - y$ if $x = \frac{1}{3}$, $y = 6$, $z = 12$

24. $xyz + yz - y$ if $x = \frac{1}{2}$, $y = 2$, $z = 8$

25. Convert 5 square feet to square inches.

26. Use unit multipliers to convert 1 mile to yards.

27. Express the perimeter of this figure in centimeters.

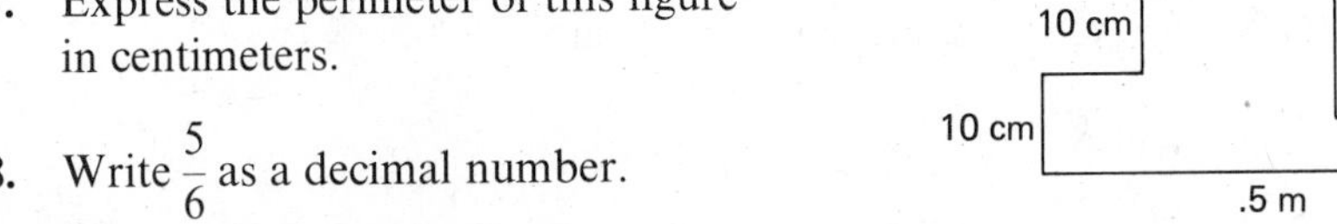

28. Write $\frac{5}{6}$ as a decimal number.

29. Round off 17,625,891.41 to the nearest million.

30. Use words to write 6256187.87621.

LESSON 44 Multiplying and dividing mixed numbers

44.A multiplying mixed numbers

The easiest way to multiply mixed numbers is to convert them to improper fractions and then multiply the improper fractions.

example 44.A.1 Multiply $4\frac{1}{3}$ by $\frac{6}{5}$.

solution We first write $4\frac{1}{3}$ as an improper fraction.

$$\frac{13}{3} \cdot \frac{6}{5}$$

We always cancel if we can. Then we multiply.

$$\frac{13}{\cancel{3}} \cdot \frac{\overset{2}{\cancel{6}}}{5} = \mathbf{\frac{26}{5}}$$

This answer can be written as $5\frac{1}{5}$ if desired.

example 44.A.2 Multiply: $2\frac{1}{4} \times 3\frac{1}{3} \times 5\frac{1}{12}$

solution First we convert each mixed number to an improper fraction.

$$\frac{9}{4} \times \frac{10}{3} \times \frac{61}{12}$$

Now we cancel where possible and then multiply.

$$\frac{\overset{3}{\cancel{9}}}{4} \times \frac{\overset{5}{\cancel{10}}}{\cancel{3}} \times \frac{61}{\underset{\underset{2}{\cancel{6}}}{\cancel{12}}} = \mathbf{\frac{305}{8}}$$

We will leave this answer as an improper fraction.

44.B dividing mixed numbers

To divide mixed numbers, we first write the mixed numbers as improper fractions. Then we invert the divisor and multiply.

example 44.B.1 Simplify: $\dfrac{2\frac{1}{8}}{3\frac{2}{3}}$

solution First we write both numbers as improper fractions.

$$\frac{\frac{17}{8}}{\frac{11}{3}}$$

Now we invert and multiply.

$$\frac{17}{8} \cdot \frac{3}{11} = \mathbf{\frac{51}{88}}$$

example 44.B.2 Simplify: $4\frac{1}{8} \div 2\frac{1}{5} \cdot 3\frac{1}{4} \div 5\frac{1}{2}$

solution First we write each number as an improper fraction.

$$\frac{33}{8} \div \frac{11}{5} \cdot \frac{13}{4} \div \frac{11}{2}$$

Next we invert the divisors and change every division sign into a multiplication sign. Then we cancel and multiply:

$$\frac{\overset{3}{\cancel{33}}}{\underset{4}{\cancel{8}}} \cdot \frac{5}{\cancel{11}} \cdot \frac{13}{4} \cdot \frac{\cancel{2}}{11} = \mathbf{\frac{195}{176}}$$

problem set 44

1. Virago bought 78 pots for \$312. What would she have had to pay for 400 pots?
2. In Graph E (Lesson 30), what was the average number of tons per month for the first 4 months of the year?
3. During the first hour, twenty-two million, forty were born. During the second hour, only fourteen million, eight hundred sixty-five thousand, nine hundred thirty-two were born. How many more were born during the first hour?
4. Henry found that each container would hold 1100 pills. If he had 4200 containers, how many pills could he ship?
5. What number is $\frac{2}{9}$ of 180?
6. What number is $\frac{1}{16}$ of 320?
7. Write $7\frac{5}{16}$ as an improper fraction.
8. Find the least common multiple of 14, 21, and 49.

Solve:

9. $\frac{7}{8}x = 4$
10. $4x = 80$
11. $\frac{x}{7} = 84$
12. $x - \frac{3}{7} = \frac{9}{14}$
13. $x + \frac{3}{11} = \frac{9}{22}$

Simplify:

14. $3 + (2 \cdot 5)3 + 4(2 \cdot 8)$
15. $5(3 - 1) + 2(6 - 5) - 2$

*16. $4\frac{1}{3} \times \frac{6}{5}$

*17. $2\frac{1}{4} \times 3\frac{1}{3} \times 5\frac{1}{12}$

*18. $\dfrac{2\frac{1}{8}}{3\frac{2}{3}}$

*19. $4\frac{1}{8} \div 2\frac{1}{5} \cdot 3\frac{1}{4} \div 5\frac{1}{2}$

20. $\frac{5}{6} + 1\frac{5}{12} - \frac{3}{4}$

21. $17\frac{7}{12} - 12\frac{3}{4}$

22. $117.89 - 112.341$

23. $14.02 \times .0015$

24. $\dfrac{7812}{.003}$

25. Evaluate: $xz + yz - xy$ if $x = 4$, $y = \frac{1}{2}$, $z = 8$

26. Use unit multipliers to convert 3 miles to yards.

27. Find the area of this figure. Dimensions are in feet.

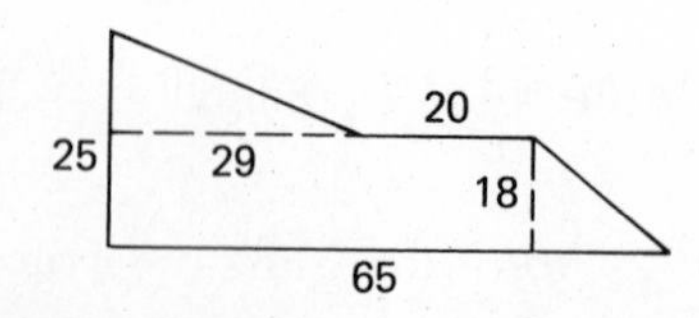

28. Draw a diagram that depicts the multiplication of 48 by $\frac{7}{12}$.

29. (a) List the prime numbers that are greater than 37 but less than 52. (b) List the multiples of 5 that are greater than 37 but less than 52.

30. Convert 5 square meters to square centimeters.

LESSON 45 Exponents and roots

45.A exponents

Often it is necessary to indicate that a number is to be multiplied by itself a given number of times. If we wish to indicate that 7 is to be used as a factor six times, we could write

$$7 \cdot 7 \cdot 7 \cdot 7 \cdot 7 \cdot 7$$

We can also designate repeated multiplication of this kind by using **exponential notation.** Exponential notation lets us express the same thought in a more concise form. To indicate the same multiplication in exponential notation, we would write

$$\text{base} \longrightarrow 7^6 \longleftarrow \text{exponent or power}$$

The lower number is called the **base,** and the upper number is called the **exponent** or **power.** The exponent tells how many times the base is used as a factor. The whole expression is called an **exponential.** We read the exponential above by saying "7 to the 6th power" or just "7 to the 6th."

$3^4 = 3 \cdot 3 \cdot 3 \cdot 3$ The base is 3 and the exponent is 4.

$4^3 = 4 \cdot 4 \cdot 4$ The base is 4 and the exponent is 3.

$\left(\frac{2}{3}\right)^3 = \frac{2}{3} \cdot \frac{2}{3} \cdot \frac{2}{3}$ The base is $\frac{2}{3}$ and the exponent is 3.

example 45.A.1 Simplify: $4^2 + 3^4 + 2^3$

solution We write each exponential in expanded form and get

$$4 \cdot 4 + 3 \cdot 3 \cdot 3 \cdot 3 + 2 \cdot 2 \cdot 2$$

Next we do the multiplications and then we add.

$$16 + 81 + 8 = \mathbf{105}$$

45.B roots of numbers

The inverse operation of raising to a power is called **taking the root.** If we use 2 as a factor four times, the answer is 16.

$$2^4 = 16$$

If we wish to undo this, we can ask "what number used as a factor four times equals 16" by writing

$$\sqrt[4]{16}$$

and since 2 used as a factor four times equals 16, the answer is 2.

$$\sqrt[4]{16} = 2$$

We read this by saying "the fourth root of 16 equals 2." We say that 16 is the **radicand,** 4 is the **index,** and 2 is the **root.** We call the symbol

$$\sqrt{}$$

the **radical sign.** If the index is not written, it is understood to be 2. We call the whole expression a **radical expression** or just a **radical.** For the present, we will restrict our attention to radicals that represent whole numbers. In later lessons, we will investigate radicals that represent decimal numbers.

example 45.B.1 Simplify: (a) $\sqrt[4]{81}$ (b) $\sqrt[3]{27}$ (c) $\sqrt{16}$ (d) $\sqrt[3]{8}$

solution
(a) Since $3 \cdot 3 \cdot 3 \cdot 3 = 81$ $\quad \sqrt[4]{81} = \mathbf{3}$

(b) Since $3 \cdot 3 \cdot 3 = 27$ $\quad \sqrt[3]{27} = \mathbf{3}$

(c) Since $4 \cdot 4 = 16$ $\quad \sqrt{16} = \mathbf{4}$

(d) Since $2 \cdot 2 \cdot 2 = 8$ $\quad \sqrt[3]{8} = \mathbf{2}$

When we simplify expressions that contain exponentials or radicals, we begin by simplifying the exponentials and radicals. Then we simplify within symbols of inclusion. Lastly, we remember to do all multiplications before we add or subtract.

example 45.B.2 Simplify: $4(3 - 2 + 8) + 2^2 - \sqrt[3]{27} + 3 \cdot 2$

solution First we simplify the exponentials and radicals.

$$4(3 - 2 + 8) + 4 - 3 + 3 \cdot 2$$

Then we simplify within the parentheses.

$$4(9) + 4 - 3 + 3 \cdot 2$$

Now we multiply where indicated.

$$36 + 4 - 3 + 6$$

We finish by adding and subtracting from left to right.

$40 - 3 + 6$ added 36 and 4

$37 + 6$ subtracted 3

43 added 6

problem set 45

1. Four large ones cost $40. How much would the customer have to pay for 120 large ones?

2. The order was for 4 bunches of asparagus at 50 cents a bunch; 9 bushels of okra at \$5.50 a bushel, and 4 pecks of beans at \$7.50 a peck. What was the total cost of the order?

3. The number of points scored by the home team for each of the first seven games was as follows: 45, 60, 50, 70, 80, 75, 70. Make a broken-line graph that presents this information.

4. William had 42 boxes and had 5040 items in all. If he divided the items evenly, how many would go in each box?

5. What number is $\frac{7}{8}$ of 400?

6. What number is $\frac{11}{6}$ of 120?

7. Write $3\frac{2}{5}$ as a improper fraction.

8. Find the least common multiple of 18, 42, and 50.

Solve:

9. $\frac{5}{3}x = 20$

10. $4x = 2$

11. $\frac{x}{4} = 7$

12. $x - 7 = 2$

Simplify:

13. $4^2 + 3^4 + 2^3$

14. $\sqrt[4]{81}$

15. $\sqrt[3]{27}$

16. $4(3 - 2 + 8) + 2^2 - \sqrt[3]{27} + 3 \cdot 2$

17. $\frac{3}{4} + 7\frac{11}{12} - \frac{5}{6}$

18. $321\frac{7}{12} - 123\frac{3}{4}$

19. $6 \cdot 3 + 4 \cdot 2 + 12$

20. $111.8 \times .007$

21. $7816.7 - 982.67$

22. $\frac{179.32}{.004}$

23. $7\frac{1}{8} \div 2\frac{1}{4} \times 3\frac{1}{6}$

24. $2\frac{1}{4} \times 6\frac{3}{4} \div 3\frac{1}{8}$

25. $\dfrac{7\frac{2}{3}}{6\frac{5}{6}}$

26. Evaluate: $xyz + y + x + z$ if $x = 3, y = 4, z = 5$

27. Express the perimeter of this figure in meters. Dimensions are in centimeters.

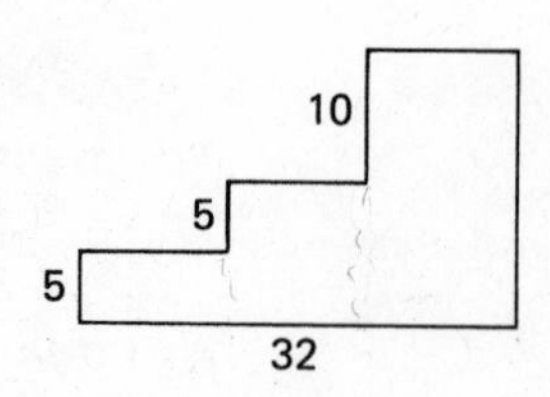

28. Use unit multipliers to convert from 5 miles to feet.

29. Draw a diagram that depicts the multiplication of 24 by $\frac{7}{12}$.

30. Convert 5 square yards to square feet.

LESSON 46 Volume

46.A volume

We use the word **area** to describe the size of a surface. When we tell how large an area is, we describe how many squares of a certain size can be drawn on the surface. The area of a surface also tells us the number of floor tiles it will take to cover the surface. Area has only length and width.

We use the word **volume** to describe a space or a solid that has depth as well as length and width. **Volume is not flat, for volume describes how many cubes of a certain size a thing will hold.** We can use sugar cubes to help us think about volume. A cube is a six-sided figure each of whose faces is square.

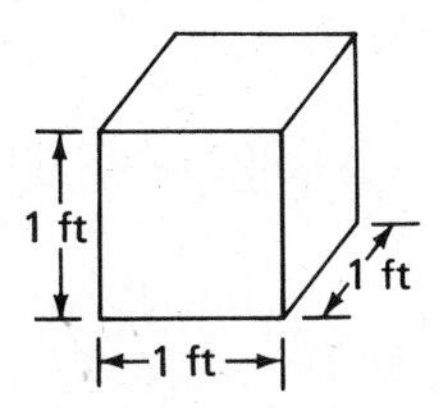

If each of the edges is 1 foot long, we say that the cube has a volume of 1 cubic foot. If each of the edges is 1 centimeter long, the volume is 1 cubic centimeter. If each of the edges is 1 mile long, the volume is 1 cubic mile, etc. We use exponents to help us abbreviate the units for volume.

$$1 \text{ cubic foot} = 1 \text{ ft}^3 \qquad 1 \text{ cubic centimeter} = 1 \text{ cm}^3$$

$$1 \text{ cubic mile} = 1 \text{ mi}^3$$

If we have a rectangular area that measures 4 feet by 2 feet, it has an area of 8 square feet.

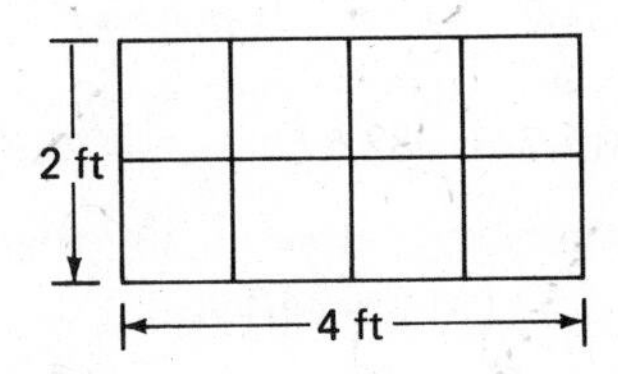

One sugar cube with a volume of 1 cubic foot can be set on each of the squares shown. If we place 1 cube on each square, we will use 8 cubes. If we stack the cubes 2 deep, we will use 16 cubes.

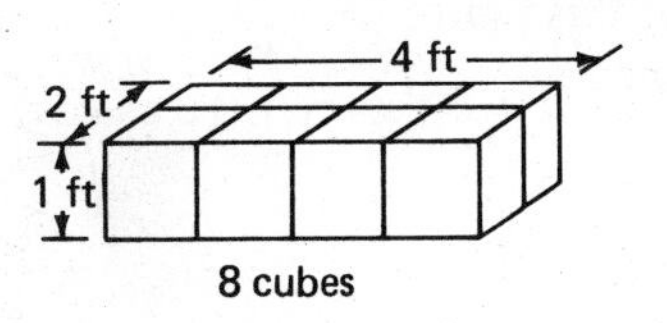

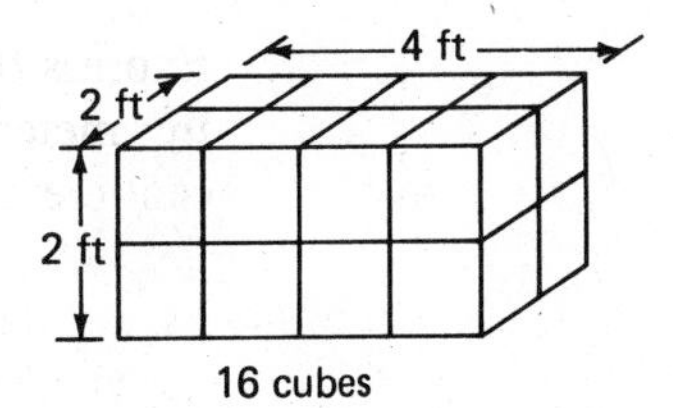

The figure on the left has a volume of 8 ft^3, and the figure on the right has a volume of 16 ft^3.

If we stack the cubes 3 deep, we would have 3 layers of 8 cubes; and if we stack them 4 deep, we would have 4 layers of 8 cubes.

$$3 \text{ layers of } 8 \text{ cubes} = 24 \text{ cubes}$$

$$4 \text{ layers of } 8 \text{ cubes} = 32 \text{ cubes}$$

Thus, the volume of a solid whose sides go straight up is the number of cubes that can be placed in the first layer times the number of layers.

example 46.A.1 Find the number of 1-cm cubes that can be placed in a box measuring 4 cm × 6 cm × 10 cm.

solution Any two measurements can be used for the bottom. We will use 4 cm and 6 cm.

$$\text{Area of bottom} = 4\text{ cm} \times 6\text{ cm} = 24\text{ cm}^2$$

Thus, 24 cubes can be placed in the first layer. If we stack them 10 cubes high, we will get

$$24\text{ cubes} \times 10 = \mathbf{240\ cubes}$$

Thus we say that the box has a volume of 240 cm^3.

example 46.A.2 How many 1-foot cubes will this prism hold? Dimensions are in feet.

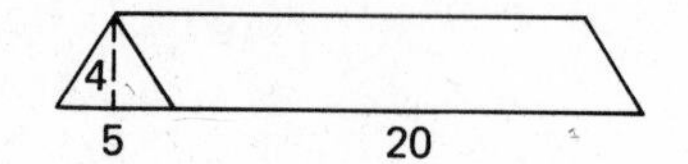

solution If we stand the prism on one end, the triangle will be the base.

$$\text{Area of triangle} = \frac{\text{base} \times \text{height}}{2} = \frac{5 \cdot 4}{2} = 10\text{ ft}^2$$

Thus we can place 10 sugar cubes on the bottom layer (if we crush them first). We will stack them 20 cubes high so the total number of cubes is

$$10\text{ cubes} \times 20 = \mathbf{200\ cubes}$$

Since the prism will hold 200 one-foot cubes, we say that the volume of the prism is 200 cubic feet.

problem set 46

1. There were three times as many dungeons as there were dragons. There were seven times as many princesses as dragons. If there were 12 dragons, how many dragons, dungeons, and princesses were there in all?

2. In Graph D (Lesson 30), what was the average income of the employees who had worked at least 10 years?

3. The number of reds exceeded the number of blues by 80. The number of greens was 20 less than the number of blues. If there were 120 blues, what was the sum of the reds, the blues, and the greens?

4. Nine thousand, six hundred forty-two exceeded the permissible number by two thousand, seventy-five. What was the permissible number?

*5. Find the number of 1-cm cubes that can be placed in a box measuring 4 cm × 6 cm × 10 cm.

*6. How many 1-foot cubes will this prism hold? Dimensions are in feet.

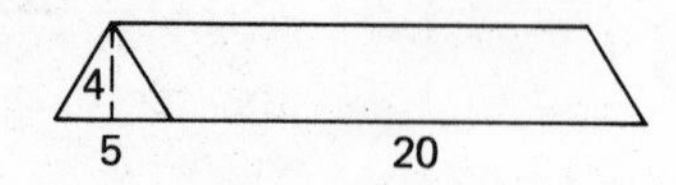

7. What number is $\frac{7}{9}$ of 1800?

8. Write $7\frac{5}{6}$ as an improper fraction.

9. Find the least common multiple of 16, 24, and 36.

Solve:

10. $\frac{4}{3}x = 160$

11. $x + 6 = 12$

12. $x - 5 = 13$

Simplify:

13. $2^2 + 3^3 + 4^2$

14. $\sqrt[4]{16}$

15. $7(2 + 6 - 4) + 2^3 - \sqrt[3]{8}$

16. $\frac{4}{5} + 6\frac{3}{10} - \frac{6}{15}$

17. $615\frac{3}{8} - 138\frac{3}{4}$

18. $7 \cdot 8 - 3 \cdot 4 + 6$

19. $134.8 \times .013$

20. $725.89 - 62.871$

21. $\frac{381.41}{.006}$

22. $3 + (6 \cdot 12)5 - 7$

23. $7(6 - 4) + 2(4 + 2) - 7 \cdot 2$

24. $3\frac{3}{5} \times 3\frac{1}{2} \div 6\frac{3}{4}$

25. $\frac{3\frac{4}{5}}{2\frac{7}{8}}$

26. Evaluate: $xyz + yz - y$ if $x = \frac{1}{6}$, $y = 3$, $z = 2$

27. Find the area of this figure. Dimensions are in feet.

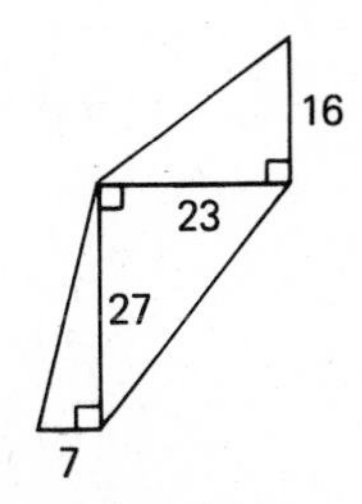

28. Use unit multipliers to convert from 6200 centimeters to kilometers.

29. Draw a diagram that depicts the multiplication of 36 by $\frac{7}{12}$.

30. Use words to write 617825.000131.

LESSON 47 *Order of operations with fractions*

47.A order of operations with fractions

Thus far, we have restricted order-of-operations problems to those containing whole numbers. The procedure is the same for fractions and mixed numbers. **Unless otherwise indicated, we always multiply before adding or subtracting.**

example 47.A.1 Simplify: $\frac{3}{20} + \frac{3}{2} \cdot \frac{2}{5}$

solution We multiply first and get

$$\frac{3}{20} + \frac{3}{5}$$

Now we find a common denominator and add.

$$\frac{3}{20} + \frac{3}{5} = \frac{3}{20} + \frac{12}{20} = \frac{15}{20} = \mathbf{\frac{3}{4}}$$

example 47.A.2 Simplify: $3\frac{1}{2} \cdot \frac{7}{8} - \frac{1}{4}$

solution We change the mixed number to an improper fraction and then we multiply.

$$\frac{7}{2} \cdot \frac{7}{8} - \frac{1}{4} = \frac{49}{16} - \frac{1}{4}$$

Now we subtract.

$$\frac{49}{16} - \frac{1}{4} = \frac{49}{16} - \frac{4}{16} = \frac{45}{16} = \mathbf{2\frac{13}{16}}$$

problem set 47

1. The pompon girls could put 44 pompons in one box. If there were 1760 pompons that had to be packed, how many boxes did the pompon girls need?

2. When the boxes came, a count revealed an overshipment. They had shipped 62 boxes. Now how many pompons were required to fill all the boxes?

3. The time for the race came down with practice. The average times were: first week, 57 seconds; second week, 55 seconds; third week, 50 seconds; fourth week, 51 seconds; fifth week, 48 seconds. Make a broken-line graph that presents this information.

4. Ronk bought 14 of them for $70. How much would he have had to pay for 42 of them?

5. Find the number of 1-foot cubes that can be placed in a box measuring 4 ft × 5 ft × 6 ft.

6. What number is $\frac{6}{7}$ of 217?

7. Write $\frac{316}{25}$ as a mixed number.

8. Find the least common multiple of 8, 21, and 24.

Solve:

9. $\frac{6}{7}x = 18$

10. $x + \frac{5}{13} = \frac{18}{26}$

11. $7x = 315$

Simplify:

12. $4 + (3 \cdot 5)6 + 4(2 \cdot 3)$

13. $6(3 - 1) + 2(3 - 1) - 4$

*14. $\frac{3}{20} + \frac{3}{2} \cdot \frac{2}{5}$

15. $3\frac{3}{4} \cdot \frac{5}{8} - \frac{3}{4}$

16. $4^2 + 3^3 - 2^3$

17. $\sqrt[3]{8} + \sqrt[3]{27}$

18. $\frac{7}{8} + 5\frac{5}{16} - \frac{3}{4}$

19. $117\frac{9}{10} - 12\frac{3}{5}$

20. $181.3 \times .012$

21. $1.825 - .981$

22. $\frac{262.15}{.003}$

23. $4\frac{3}{4} \div 2\frac{1}{3} \times 3\frac{1}{3} \div 1\frac{1}{4}$

24. $6\frac{3}{8} \times 2\frac{4}{5} \div 2\frac{1}{2}$

25. $\dfrac{8\frac{6}{7}}{3\frac{5}{14}}$

26. Evaluate: $xy + xyz - z$ if $x = 24, y = \frac{1}{4}, z = 2$

27. Find the perimeter of this figure. Dimensions are in centimeters.

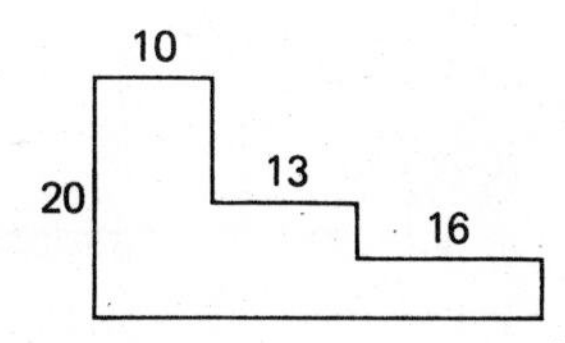

28. Use unit multipliers to convert from 360 yards to inches.

29. Draw a diagram that depicts the multiplication of 25 by $\frac{3}{5}$.

30. Round off $621{,}721.8\overline{17}$ to four decimal places.

LESSON 48 Evaluation of exponentials and radicals

48.A evaluation of exponentials and radicals

Exponentials designate a base and tell us how many times the base is to be used as a factor. The expression 5^3 means

$$5 \cdot 5 \cdot 5$$

When the base is a letter, the exponent tells us how many times the letter is to be used as a factor. Thus the expression x^3 means

$$x \cdot x \cdot x$$

When we evaluate an exponential that contains letters, we replace the letters with the proper numbers.

example 48.A.1 Evaluate m^x if $m = 3$ and $x = 5$.

solution First we replace m with 3 and x with 5 and get

$$3^5$$

This indicates that 3 is to be a factor 5 times so the answer is 243.

$$3 \cdot 3 \cdot 3 \cdot 3 \cdot 3 = \mathbf{243}$$

example 48.A.2 Evaluate $4p^k$ if $p = 2$ and $k = 7$.

solution This indicates that 2 is to be used as a factor 7 times, and this product is to be multiplied by 4.

$$4 \times 2^7 = 4 \times 2 \cdot 2 \cdot 2 \cdot 2 \cdot 2 \cdot 2 \cdot 2 = \mathbf{512}$$

Radicals are also evaluated by replacing the letters with the designated numbers.

example 48.A.3 Evaluate $\sqrt[p]{k}$ if $p = 2$ and $k = 16$.

solution We first replace p with 2 and k with 16 and get

$$\sqrt[2]{16} \quad \text{or} \quad \sqrt{16}$$

It was not necessary to write the 2 because if no index is written, the index is understood to be 2. Now we finish.

$$\sqrt{16} = \mathbf{4}$$

problem set 48

1. Sarah sold whortleberries after school. If she sold 40 pecks for $640, how much money would she get for 100 pecks?
2. Irwin stretched but could only reach seventy-one and fourteen hundred, three ten-millionths inches. He lost because Eisel could reach seventy-two and sixteen hundred, forty-two hundred-thousandths inches. How much farther could Eisel reach?
3. Knapp hid them in discrete bunches of 48. Some were under the front porch. In all he hid 1305 discrete bunches. What was the total of his hiding efforts?
4. In Graph B (Lesson 30), how many more type C cars were sold in 1980 than type B cars sold in 1970?
5. Find the number of 1-foot cubes this prism will hold. Dimensions are in feet.

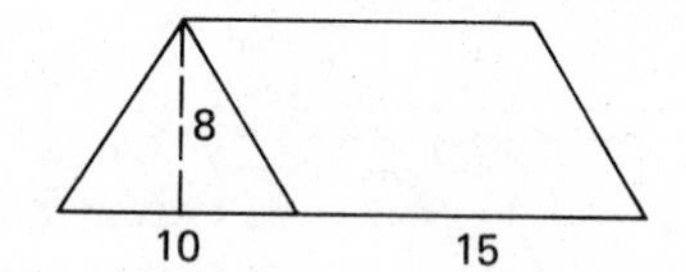

6. What number is $\frac{3}{8}$ of 256?
7. Write $\frac{213}{8}$ as a mixed number.
8. Find the least common multiple of 16, 24, and 30.

Solve:

9. $\frac{7}{8}x = 14$
10. $x - \frac{1}{8} = \frac{3}{16}$
11. $x + \frac{5}{8} = \frac{14}{5}$

Simplify:

12. $6 \cdot 3 + (3 \cdot 2)5 - 4(2 \cdot 2)$
13. $4(6 - 2 + 1) + 3^2 - \sqrt[3]{8}$
14. $\frac{7}{20} + \frac{4}{5} \cdot \frac{2}{3}$
15. $3\frac{1}{6} \cdot \frac{1}{8} - \frac{4}{12}$
16. $3^2 + 4^2 - 5^2$
17. $\frac{3}{4} + 7\frac{1}{16} - \frac{5}{8}$
18. $113\frac{4}{7} - 32\frac{1}{14}$
19. $31.62 \times .08$
20. $89.265 - 6.898$
21. $\frac{261.82}{.004}$
22. $3\frac{2}{3} \times 2\frac{3}{4} \div 3\frac{1}{2}$
23. $\dfrac{6\frac{3}{4}}{7\frac{11}{12}}$

Evaluate:

24. $xyz - x + xy \quad \text{if } x = 12,\ y = \frac{1}{4},\ z = 6$

*25. m^x if $m = 3$ and $x = 5$

26. $\sqrt[p]{k}$ if $p = 3$ and $k = 8$

27. Find the area of this figure. Dimensions are in meters.

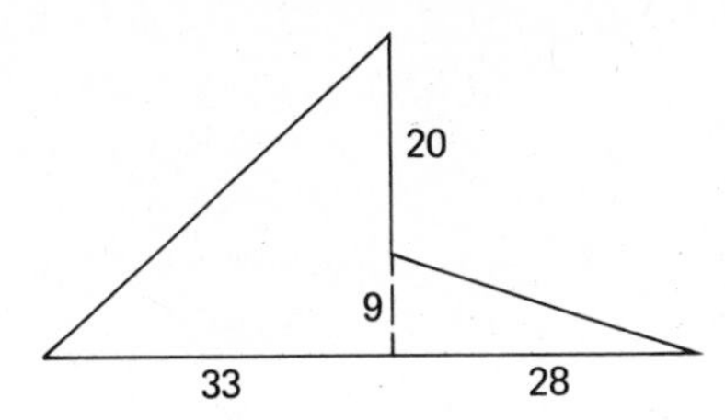

28. Use unit multipliers to convert 250,001 centimeters to kilometers.

29. Draw a diagram that depicts the multiplication of 20 by $\frac{3}{5}$.

30. Write 321,621.32 in words.

LESSON 49 Fractional part of a number · Fractional equations

49.A fractional part of a number

We remember that a fraction designates a part of a whole. On the left we show a rectangle that represents the number 1 or the whole thing. On the right, we divide the whole thing into 5 parts and shade in 2 of the parts. This is what we mean by the fraction $\frac{2}{5}$.

1 =

1

Whole thing

$\frac{1}{5}$	$\frac{1}{5}$	$\frac{1}{5}$	$\frac{1}{5}$	$\frac{1}{5}$

$\frac{2}{5}$

If the whole is a number greater than 1, then $\frac{2}{5}$ of the number means 2 of the 5 equal parts of the number. For example, if the number is 150, each of the 5 equal parts is 30.

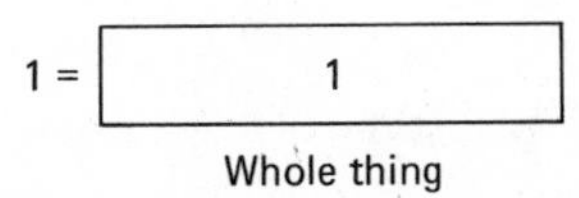

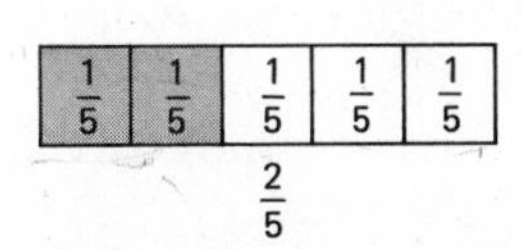

Then 2 of these parts equals $30 + 30 = 60$.

$$\frac{2}{5} \times 150 = 60$$

49.B fractional equations

The general equation for the fractional part of a number is very important. The equation above written in words is

Two-fifths of 150 is 60

This statement contains a fraction and two numbers. One of the numbers associates with the word **of** and the other associates with the word **is.** We can write the general form of this equation as

$$F \times of = is$$

The letter **F** stands for **fraction** and the words **of** and **is** designate the two numbers.

example 49.B.1 What fraction of 75 is 25?

solution First we write the general equation.

$$F \times of = is$$

Now we replace **F** with WF for "what fraction." We replace **of** with 75 and replace **is** with 25.

$$WF \times 75 = 25$$

We can undo multiplication by 75 by dividing by 75. Thus, we divide both sides by 75.

$$\frac{WF \times 75}{75} = \frac{25}{75} \longrightarrow WF = \mathbf{\frac{1}{3}}$$

Thus we find that $\frac{1}{3}$ of 75 is 25.

example 49.B.2 Five-sevenths of what number is $\frac{5}{2}$?

solution We will use the equation

$$F \times of = is$$

We replace **F** with $\frac{5}{7}$, replace **of** with WN and replace **is** with $\frac{5}{2}$.

$$\frac{5}{7} \times WN = \frac{5}{2}$$

Then we solve by multiplying both sides by $\frac{7}{5}$.

$$\frac{7}{5} \cdot \frac{5}{7} \times WN = \frac{5}{2} \cdot \frac{7}{5} \longrightarrow \mathbf{WN = \frac{7}{2}}$$

Thus we find that $\frac{5}{7}$ of $\frac{7}{2}$ is $\frac{5}{2}$.

example 49.B.3 Five-sevenths of $\frac{5}{2}$ is what number?

solution Again we use the equation

$$F \times of = is$$

We replace **F** with $\frac{5}{7}$, **of** with $\frac{5}{2}$, and **is** with WN.

$$\frac{5}{7} \cdot \frac{5}{2} = WN$$

Then we multiply.

$$\mathbf{\frac{25}{14} = WN}$$

Thus we find that five-sevenths of $\frac{5}{2}$ is $\frac{25}{14}$.

problem set 49

1. Big Daddy put all his money in gilt-edged stock certificates. If he invested \$64,000 and paid \$40 for each certificate, how many certificates did Big Daddy buy?

2. Marsha guessed forty-one thousand, fifteen millionths. Then Kevin the Pipe Man guessed twenty-seven ten-thousandths. What was the sum of their guesses?

3. The numbers of dogs entered in the contest were as follows: hounds, 140; collies, 80; terriers, 110; mastiffs, 60. Make a bar graph that presents this information.

4. The weights of the first four contestants were 410 lb, 380 lb, 476 lb, and 502 lb. What was the average weight of the contestants?

*5. Five-sevenths of what number is $\frac{5}{2}$?

6. Three-fifths of what number is $\frac{10}{3}$?

7. Find the number of 1-ft cubes that a box which measures 3 ft × 4 ft × 6 ft will hold.

8. What number is $\frac{4}{7}$ of 210?

9. Write $6\frac{27}{31}$ as an improper fraction.

10. Find the least common multiple of 18, 20, and 24.

Solve:

11. $\frac{8}{9}x = 16$

12. $x - 3\frac{1}{4} = \frac{1}{2}$

Simplify:

13. $7 \cdot 2 + (3 \cdot 2)4 - 3(2 \cdot 1)$

14. $6(3 - 2 + 1) + 3^2 - \sqrt[3]{8}$

15. $\frac{3}{20} + \frac{4}{5} \cdot \frac{3}{4}$

16. $4\frac{2}{3} \cdot \frac{3}{4} - \frac{7}{12}$

17. $137\frac{5}{7} - 16\frac{1}{14}$

18. $\frac{4}{5} + 3\frac{1}{10} - \frac{14}{25}$

19. $3^2 - \sqrt{9} + \sqrt[3]{8}$

20. $21.32 \times .06$

21. $213.81 - 11.713$

22. $\frac{6111.2}{.005}$

23. $\frac{6\frac{4}{5}}{5\frac{3}{4}}$

Evaluate:

24. $xyz + yz - xy$ if $x = 20$, $y = \frac{1}{5}$, $z = 5$

25. y^x if $y = 3$ and $x = 3$

26. $\sqrt[p]{q}$ if $p = 3$ and $q = 27$

27. Find the volume of this figure. Dimensions are in feet.

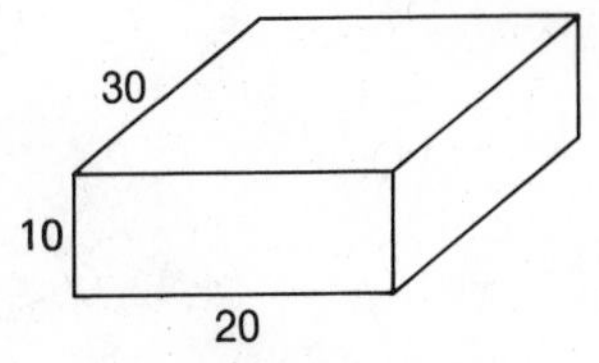

28. Use unit multipliers to convert 2 miles to inches.

29. Draw a diagram that depicts the multiplication of 30 by $\frac{5}{6}$.

30. (a) List the prime numbers between 41 and 51. (b) List the multiples of 12 between 41 and 51.

LESSON 50 Surface area

50.A surface area

The surface area of a solid is the total area of all the exposed surfaces of the solid. A rectangular box has six surfaces. It has a top and a bottom, a front and a back, and two sides. The surface area is the sum of these six areas.

example 50.A.1 Find the surface area of this box. All dimensions are in meters.

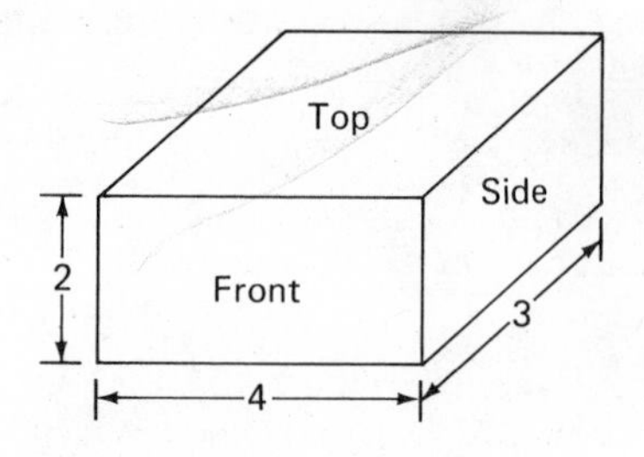

solution Since the box is rectangular, the areas of the top and the bottom are equal. The area of the front equals the area of the back, and the areas of the two sides are equal.

$$\begin{aligned}
\text{Area of front} &= 4\text{ m} \times 2\text{ m} = 8\text{ m}^2\\
\text{Area of back} &= 4\text{ m} \times 2\text{ m} = 8\text{ m}^2\\
\text{Area of top} &= 4\text{ m} \times 3\text{ m} = 12\text{ m}^2\\
\text{Area of bottom} &= 4\text{ m} \times 3\text{ m} = 12\text{ m}^2\\
\text{Area of side} &= 3\text{ m} \times 2\text{ m} = 6\text{ m}^2\\
\text{Area of side} &= 3\text{ m} \times 2\text{ m} = 6\text{ m}^2\\
\text{Surface area} &= \text{total} = \mathbf{52\text{ m}^2}
\end{aligned}$$

example 50.A.2 Find the surface area of this prism. All dimensions are in centimeters.

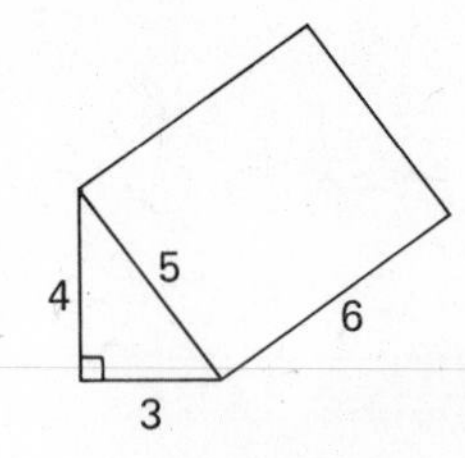

solution The prism has two ends that are triangles. It has three faces that are rectangles.

$$\begin{aligned}
\text{Area of one end} &= \frac{4\text{ cm} \times 3\text{ cm}}{2} = 6\text{ cm}^2\\
\text{Area of one end} &= \frac{4\text{ cm} \times 3\text{ cm}}{2} = 6\text{ cm}^2\\
\text{Area of bottom} &= 3\text{ cm} \times 6\text{ cm} = 18\text{ cm}^2\\
\text{Area of back} &= 4\text{ cm} \times 6\text{ cm} = 24\text{ cm}^2\\
\text{Area of front} &= 5\text{ cm} \times 6\text{ cm} = 30\text{ cm}^2\\
\text{Surface area} &= \text{total} = \mathbf{84\text{ cm}^2}
\end{aligned}$$

problem set 50

1. On one hand there were one hundred forty-two thousand, seven hundred sixty-three. On the other hand there were two hundred twenty-eight thousand, fourteen. How many more were there on the other hand?

2. The whole batch cost $28,000 and contained 140 items. What would 200 items cost?

3. In Graph E (Lesson 30), what was the average number of tons for the 4-month period of March, April, May, and June?

4. The times for the 100-meter dash in seconds were 10.1, 10.2, 10.4, 10.6, and 10.3 seconds. What was the average time of the racers?

5. Find the surface area of this box. All dimensions are in meters.

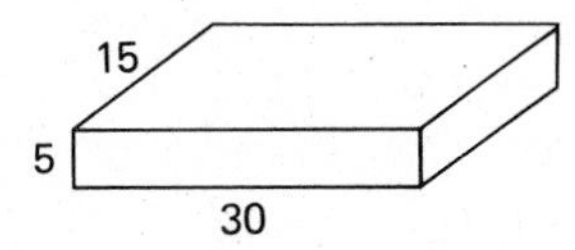

6. Find the number of 1-centimeter cubes this prism will hold. Dimensions are in centimeters.

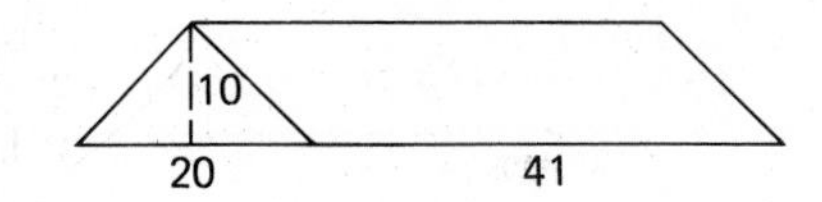

7. What number is $\frac{5}{8}$ of 168?

8. Four-fifths of what number is $\frac{7}{2}$?

9. Find the least common multiple of 18, 24, and 36.

10. Five-sixths of what number is $4\frac{1}{3}$?

11. Write $\frac{168}{7}$ as a mixed number.

Solve:

12. $\frac{8}{9}x = 16$

13. $x - \frac{3}{5} = 2\frac{3}{10}$

Simplify:

14. $6 \cdot 2 + (6 \cdot 3)7 - 4(3 \cdot 1)$

15. $5(6 - 3 + 1) + 6^2 - \sqrt{100}$

16. $\frac{11}{14} + \frac{3}{7} \cdot \frac{3}{4}$

17. $6\frac{2}{3} \cdot \frac{1}{4} - \frac{3}{4}$

18. $3^2 + \sqrt[3]{27} - \sqrt[3]{8}$

19. $\frac{3}{5} + 3\frac{4}{15} - \frac{7}{30}$

20. $117\frac{3}{8} - 14\frac{7}{16}$

21. $16.82 \times .016$

22. $118.321 - 81.34$

23. $\frac{16.25}{.03}$

24. $4\frac{2}{3} \times 2\frac{3}{4} \div 1\frac{5}{12}$

25. $\frac{7\frac{2}{3}}{2\frac{5}{6}}$

Evaluate:

26. $x + xy + xyz$ if $x = 1$, $y = 10$, $z = \frac{1}{5}$

27. p^q if $p = 2$ and $q = 3$

28. Use unit multipliers to convert 6 miles to feet.

29. Draw a diagram that depicts the multiplication of 27 by $\frac{7}{9}$.

30. Convert 12 square kilometers to square meters.

LESSON 51 *Multiple symbols of inclusion*

51.A multiple symbols of inclusion

When an expression contains more than one symbol of inclusion, we begin by simplifying within the innermost symbol of inclusion.

example 51.A.1 Simplify: $24 - 2[(5 - 2)(14 - 12) + 3]$

solution We begin by simplifying within the parentheses.

$$24 - 2[(3)(2) + 3]$$

Now we simplify within the braces. We remember to multiply before we add.

$$24 - 2[9]$$

Again we multiply first and then subtract.

$$24 - 18 = \mathbf{6}$$

example 51.A.2 Simplify: $33 - 2[3(3 + 12) - (5 \cdot 2)3]$

solution We begin by simplifying within the parentheses.

$$33 - 2[3(15) - (10)3]$$

Now we simplify within the brackets.

$$33 - 2[45 - 30]$$

$$33 - 2[15]$$

Next we multiply and then we subtract.

$$33 - 30 = \mathbf{3}$$

problem set 51

1. On the first day the attendance was forty-seven thousand, three hundred sixty-four. On the second day the attendance was fifty-three thousand, seven. How many more attended on the second day?

2. Each bin held 14 uniforms. If there were 140 bins in all, how many uniforms did they hold?

3. Fifteen could be purchased for only $315. How much would 140 cost?

4. The bulls weighed 2153 lb, 1491 lb, and 1840 lb. What was the average weight of the bulls?

5. Find the volume of the prism. Dimensions are in feet.

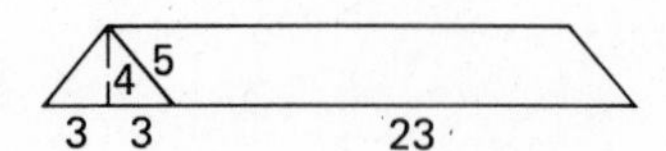

6. Find the surface area of the prism given in Problem 5.

7. Find the least common multiple of 14, 21, and 27.

8. Five-eighths of what number is 100?

9. What fraction of 64 is 56?

10. Seven-elevenths of 88 is what number?

11. Write $\frac{131}{15}$ as a mixed number.

Solve:

12. $\frac{4}{7}x = 112$

13. $x + \frac{7}{12} = 3\frac{5}{12}$

Simplify:

*__14.__ $24 - 2[(5 - 2)(14 - 12) + 3]$

15. $15 + 3[(6 - 4)(7 - 4) - 2]$

16. $4[(3 - 1)(3 + 2) - 1] + 25$

17. $7\frac{3}{4} \cdot \frac{2}{3} - \frac{5}{12}$

18. $\frac{13}{14} + \frac{2}{7} \cdot \frac{3}{4}$

19. $\frac{4}{5} + 2\frac{7}{10} - \frac{3}{20}$

20. $2^3 + 3^3 - \sqrt[3]{8}$

21. $36\frac{6}{7} - 14\frac{3}{14}$

22. $171.3 \times .012$

23. $62.891 - 18.812$

24. $\frac{17.025}{.003}$

25. $4\frac{1}{8} \div 2\frac{1}{4} \times 3\frac{1}{2} \div 1\frac{1}{16}$

26. $\frac{8\frac{1}{2}}{4\frac{1}{7}}$

Evaluate:

27. $xyz + yz - x$ if $x = 16, y = \frac{1}{8}, z = 24$

28. $p^q + p + q$ if $p = 2, q = 3$

29. Use unit mulitpliers to convert 2162.18 centimeters to kilometers.

30. Write 16821621.00321 in words.

LESSON 52 Complex decimal numbers

52.A complex decimal numbers

A complex decimal number is a number that is part decimal and part fraction such as

$$.043\frac{2}{3}$$

If we try to write this number as a decimal number, we will find that it requires an infinite number of digits. But we can write it as a fraction of whole numbers. The trick is to realize that both the 3 and the $\frac{2}{3}$ are in the thousandths' place.

$$\underset{\text{tenths' place}}{.0}\ \overset{\text{hundredths' place}}{4}\ \underset{\text{thousandths' place}}{\left(3\frac{2}{3}\right)}$$

We read this number as "forty-three and two-thirds thousandths." We can write it as

$$\frac{43\frac{2}{3}}{1000}$$

Now we write $43\frac{2}{3}$ as an improper fraction.

$$\frac{\frac{131}{3}}{1000}$$

We finish by multiplying above and below by $\frac{1}{1000}$.

$$\frac{\frac{131}{3} \times \frac{1}{1000}}{1000 \times \frac{1}{1000}} = \mathbf{\frac{131}{3000}}$$

example 52.A.1 Write $.0016\frac{3}{8}$ as a fraction of whole numbers.

solution We note that $6\frac{3}{8}$ is in the ten-thousandths' place. Thus we can read the number as "sixteen and three-eighths ten-thousandths."

We can write this as a fraction whose bottom is 10,000.

$$\frac{16\frac{3}{8}}{10{,}000}$$

Next we write $16\frac{3}{8}$ as an improper fraction.

$$\frac{\frac{131}{8}}{10{,}000}$$

Now we multiply above and below by $\frac{1}{10{,}000}$.

$$\frac{\frac{131}{8} \times \frac{1}{10{,}000}}{10{,}000 \times \frac{1}{10{,}000}} = \mathbf{\frac{131}{80{,}000}}$$

problem set 52

1. At the sock hop there were 420 pairs of red socks, 375 pairs of black socks, and 835 pairs that were mismatched. By how many pairs did the mismatched socks outnumber the sum of the reds and blacks?

2. The original diameter of the microbe was one thousand, four hundred seventy-five millionths inches. Later the diameter increased to one hundred three ten-thousandths inches. By how much did the diameter increase?

3. Jimmy bought 1900 oscillators for $38,000. How much would he have to pay for 5000 oscillators?

4. In the first five games the team scored 86 points, 92 points, 80 points, 70 points, and 104 points. What was their average score?

*5. Write $.0016\frac{3}{8}$ as a fraction of whole numbers.

6. Write $.013\frac{2}{3}$ as a fraction of whole numbers.

7. Find the volume of the rectangular box. Dimensions are in yards.

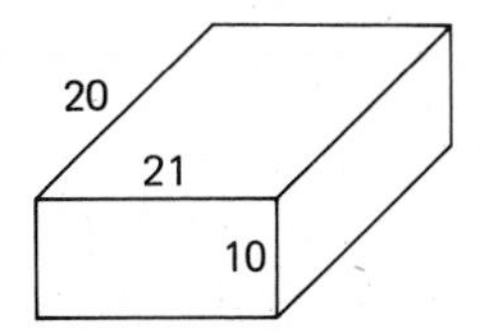

8. Find the surface area of the rectangular box shown in Problem 7.

9. Find the least common multiple of 20, 30, and 36.

10. Three-fourths of what number is 51?

11. Three-sevenths of 91 is what number?

12. Write $7\frac{3}{22}$ as an improper fraction.

Solve:

13. $\frac{5}{4}x = 120$

14. $x + \frac{3}{11} = \frac{51}{22}$

Simplify:

15. $17 - 2[(6 - 4)(7 - 3) - 7]$

16. $3^2 + 2[(7 + 1)(7 - 5) - 12]$

17. $6\frac{2}{3} \cdot \frac{1}{4} - \frac{5}{12}$

18. $\frac{5}{6} + 1\frac{7}{12} - \frac{2}{3}$

19. $2 + 2^2 + 2^3$

20. $\sqrt[3]{27} - \sqrt[2]{9} + 3$

21. $16\frac{3}{5} - 5\frac{1}{10}$

22. $181.4 \times .0012$

23. $1189.26 - 91.872$

24. $\frac{181.02}{.006}$

25. $7\frac{1}{4} \div 3\frac{1}{3} \times 1\frac{2}{3} \times 1\frac{1}{6}$

26. $\dfrac{3\frac{2}{7}}{2\frac{1}{14}}$

Evaluate:

27. $xyzt + xyz + yzt$ if $x = \frac{1}{6}, y = 3, z = 4, t = 5$

28. $\sqrt[x]{y} + x$ if $x = 3, y = 8$

29. Use unit multipliers to convert 625 yards to inches.

30. Round off 621.5621812 to the thousandths' place.

LESSON 53 Decimal part of a number

53.A decimal part of a number

The equation for the fractional part of a number is

$$F \times of = is$$

The equation for the decimal part of a number is exactly the same except that we use D for decimal in place of F for fraction.

$$D \times of = is$$

example 53.A.1 .4 of what number is 72?

solution We will use the equation for the decimal part of a number

$$D \times of = is$$

and replace **D** with .4, **of** with WN and **is** with 72

$$.4 \times WN = 72$$

We divide by .4 to solve.

$$\frac{\cancel{.4} \times WN}{\cancel{.4}} = \frac{72}{.4}$$

$$\mathbf{WN = 180}$$

Thus, .4 of 180 is 72.

example 53.A.2 What decimal of 240 is 90?

solution We use the decimal form of the equation

$$D \times of = is$$

and replace **of** with 240 and **is** with 90.

$$D \times 240 = 90$$

Then we divide both sides by 240.

$$\frac{D \times 240}{240} = \frac{90}{240} \quad \text{divided by 240}$$

$$\mathbf{D = .375} \quad \text{simplified}$$

problem set 53

1. Willikin hid one hundred forty million, fourteen in the boscage of the weald. The next week he hid an additional fifteen million, nine hundred eighty-two thousand. How many did Willikin hide in all?
2. Bad King John bought 80 horses at the auction. If he paid 320 guineas for them, what would he have had to pay for 320 horses?
3. Richard the Lion Hearted counted his troops on the outskirts of Accra. He counted seven thousand, nine hundred forty-two. If this was two hundred forty-two more than he counted yesterday, what was yesterday's count?
4. The first number was 186,925. The second number was 905,106. The third number was 2,105,061. How much greater was the third number than the sum of the first two?

5. .6 of what number is 72?

6. What decimal of 360 is 60?

7. Write $.011\frac{3}{5}$ as a fraction of whole numbers.

8. Find the volume of the prism. Dimensions are in meters.

9. Find the surface area of the prism given in Problem 8.

10. Find the least common multiple of 15, 20, and 30.

11. What fraction of 56 is 21?

12. Eight-thirteenths of what number is 16?

13. Write $\frac{181}{20}$ as a mixed number.

Solve:

14. $\frac{8}{7}x = 104$

15. $x - \frac{5}{12} = \frac{15}{4}$

Simplify:

16. $24 + 2[6(3 - 1)(7 - 3) - 6]$

17. $2^3 + 2[(8 + 1)(6 - 5) - 3]$

18. $5\frac{6}{7} \cdot \frac{3}{2} - 1\frac{1}{14}$

19. $\frac{5}{6} + 1\frac{5}{12} - 1\frac{1}{3}$

20. $1 + 2^2 + 4^2 - 3^2$

21. $\sqrt[3]{8} + \sqrt[3]{27} - 5$

22. $23\frac{4}{5} - 6\frac{1}{15}$

23. $181.4 \times .0013$

24. $\frac{182.101}{.0006}$

25. $2811.62 - 13.981$

26. $8\frac{3}{4} \div 1\frac{1}{3} \times 1\frac{3}{4} \div \frac{1}{12}$

27. $\dfrac{2\frac{3}{7}}{3\frac{3}{14}}$

Evaluate:

28. $xyt - yt + x$ if $x = 6$, $y = \frac{1}{3}$, $t = 12$

29. $\sqrt[x]{y} + xy$ if $x = 3$, $y = 8$

30. Draw a diagram that depicts multiplication of 20 by $\frac{3}{5}$.

LESSON 54 *Fractions and symbols of inclusion*

54.A fractions and symbols of inclusion

We have restricted our practice with symbols of inclusion to problems that have whole numbers. Symbols of inclusion can also be used to practice the simplification of fractional expressions and mixed numbers.

example 54.A.1 Simplify: $\frac{3}{2}\left(\frac{1}{4} + \frac{7}{16}\right) - \frac{1}{8}$

solution First we simplify within the parentheses.

$$\frac{3}{2}\left(\frac{4}{16} + \frac{7}{16}\right) - \frac{1}{8} = \frac{3}{2}\left(\frac{11}{16}\right) - \frac{1}{8}$$

Next we multiply

$$\frac{33}{32} - \frac{1}{8}$$

and now we subtract

$$\frac{33}{32} - \frac{4}{32} = \mathbf{\frac{29}{32}}$$

example 54.A.2 Simplify: $\frac{1}{3}\left(3\frac{1}{4} - \frac{1}{8}\right) + \frac{1}{48}$

solution First we simplify within the parentheses.

$$\frac{1}{3}\left(\frac{26}{8} - \frac{1}{8}\right) + \frac{1}{48} = \frac{1}{3}\left(\frac{25}{8}\right) + \frac{1}{48}$$

Then we multiply. Then we add and simplify the result.

$$\frac{25}{24} + \frac{1}{48} = \frac{50}{48} + \frac{1}{48} = \frac{51}{48} = \mathbf{\frac{17}{16}}$$

example 54.A.3 Simplify: $\frac{2}{5}\left[\left(\frac{2}{3} - \frac{1}{6}\right) + \left(\frac{1}{3} + \frac{5}{6}\right)\right]$

solution First we simplify within the parentheses.

$$\frac{2}{5}\left[\left(\frac{1}{2}\right) + \left(\frac{7}{6}\right)\right]$$

Now we simplify within the brackets and get

$$\frac{2}{5}\left[\frac{10}{6}\right]$$

and then we multiply

$$\frac{2}{5}\left[\frac{10}{6}\right] = \mathbf{\frac{2}{3}}$$

problem set 54

1. He was a covetous little beggar as he wanted everything he saw. But his funds were limited, and he bought only 70 of them for which he paid $3500. What would he have had to pay for 720 of them?

2. The fishmonger sold 40 codfish at 10 shillings each, 59 squid at 30 shillings each, and 1 grouper for 300 shillings. What was the average price paid for each fish?

3. The crones and the curmudgeons forced their way in until 900,062 had arrived. This was 202,020 more than had come last time. How many came last time?

4. The number of telephone calls varied from month to month. The number of calls for the first 4 months were: January, 5000; February, 7500; March, 4000; and April, 9000. Make a broken-line graph that displays this information.

5. .3 of what number is 36?

6. What decimal of 480 is 60?

7. Write $.023\frac{4}{5}$ as a fraction of whole numbers.

8. Find the area of the following figure. Dimensions are in centimeters.

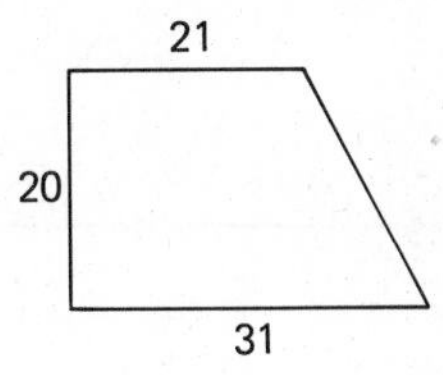

9. What is the volume of a solid whose base is the figure shown in Problem 8 and whose height is 10 centimeters?

10. Find the least common multiple of 10, 15, and 25.

11. Six-sevenths of 98 is what number?

12. What fraction of 64 is 48?

13. Write $7\frac{21}{23}$ as an improper fraction.

Solve:

14. $\frac{12}{13}x = 60$

15. $x + \frac{3}{7} = \frac{29}{14}$

Simplify:

16. $36 - 3[(6 - 3)(3 - 1) - 4]$

17. $2^2 + 2[2(3 + 1)(3 - 2) - 1]$

18. $4\frac{3}{5} \cdot \frac{2}{3} - \frac{14}{15}$

19. $5\frac{2}{3} - 3\frac{5}{6} + \frac{5}{18}$

20. $3^2 + 2^2 - 3 + \sqrt[3]{8}$

21. $197.3 \times .013$

22. $6211.89 - 8.987$

23. $\frac{192.03}{.05}$

24. $\dfrac{1\frac{3}{4}}{2\frac{1}{5}}$

***25.** $\frac{3}{2}\left(\frac{1}{4} + \frac{7}{16}\right) - \frac{1}{8}$

26. $\frac{1}{4}\left(2\frac{1}{4} - \frac{1}{8}\right) + \frac{3}{16}$

27. $3\frac{2}{3} \div 1\frac{1}{4} \times 2\frac{1}{6} \times 1\frac{1}{3}$

Evaluate:

28. $xyz + xz - z$ if $x = \frac{1}{3}$, $y = 9$, $z = 12$

29. $\sqrt[x]{y} + x^y$ if $x = 2$, $y = 4$

30. Round off $18.0\overline{618}$ to the nearest millionth.

LESSON 55 Percent

55.A percent

The language spoken by the ancient Romans was Latin. The Latin word for *by* was *per* and the word for *one hundred* was *centum*. We put these words together in English to form the word **percent.** From this we see that percent means by the 100.

One percent of a number is one-hundredth of the number. To demonstrate we will use the number 256.

2.56	2.56	2.56	2.56	2.56	2.56	2.56	2.56	2.56	2.56
2.56	2.56	2.56	2.56	2.56	2.56	2.56	2.56	2.56	2.56
2.56	2.56	2.56	2.56	2.56	2.56	2.56	2.56	2.56	2.56
2.56	2.56	2.56	2.56	2.56	2.56	2.56	2.56	2.56	2.56
2.56	2.56	2.56	2.56	2.56	2.56	2.56	2.56	2.56	2.56
2.56	2.56	2.56	2.56	2.56	2.56	2.56	2.56	2.56	2.56
2.56	2.56	2.56	2.56	2.56	2.56	2.56	2.56	2.56	2.56
2.56	2.56	2.56	2.56	2.56	2.56	2.56	2.56	2.56	2.56
2.56	2.56	2.56	2.56	2.56	2.56	2.56	2.56	2.56	2.56
2.56	2.56	2.56	2.56	2.56	2.56	2.56	2.56	2.56	2.56

If we divide 256 into 100 parts, each part is 2.56. Thus 2.56 is 1 percent of 256 because 1 percent is exactly the same thing as one-hundredth. If we want 5 percent of 256, we must use 5 of these parts.

$$5 \text{ percent} = 5(2.56) = 12.8$$

One hundred and sixty percent of 256 would be 160 of these parts.

$$160 \text{ percent} = 160(2.56) = 409.6$$

From this we see that we can find a given percent of a number if we first divide the number into 100 equal parts. Each of these parts is 1 percent of the number.

example 55.A.1 (a) What is 1 percent of 7500? (b) What is 34 percent of 7500?

solution (a) To find 1 percent of 7500, we divide 7500 by 100.

$$1 \text{ percent of } 7500 = \frac{7500}{100} = \mathbf{75}$$

(b) 34 percent is 34 times the value of 1 percent.

$$34 \times 75 = \mathbf{2550}$$

example 55.A.2 (a) What is 1 percent of 67? (b) What is 140 percent of 67?

solution (a) To find 1 percent of 67 we divide by 100.

$$1 \text{ percent of } 67 = \frac{67}{100} = \mathbf{.67}$$

(b) 140 percent is 140 of these parts.

$$140 \times .67 = \mathbf{93.8}$$

problem set 55

1. The new tires cost \$780, and for this price the dealer got 10 tires. What would he have had to pay for 58 tires?

2. Two numbers were proposed. The first number was nine million, forty-seven. The second number was eight million, seven hundred ninety-three thousand, two hundred fifteen. By how much was the first number greater?

3. The first four cows through the chute were nice and fat. The first one weighed 1200 lb, the second one weighed 900 lb, the third one weighed 840 lb, and the last one weighed 1050 lb. What was the average weight of the cows?

4. In Graph A (Problem Set 30), how many more (or fewer) inches did it rain in February than in April?

5. (a) What is 1 percent of 6200? (b) What is 24 percent of 6200?

6. (a) What is 1 percent of 52? (b) What is 140 percent of 52?

7. Write $.016\frac{1}{4}$ as a fraction of whole numbers.

8. .6 of what number is 42?

9. What fraction of 57 is 45?

10. What decimal of 280 is 70?

11. Four-ninths of 99 is what number?

12. What is the volume of a solid whose base is the figure shown and whose height is 5 centimeters? Dimensions shown are in centimeters.

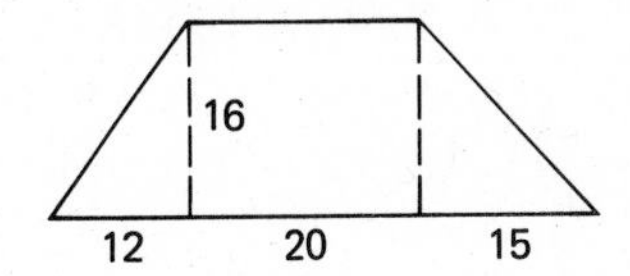

13. Find the least common multiple of 18, 24, and 30.

14. Write $6\frac{16}{17}$ as an improper fraction.

Solve:

15. $\frac{4}{15}x = 8$

16. $x + 1\frac{3}{5} = 3\frac{7}{10}$

Simplify:

17. $72 - 3[(7 - 4)(3 - 1) - 4]$

18. $3^3 + 2^2[2(2 + 1)(2 - 1) - 5]$

19. $4\frac{4}{5} - 3\frac{2}{3} + \frac{7}{15}$

20. $2^3 + 3^2 - \sqrt[3]{27}$

21. $132.7 \times .012$

22. $18{,}251.3 - 62.982$

23. $\frac{135.06}{.003}$

24. $\frac{1\frac{4}{5}}{2\frac{1}{3}}$

25. $\frac{1}{3}\left(\frac{1}{6} + \frac{5}{12}\right) - \frac{1}{36}$

26. $\frac{1}{4}\left(3\frac{1}{3} - \frac{1}{6}\right) + \frac{1}{12}$

27. $2\frac{1}{3} \div 1\frac{1}{6} \times 3\frac{1}{4} \div 1\frac{1}{3}$

Evaluate:

28. $xy + x + xyz - z$ if $x = 24$, $y = \frac{1}{6}$, $z = 3$

29. $\sqrt[p]{q} + pq$ if $p = 3, q = 8$

30. Write 187625113.71 in words.

LESSON 56 Ratio

56.A ratio

Ratio is another word for fraction. If we write

$$\frac{3}{4}$$

we can say that we have written the fraction three-fourths. We can also say that we have written the ratio of 3 to 4. There are many equivalent forms of this ratio (fraction).

$$\frac{3}{4} \quad \frac{6}{8} \quad \frac{15}{20} \quad \frac{90}{120} \quad \frac{150}{200} \quad \frac{450}{600}$$

All these ratios have the same value.

When we write an equation that consists of two ratios connected by an equals sign such as

$$\frac{3}{4} = \frac{6}{8}$$

we say that we have written a **proportion.** Proportions can be either true proportions or false proportions.

$$\frac{1}{2} = \frac{5}{10} \quad \text{true} \qquad \frac{1}{2} = \frac{7}{10} \quad \text{false}$$

The proportion on the left is a true proportion and the proportion on the right is a false proportion.

We find it helpful to note that the cross products of a true proportion are equal to each other.

$$\frac{3}{4} \times \frac{6}{8} \qquad 4 \cdot 6 = 24 \qquad 3 \cdot 8 = 24$$

Here one cross product is 4 times 6, and the other cross product is 3 times 8. Both cross products equal 24.

When one part of a proportion is a variable, the proportion is a **conditional proportion.** Conditional proportions can also be called **conditional equations.**

$$\frac{20}{15} = \frac{4}{x}$$

This proportion is a conditional proportion and is neither true nor false. There is a value of x that will make this proportion a true proportion. To find this value of x, we begin by setting the cross products equal to each other.

example 56.A.1 Solve: $\frac{20}{15} = \frac{4}{x}$

solution We begin by setting the cross products equal to each other.

$$20x = 4 \cdot 15$$

To solve we multiply 4 by 15 and divide both sides by 20.

$$\frac{20x}{20} = \frac{60}{20}$$

$$\mathbf{x = 3}$$

example 56.A.2 Solve: $\frac{4}{5} = \frac{p}{7}$

solution First we set the cross products equal to each other.

$$4 \cdot 7 = 5p$$

And then we divide both sides by 5.

$$\frac{4 \cdot 7}{5} = \frac{5p}{5}$$

$$\mathbf{\frac{28}{5} = p}$$

We decide to leave the answer in the form of an improper fraction.

problem set 56

1. If 440 of the red ones could be purchased for only \$3520, what would be the cost if only 80 were purchased?
2. The first five numbers were 7, 14, 8, 32, and 14. What was the average of these numbers?
3. The first try resulted in 1436. The second try resulted in 1892. The third try was the big one as it came to 4400. By how much did the third try exceed the sum of the first two?
4. Ninety-seven was a good guess. However, 5 times this number increased by six hundred forty-two was the correct number. What was the correct number?
5. (a) What is 1 percent of 3600? (b) What is 16 percent of 3600?
6. (a) What is 1 percent of 25? (b) What is 150 percent of 25?
7. Write $.013\frac{4}{5}$ as a fraction of whole numbers.
8. .8 of what number is 48?
9. What fraction of 60 is 48?
10. What decimal of 350 is 70?
11. What is the volume of a solid whose base is the figure shown and whose height is 2 feet? Dimensions are in feet.

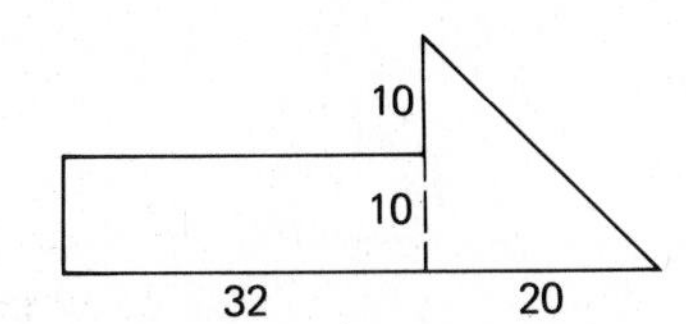

12. Find the least common multiple of 12, 14, and 18.
13. Write $\frac{192}{7}$ as a mixed number.

Solve:

14. $\frac{15}{7} = \frac{4}{x}$ **15.** $\frac{5}{6} = \frac{p}{4}$ **16.** $\frac{7}{15}x = 3$

17. $x + 3\frac{3}{14} = 4\frac{5}{28}$

Simplify:

18. $64 - 2[(3 - 1)(5 - 2) + 1]$ **19.** $2^3 + 2^2[3(2 - 1)(2 + 2) - 7]$

20. $2\frac{1}{3} \cdot 1\frac{3}{4} - \frac{7}{12}$ **21.** $6\frac{2}{5} - 2\frac{1}{4} + \frac{3}{40}$

22. $2^4 + 3^3 - 2^3 + \sqrt[3]{8}$ **23.** $1921 \times .0016$

24. $9218.821 - 61.872$ **25.** $\frac{1721.2}{.02}$

26. $\frac{1}{4}\left(\frac{1}{6} + \frac{1}{3}\right) - \frac{3}{24}$ **27.** $\frac{1}{5}\left(3\frac{1}{4} - 2\frac{1}{3}\right) + \frac{7}{15}$

28. $\dfrac{2\frac{3}{4}}{1\frac{7}{8}}$ **29.** $6\frac{2}{3} \div 1\frac{1}{6} \times 3\frac{1}{3} \div 2\frac{4}{5}$

30. (a) List the prime numbers between 50 and 59. (b) List the multiples of 3 between 50 and 60.

LESSON 57 Decimals, fractions, and percent

57.A decimals, fractions, and percent

We can designate the same part of a whole by using a fraction or a decimal number or by using percent.

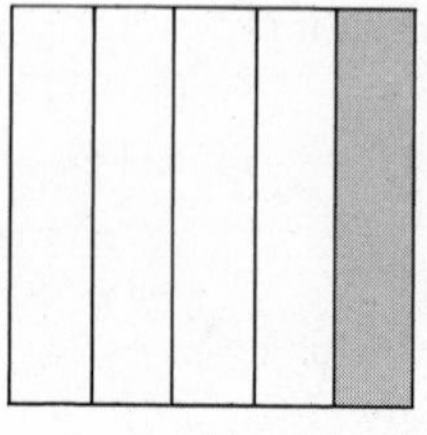

$\frac{1}{5}$

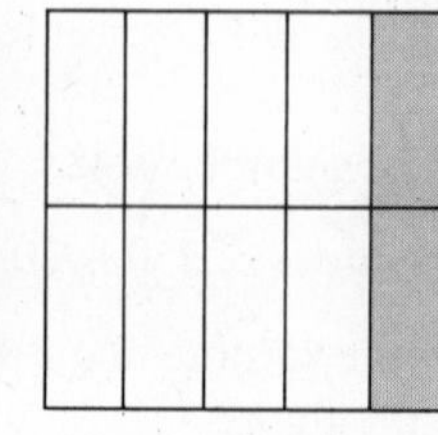

.2

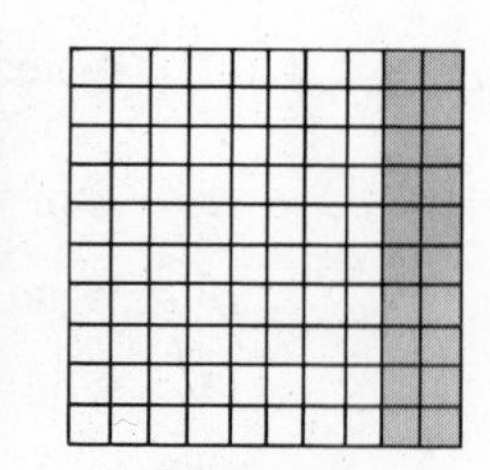

20 percent

In the left figure we have shaded 1 of the 5 parts to represent $\frac{1}{5}$. The center figure has 10 parts, and we have shaded 2 of them to represent two-tenths, which is .2. The right figure has been divided into 100 parts; 20 of these parts have been shaded to represent

20 percent. All three shaded areas are equal. From these diagrams we see that

$$\frac{1}{5} \quad \text{and} \quad .2 \quad \text{and} \quad 20 \text{ percent}$$

are three ways of saying the same thing.

example 57.A.1 What are the decimal and percent equivalents of $\frac{51}{100}$?

solution (a) To find the decimal equivalent of $\frac{51}{100}$, we divide.

$$\frac{51}{100} = \mathbf{.51}$$

(b) To change a decimal to percent, we move the decimal point two places to the right.

$$.51 = \mathbf{51\ percent}$$

57.B reference numbers

Many people understand how to change from one of the three forms to the others, but still they make mistakes. The mistake made most often is moving the decimal point the wrong way. To prevent mistakes, it is helpful to memorize one set of numbers that is correct. Then we use these numbers as reference numbers. In this book we will use the numbers from the last problem as reference numbers.

example 57.B.1 Complete the following table.

Fraction	*Decimal*	*Percent*
$\frac{1}{16}$	(a)	(b)

solution The three blank spaces are provided for the reference numbers. We write them in these spaces.

Fraction	*Decimal*	*Percent*
$\frac{51}{100}$	.51	51
$\frac{1}{16}$	(a)	(b)

(a) To find the decimal form, we divide.

$$\begin{array}{r} .0625 \\ 16\overline{)1.000} \\ \underline{96} \\ 40 \\ \underline{32} \\ 80 \\ \underline{80} \end{array}$$

(b) The reference numbers show that we move the decimal point two places to the right to find the percent form. Thus

.0625 is equivalent to 6.25 percent

Fraction	*Decimal*	*Percent*
$\frac{51}{100}$	.51	51
$\frac{1}{16}$	**.0625**	**6.25**

example 57.B.2 Complete the following table.

Fraction	*Decimal*	*Percent*
(b)	(a)	25

solution First we write the reference numbers.

Fraction	*Decimal*	*Percent*
$\frac{51}{100}$	.51	51
(b)	(a)	25

(a) The reference numbers remind us to move the decimal point two places to the left to get the decimal form. Thus

25 percent is equivalent to the decimal number .25

(b) To change .25 to a fraction, we write it as twenty-five hundredths and then simplify.

$$\frac{25}{100} = \frac{1}{4}$$

Thus we have

Fraction	*Decimal*	*Percent*
$\frac{51}{100}$	.51	51
$\mathbf{\frac{1}{4}}$	**.25**	25

problem set 57

1. There was a plethora of new ideas. The first day there were 151,061 new ideas. The second day there were 69,085 new ideas. How many new ideas were there in all?

2. While new ideas abounded, there was a dearth of suggestions as to how they were to be used. One day there were 143,000 new ideas but only 9614 proposals for their use. How many more new ideas were there than proposals?

3. If 53 of them cost $742, how much would 25 of them cost?

4. One hundred forty-five thousand came on the first day. On the second day 5 times this many came. How many came in all?

*5. Complete the following table. Begin by inserting reference numbers.

Fraction	*Decimal*	*Percent*
(a)	(b)	25
$\frac{1}{16}$	(c)	(d)

6. Write $.001\frac{2}{5}$ as a fraction of whole numbers.

7. .7 of what number is 42?

8. What fraction of 72 is 64?

9. What decimal of 420 is 273?

10. What is the volume of a solid whose base is the figure shown and whose height is 3 inches? Dimensions are in inches.

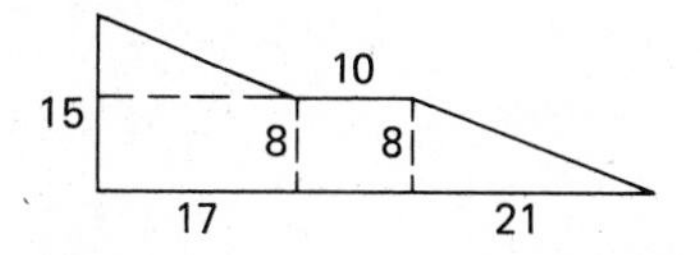

11. Find the least common multiple of 15, 18, and 20.

12. Write $17\frac{1}{12}$ as an improper fraction.

Solve:

13. $\frac{16}{5} = \frac{3}{x}$

14. $\frac{6}{7} = \frac{p}{14}$

15. $\frac{5}{12}x = 9$

16. $x - 2\frac{3}{14} = 1\frac{1}{21}$

Simplify:

17. $38 - 2[(6 - 5)(4 - 1) + 1]$

18. $2^3 + 3^2[(7 - 2)(3 - 1)3 - 25]$

19. $1\frac{1}{3} \cdot 2\frac{3}{4} - \frac{7}{12}$

20. $5\frac{3}{5} - 1\frac{3}{4} + \frac{7}{20}$

21. $2^3 + 3^2 - 2^2 + \sqrt{16}$

22. $16.82 \times .013$

23. $1262.81 - 12.981$

24. $\frac{618.21}{.004}$

25. $\frac{1}{5}\left(\frac{2}{3} + \frac{1}{2}\right) - \frac{2}{15}$

26. $\frac{1}{4}\left(2\frac{3}{4} - 1\frac{1}{8}\right) + \frac{3}{16}$

27. $\dfrac{1\frac{4}{7}}{2\frac{3}{4}}$

28. $5\frac{1}{3} \div 1\frac{1}{6} \times 3\frac{1}{3} \div \frac{1}{6}$

29. Evaluate: $x^2 + 2xy + y^2$ if $x = 2, y = 3$

30. Write 16828732.013 in words.

LESSON 58 Equations with mixed numbers

58.A equations with mixed numbers

When equations contain mixed numbers, it is often helpful if we convert the mixed numbers to improper fractions as the first step.

example 58.A.1 Solve: $2\frac{1}{3}m = 5$

solution As the first step, we write $2\frac{1}{3}$ as an improper fraction.

$$\frac{7}{3}m = 5$$

Now we solve by multiplying both sides by $\frac{3}{7}$.

$$\frac{3}{7} \cdot \frac{7}{3}m = 5 \cdot \frac{3}{7} \qquad \text{multiplied by } \frac{3}{7}$$

$$\mathbf{m = \frac{15}{7}} \qquad \text{simplified}$$

example 58.A.2 Solve: $3\frac{1}{2}k = 4\frac{1}{5}$

solution This time there are two mixed numbers. As the first step, we write them both as improper fractions.

$$\frac{7}{2}k = \frac{21}{5}$$

To solve, we multiply both sides by $\frac{2}{7}$.

$$\frac{2}{7} \cdot \frac{7}{2}k = \frac{21}{5} \cdot \frac{2}{7} \qquad \text{multiplied by } \frac{2}{7}$$

$$\mathbf{k = \frac{6}{5}} \qquad \text{simplified}$$

58.B fractional parts of a number

Mixed numbers can also be used in fractional parts of number problems. Again the first step is to write the mixed number as an improper fraction.

example 58.B.1 $2\frac{1}{2}$ of what number is $7\frac{1}{3}$?

solution We will use the general equation

$$F \times of = is$$

We replace the mixed numbers with their improper fraction equivalents.

$$\frac{5}{2} \times WN = \frac{22}{3}$$

We solve by multiplying both sides by $\frac{2}{5}$.

$$\frac{2}{5} \cdot \frac{5}{2} \times WN = \frac{22}{3} \cdot \frac{2}{5} \qquad \text{multiplied by } \frac{2}{5}$$

$$\mathbf{WN = \frac{44}{15}} \qquad \text{simplified}$$

example 58.B.2 $2\frac{1}{5}$ of $8\frac{1}{8}$ is what number?

solution Our basic equation is

$$F \times of = is$$

We replace both numbers with their improper-fraction equivalents.

$$\frac{11}{5} \cdot \frac{65}{8} = WN \qquad \text{substituted}$$

Now we cancel and then multiply.

$$\frac{11}{1} \cdot \frac{13}{8} = WN \qquad \text{canceled}$$

$$\mathbf{\frac{143}{8} = WN} \qquad \text{multiplied}$$

problem set 58

1. The first three numbers were 181,632; 407,815; and 50,219. What was the average of these numbers?

2. Grand hamadas did not come cheap as 7 of them cost $280,000. What would 30 grand hamadas have cost?

3. The first weighing came to 140,006 pounds. The second weighing only came to 132,468 pounds. By how much was the first weighing greater?

4. The two weighings above were both surpassed by the third weighing, which was 296,475 pounds. What was the total of all three weighings?

5. Complete the following table. Begin by inserting reference numbers.

Fraction	*Decimal*	*Percent*
(a)	(b)	16
$\frac{1}{8}$	(c)	(d)

6. Write $.0013\frac{3}{4}$ as a fraction of whole numbers.

7. .9 of what number is 72?

8. What fraction of 48 is 40?

9. What decimal of 630 is 441?

10. $3\frac{1}{4}$ of what number is $2\frac{1}{3}$?

11. $1\frac{1}{4}$ of $6\frac{1}{3}$ is what number?

12. What is the volume of a solid whose base is the figure shown and whose height is 4 inches? Dimensions are in inches.

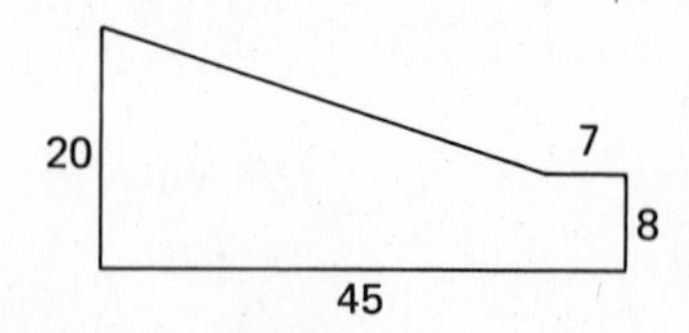

13. Find the least common multiple of 8, 18, and 20.

14. Write $\frac{161}{8}$ as a mixed number.

Solve:

15. $\frac{15}{4} = \frac{10}{x}$

16. $\frac{3}{5} = \frac{p}{12}$

17. $1\frac{3}{5}x = 6$

18. $3\frac{1}{4}p = 5$

Simplify:

19. $49 - 2[(5 - 2^2)(4 + 2) - 5]$

20. $\sqrt[3]{8} + 2^3[2^2(2^3 - 5) - 4]$

21. $2\frac{1}{3} \cdot 3\frac{1}{4} - \frac{11}{12}$

22. $14\frac{2}{3} - 12\frac{7}{8} + \frac{11}{48}$

23. $171.6 \times .007$

24. $1171.61 - 13.321$

25. $\frac{611.51}{.03}$

26. $\frac{1}{6}\left(\frac{1}{3} + \frac{1}{2}\right) - \frac{5}{36}$

27. $\frac{1}{5}\left(\frac{1}{4} - \frac{1}{8}\right) + 2\frac{7}{8}$

28. $\dfrac{6\frac{2}{3}}{2\frac{1}{4}}$

29. $6\frac{1}{4} \div 3\frac{2}{3} \times 2\frac{1}{4} \div \frac{1}{8}$

30. Evaluate: $x^3 + 3xy^2 + 3x^2y + y^3$ if $x = 1, y = 2$

LESSON 59 *Rate*

59.A

rate Rate problems involve three variables. One variable is the rate, and is the result of dividing the other two variables. As an example, consider the statement that Obidiah can wash 4 cars in 2 hours. We can find two rates from this statement. To find one of the rates, we divide 4 cars by 2 hours. To find the other rate, we divide 2 hours by 4 cars.

$$\frac{4 \text{ cars}}{2 \text{ hours}} = 2 \text{ cars per hour} \qquad \frac{2 \text{ hours}}{4 \text{ cars}} = \frac{1}{2} \text{ hour per car}$$

Rates do not have to involve time. If there were 200 redheads in a student body of 4000, we could convey this information as a rate by dividing 200 redheads by 4000 students. Also we could divide 4000 students by 200 redheads.

$$\frac{200 \text{ redheads}}{4000 \text{ students}} = \frac{1}{20} \text{ redhead per 1 student}$$

$$\frac{4000 \text{ students}}{200 \text{ redheads}} = 20 \text{ students per 1 redhead}$$

From these four examples, we see that the word **per** means **divided by.** Thus, if someone says 400 miles per hour, they are talking about the fraction

$$\frac{400 \text{ miles}}{1 \text{ hour}}$$

and if they say 580 bushels per picker, they are describing the fraction

$$\frac{580 \text{ bushels}}{1 \text{ picker}}$$

Of course, both of these can be inverted by using *per* with the other word. That is, 1 hour per 400 miles and 1 picker per 460 bushels would describe the fractions

$$\frac{1 \text{ hour}}{400 \text{ miles}} = \frac{1}{400}\,\frac{\text{hour}}{\text{mile}} \qquad \frac{1 \text{ picker}}{460 \text{ bushels}} = \frac{1}{460}\,\frac{\text{picker}}{\text{bushel}}$$

59.B distance equals rate times time

Thus, we see that there are rates that do not have time as one of the variables. If time is one of the variables, time does not have to be in the denominator. But rates with time in the denominator seem to be encountered more often than other rates. Thus, we will concentrate on these rates. The most common rate is distance per unit time, such as

$$60\,\frac{\text{miles}}{\text{hour}} \qquad 40\,\frac{\text{feet}}{\text{second}} \qquad 30\,\frac{\text{yards}}{\text{minute}}$$

If we multiply one of these rates by the proper time, the answer is distance. Thus we say that

$$\text{Rate} \times \text{time} = \text{distance}$$

We can see this if we multiply 40 miles per hour by 2 hours.

$$\frac{40 \text{ mi}}{1 \cancel{\text{hr}}} \times 2 \cancel{\text{hr}} = 80 \text{ mi}$$

We see that we can cancel the "hr" on top with the "hr" on the bottom.

example 59.B.1 Eudemonia covered the 560 yards in 7 seconds. (a) What was her rate? (b) How far would she go in 60 seconds?

solution (a) To find her rate, we divide 560 yards by 7 seconds.

$$\frac{560 \text{ yards}}{7 \text{ seconds}} = \mathbf{80\,\frac{yards}{second}}$$

(b) To find the distance, we multiply rate by time.

$$\text{Distance} = \text{rate} \times \text{time} \longrightarrow \text{Distance} = \frac{80 \text{ yd}}{1 \cancel{\text{sec}}} \times 60 \cancel{\text{sec}} = \mathbf{4800\ yd}$$

problem set 59

*1. Eudemonia covered the 560 yards in 7 seconds. (a) What was her rate? (b) How far did she go in 60 seconds?

2. The Black Prince reined in after traveling 56 miles in 8 hours. If he travelled at the same rate on the next day, how far could he go in 14 hours?

3. Eau de Vapid meandered 40 miles in 5 hours. Then he sauntered 60 miles in 10 hours. By how much did his first rate exceed his second rate?

4. On the next leg of the trip, Eau covered 40 miles in 20 hours. What was his average speed for all three legs of the trip? (*Note:* Averages cannot be averaged; thus, to find the average speed for all three legs of the trip, the total distance must be divided by the total time.)

5. Complete the following table. Begin by inserting reference numbers.

Fraction	*Decimal*	*Percent*
(a)	.24	(b)
$\frac{3}{5}$	(c)	(d)

6. Write $.0017\frac{2}{3}$ as a fraction of whole numbers.

7. .7 of what number is 490?

8. What fraction of 36 is 24?

9. What decimal of 720 is 420?

10. $3\frac{1}{3}$ of what number is $4\frac{1}{2}$?

11. $2\frac{1}{4}$ of $2\frac{1}{3}$ is what number?

12. What is the volume of a solid whose base is a figure shown and whose height is 6 feet? Dimensions are in feet.

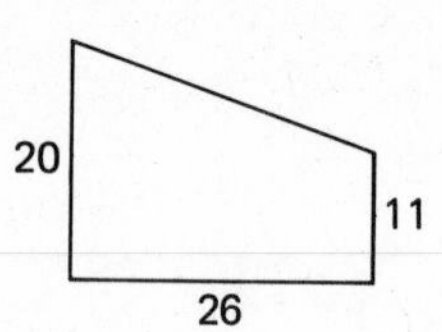

13. Find the least common multiple of 16, 24, and 36.

14. Write $18\frac{17}{18}$ as an improper fraction.

Solve:

15. $\frac{16}{5} = \frac{8}{x}$

16. $\frac{3}{4} = \frac{5}{p}$

17. $2\frac{4}{5}x = 5$

18. $4\frac{1}{5}p = 6$

Simplify:

19. $54 - 2[(6 - 2^2)(3 + 1) - 5]$

20. $\sqrt{16} + 2^2[2(3^2 - 2^2) - 5]$

21. $3\frac{1}{3} \cdot 2\frac{1}{4} - \frac{5}{12}$

22. $15\frac{6}{7} - 3\frac{3}{14} + \frac{9}{14}$

23. $621.8 \times .018$

24. $2612.81 - 14.313$

25. $\dfrac{1821.5}{.7}$

26. $\dfrac{1}{5}\left(\dfrac{1}{2} + 2\dfrac{1}{3}\right) - \dfrac{4}{15}$

27. $\dfrac{1}{4}\left(\dfrac{1}{6} + 1\dfrac{1}{4}\right) - \dfrac{1}{4}$

28. $\dfrac{3\frac{1}{3}}{1\frac{1}{6}}$

29. $3\dfrac{1}{2} \times 6\dfrac{1}{3} \div 2\dfrac{1}{3} \times 1\dfrac{1}{3}$

30. Evaluate: $xyz + z^2 + y^2 - y^z$ if $x = \dfrac{1}{3}, y = 6, z = 2$

LESSON 60 Proportions with fractions

60.A proportions with fractions

There is no change in the method of solving conditional proportions when they contain fractions or mixed numbers. The first step is to cross multiply. Then we divide or multiply as required to complete the solution.

example 60.A.1 Solve: $\dfrac{\frac{2}{3}}{x} = \dfrac{\frac{5}{8}}{\frac{1}{5}}$

solution As the first step we cross multiply.

$$\frac{2}{3} \cdot \frac{1}{5} = \frac{5}{8}x \qquad \text{cross multiplied}$$

$$\frac{2}{15} = \frac{5}{8}x \qquad \text{simplified}$$

We finish by multiplying both sides by $\dfrac{8}{5}$:

$$\frac{8}{5} \cdot \frac{2}{15} = \frac{8}{5} \cdot \frac{5}{8}x \qquad \text{multiplied by } \frac{8}{5}$$

$$\mathbf{\frac{16}{75} = x} \qquad \text{simplified}$$

example 60.A.2 Solve: $\dfrac{1\frac{1}{4}}{2\frac{1}{3}} = \dfrac{5}{x}$

solution As the first step, we change the mixed numbers to improper fractions. Then we cross multiply.

$$\frac{\frac{5}{4}}{\frac{7}{3}} = \frac{5}{x} \qquad \text{improper fractions}$$

$$\frac{5}{4}x = \frac{35}{3} \qquad \text{cross multiplied}$$

Now we finish by multiplying both sides by $\frac{4}{5}$.

$$\frac{4}{5} \cdot \frac{5}{4}x = \frac{35}{3} \cdot \frac{4}{5} \qquad \text{multiplied by } \frac{4}{5}$$

$$x = \frac{28}{3} = \mathbf{9\frac{1}{3}} \qquad \text{simplified}$$

problem set 60

1. Mandolin traveled 400 miles in 20 hours. If he increased his speed 10 miles per hour, how far could he go in the next 20 hours?

2. The magic stage traveled 1600 miles in 40 hours. If the stage reduced speed by 10 miles per hour, how far could it go in the next 40 hours?

3. The first was 14,058 and the second was 21,052. The third was 44,044. How much greater was the third than the sum of the first two?

4. The first measurement was one hundred forty-one ten-thousandths. The second measurement was one thousand, three hundred forty-two hundred-thousandths. What was the sum of the measurements?

5. Complete the following table. Begin by inserting reference numbers.

Fraction	*Decimal*	*Percent*
(a)	.12	(b)
$\frac{5}{6}$	(c)	(d)

6. Write $.0015\frac{6}{11}$ as a fraction of whole numbers.

7. .4 of what number is 316?

8. What fraction of 45 is 40?

9. What decimal of 640 is 560?

10. $2\frac{1}{4}$ of what number is $6\frac{1}{3}$?

11. $3\frac{1}{2}$ of $1\frac{1}{10}$ is what number?

12. What is the volume of a solid whose base is the figure shown and whose height is 2 meters? Dimensions are in meters. (Be very careful.)

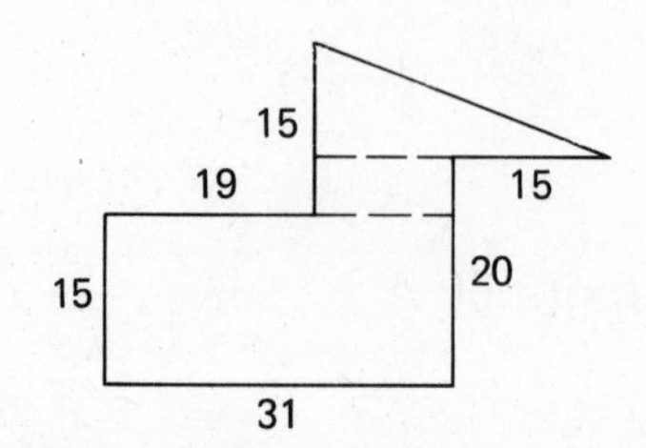

13. Find the least common multiple of 12, 16, and 30.

14. Write $\frac{831}{12}$ as a mixed number.

Solve:

15. $\frac{1\frac{1}{3}}{2\frac{2}{5}} = \frac{6}{x}$

16. $\frac{\frac{2}{3}}{\frac{3}{4}} = \frac{\frac{5}{12}}{p}$

17. $3\frac{5}{6}x = 7\frac{1}{2}$

18. $3\frac{2}{5}p = 1\frac{2}{5}$

Simplify:

19. $64 - 3[(6 - 2^2)(3^2 - 2^2) + 1]$

20. $\sqrt[3]{8} + \sqrt{16}[2(3 - 1) + 7]$

21. $2\frac{1}{4} \cdot 3\frac{2}{3} - \frac{5}{6}$

22. $17\frac{3}{4} - 4\frac{1}{5} + \frac{7}{20}$

23. $361.4 \times .0012$

24. $18{,}191.8 - 19.762$

25. $\frac{1762.3}{.6}$

26. $\frac{1}{4}\left(\frac{1}{2} + 3\frac{1}{4}\right) - \frac{5}{8}$

27. $\frac{1}{3}\left(\frac{1}{2} + 3\frac{1}{3}\right) - \frac{5}{6}$

28. $\frac{4\frac{2}{3}}{2\frac{1}{6}}$

29. $1\frac{1}{2} \times 6\frac{2}{3} \div 3\frac{1}{6} \times 1\frac{2}{3}$

30. Evaluate: $x^2 + y^2 + 2xy + 3x$ if $x = 3, y = 5$

LESSON 61 *Dividing fractional units · Three forms of the distance equation*

61.A dividing fractional units

To divide by a fraction, we invert the fraction and multiply.

$$\frac{4}{\frac{3}{8}} \quad \text{equals} \quad 4 \times \frac{8}{3} = \frac{32}{3}$$

Here to divide 4 by $\frac{3}{8}$, we multiplied 4 by the reciprocal of $\frac{3}{8}$, which is $\frac{8}{3}$. When we divide by fractional units such as miles per hour, we will use a similar procedure. We will divide the numbers and invert the units. To perform the division

$$\frac{40 \text{ mi}}{10 \frac{\text{mi}}{\text{hr}}}$$

we divide 40 by 10 and invert "miles per hour" and multiply by "hour" over "miles."

$$\frac{40 \text{ mi}}{10 \frac{\text{mi}}{\text{hr}}} = 4 \cancel{\text{mi}} \times \frac{\text{hr}}{\cancel{\text{mi}}} = 4 \text{ hr}$$

We will use this procedure whenever we solve distance equations for the time variable.

61.B three forms of the distance equation

To solve the equation

$$4x = 12$$

we divide both sides by 4.

$$\frac{4x}{4} = \frac{12}{4} \longrightarrow x = 3$$

On the left, the 4s cancel because 4 over 4 has a value of 1. If we use D for distance, R for rate, and T for time, we find that the distance equation has the same form as this equation. The distance equation is

$$R \cdot T = D$$

We can solve this equation for R by dividing both sides by T,

$$\frac{R \cdot T}{T} = \frac{D}{T} \longrightarrow R = \frac{D}{T}$$

and we can also solve for T by dividing both sides by R.

$$\frac{R \cdot T}{R} = \frac{D}{R} \longrightarrow T = \frac{D}{R}$$

Thus we see that the distance equation has three forms:

$$\text{(a)} \quad R \cdot T = D \qquad \text{(b)} \quad R = \frac{D}{T} \qquad \text{(c)} \quad T = \frac{D}{R}$$

Some things in mathematics must be memorized. We could memorize all three of these equations, but we usually try to memorize as few things as possible. Thus, it is better to memorize only one form and change this form to another form when we need the other form.

example 61.B.1 The last of the goblins ran the 4600 feet at 230 feet per second. How long did it take him?

solution We begin with the basic equation

$$R \cdot T = D$$

and solve for time by dividing both sides by R.

$$\frac{R \cdot T}{R} = \frac{D}{R} \longrightarrow T = \frac{D}{R}$$

Thus we can find time by dividing the distance by the rate.

$$\frac{4600 \text{ ft}}{230 \frac{\text{ft}}{\text{sec}}} = \frac{4600}{230} \cancel{\text{ft}} \times \frac{\text{sec}}{\cancel{\text{ft}}} = \textbf{20 sec}$$

example 61.B.2 Ethelred the Unready was not ready. Yet he covered the 480 miles in 4 hours. (a) What was his rate? (b) How long would it take him to go 1440 miles?

solution (a) First we solve the basic equation for rate and then we substitute.

$$R \cdot T = D \longrightarrow \frac{R \cdot T}{T} = \frac{D}{T} \longrightarrow R = \frac{D}{T} \longrightarrow R = \frac{480 \text{ mi}}{4 \text{ hr}} = \mathbf{120 \frac{mi}{hr}}$$

(b) Next we solve the basic equation for time.

$$R \cdot T = D \longrightarrow \frac{R \cdot T}{R} = \frac{D}{R} \longrightarrow T = \frac{D}{R}$$

Now we find the time by dividing the distance by the rate.

$$T = \frac{1440 \text{ mi}}{120 \dfrac{\text{mi}}{\text{hr}}} = 12 \cancel{\text{mi}} \cdot \frac{\text{hr}}{\cancel{\text{mi}}} = \mathbf{12 \text{ hr}}$$

problem set 61

*1. The last of the goblins ran the 4600 feet at 230 feet per second. How long did it take him?

*2. Ethelred the Unready was not ready. Yet he covered the 480 miles in 4 hours. (a) What was his rate? (b) How long would it take him to go 1440 miles?

3. The last leg of the journey was to be 1200 miles, so the tour group got an early start. If their rate was 60 miles per hour, how long would it take to get there?

4. Hadrian's men increased their efforts and built 430 feet of wall in 1 day. At this rate, how long would it take to build 7310 feet of wall?

5. Complete the following table. Begin by inserting the reference numbers.

Fraction	*Decimal*	*Percent*
(a)	.22	(b)
$\frac{21}{25}$	(c)	(d)

6. Write $.014\frac{5}{6}$ as a fraction of whole numbers.

7. .8 of what number is 96?

8. What fraction of 52 is 30?

9. What decimal of 700 is 581?

10. $3\frac{1}{5}$ of what number is $7\frac{1}{3}$?

11. $2\frac{1}{10}$ of $1\frac{3}{4}$ is what number?

12. What is the volume of a solid whose base is the figure shown and whose height is 3 feet? Dimensions are in feet.

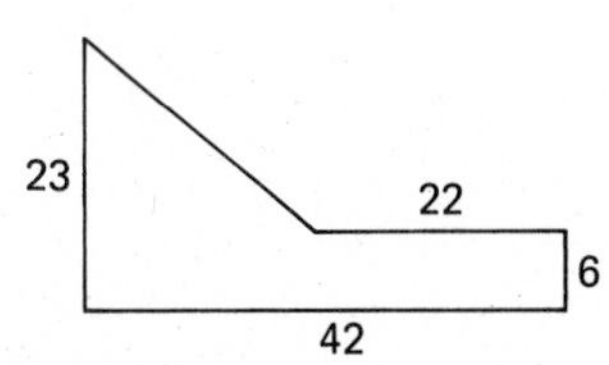

13. Find the least common multiple of 12, 15, and 25.

14. Write $\frac{622}{7}$ as a mixed number.

Solve:

15. $\dfrac{2\frac{1}{4}}{3\frac{1}{3}} = \dfrac{12}{x}$

16. $\dfrac{\frac{7}{12}}{\frac{14}{6}} = \dfrac{\frac{4}{5}}{p}$

17. $4\frac{3}{7}x = 7\frac{1}{4}$

18. $x + 6\frac{3}{14} = 12\frac{2}{7}$

Simplify:

19. $96 - 4[(7 - 2^2)(1 + 2) - 2]$

20. $\sqrt{9} + \sqrt{25}[3(3 - 1) - 2]$

21. $2\frac{1}{3} \cdot 3\frac{1}{4} - \frac{5}{6}$

22. $14\frac{4}{5} - 4\frac{3}{4} + \frac{9}{10}$

23. $6111 \times .0013$

24. $22.8971 - 9.89121$

25. $\dfrac{187.61}{.005}$

26. $\frac{1}{3}\left(\frac{1}{2} + 2\frac{1}{3}\right) - \frac{7}{18}$

27. $\frac{1}{4}\left(\frac{1}{6} + 3\frac{1}{2}\right) - \frac{4}{5}$

28. $\dfrac{4\frac{1}{3}}{5\frac{5}{6}}$

29. $2\frac{1}{2} \times 3\frac{2}{3} \div 1\frac{5}{6} \times \frac{1}{3}$

30. Evaluate: $x^2 + xy + xyz + x^z$ if $x = 6$, $y = \frac{1}{3}$, $z = 2$

LESSON 62 *Circles · Circumference and pi*

62.A circles

Every point on a circle is the same distance from the center of the circle. We call this distance the **radius** of the circle. We call a line that connects two points on a circle a **chord** of the circle. A **diameter** of a circle is a chord that passes through the center of the circle. From the circles shown here, we see that a diameter of a circle is twice the radius of the same circle.

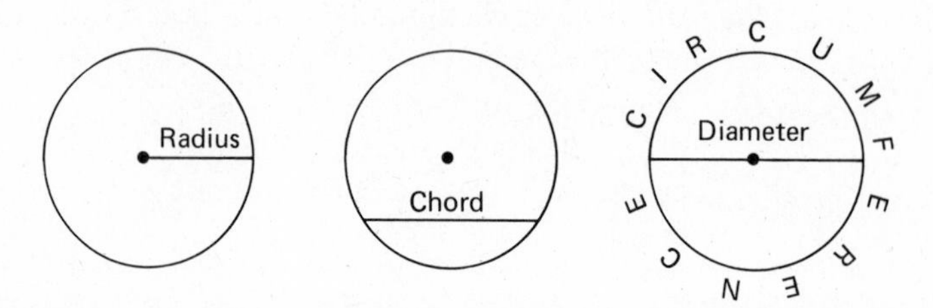

The circumference of a circle is the total distance around the circle. Many ancients thought that the circle was the perfect geometric figure. They were especially interested in the relationship between the diameter of a circle and the circumference of the same circle. They found that three diameters would not go all the way around the circle. There is a little extra left over no matter how large or how small the circle is.

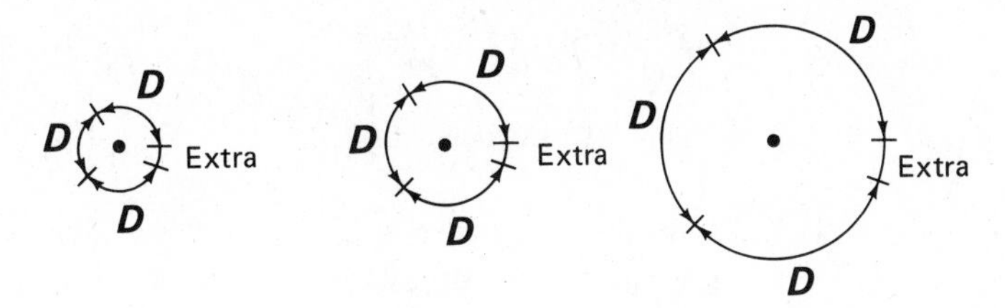

Now we know that the number of times the diameter will go around a circle is approximately

3.14 times

This is not exact because the exact numeral for this number has more digits than could ever be counted. We use the symbol π to represent this number and write it as *pi* (pronounced "pie"). We will use 3.14 as an approximation of π. If we multiply the diameter D by π, we can find the circumference of the circle. If we multiply the square of the radius R by π, we can find the area of the circle.

$$\text{Circumference} = \pi D \qquad \text{Area} = \pi R^2$$

example 62.A.1 The radius of a circle is 5 centimeters (cm). What is the (a) circumference and (b) the area of the circle?

solution (a) If the radius is 5 cm, the diameter is 10 cm. So

$$\begin{aligned}\text{Circumference} &= \pi D \\ &= 3.14(10) \\ &= \mathbf{31.4\ cm}\end{aligned}$$

(b) The area is π times the radius squared. So

$$\begin{aligned}\text{Area} &= \pi R^2 \\ &= 3.14(5)^2 \\ &= \mathbf{78.5\ cm^2}\end{aligned}$$

problem set 62

1. The bus labored up the hill at 400 yards per minute. If it was 28,000 yards to the top, how long would it take the bus to get there?

2. The skiers could traverse 7000 feet of rough terrain in 20 minutes. How long would it take them to traverse 26,950 feet of rough terrain?

3. The new machine bottled 8000 bottles in 4 hours. How long would it take the machine to bottle 40,000 bottles?

4. The tram could cover 40 centimeters in 1 second. How far could the tram go in 20,000 seconds?

5. Complete the following table. Begin by inserting the reference numbers.

Fraction	*Decimal*	*Percent*
(a)	.16	(b)
$\frac{2}{5}$	(c)	(d)

6. Write $.001\frac{12}{13}$ as a fraction of whole numbers.

7. What decimal of 720 is 80?

8. What fraction of 625 is 75?

9. $6\frac{2}{5}$ of what number is $6\frac{1}{4}$?

10. $3\frac{1}{5}$ of $1\frac{1}{2}$ is what number?

11. The radius of a circle is 10 centimeters. What is the (a) circumference of the circle and (b) what is the area of the circle?

12. The base of a solid is shown. If the walls of the solid are 5 inches high, what is the volume of the solid? Dimensions are in inches.

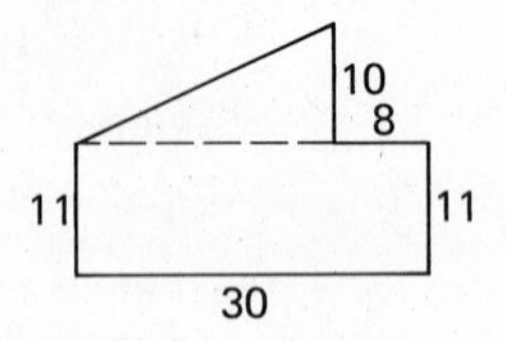

13. Find the least common multiple of 16, 18, and 40.

14. Write $12\frac{11}{12}$ as an improper fraction.

Solve:

15. $\dfrac{2\frac{3}{4}}{4\frac{2}{5}} = \dfrac{6}{x}$

16. $\dfrac{\frac{5}{12}}{\frac{25}{6}} = \dfrac{\frac{5}{24}}{k}$

17. $6\frac{2}{7}x = 2\frac{1}{3}$

18. $x - 3\frac{6}{7} = 7\frac{16}{21}$

Simplify:

19. $56 - 3[3(3 - 1)(2^2 - 2) - 5]$

20. $\sqrt{4} + 5[3(2 - 1) + \sqrt{9}]$

21. $\frac{4}{5} + 2\frac{1}{5} \cdot 1\frac{3}{4}$

22. $13\frac{7}{10} - 3\frac{2}{5} + \frac{4}{15}$

23. $621.3 \times .0014$

24. $21.62 - 18.9261$

25. $\dfrac{123.45}{.006}$

26. $\frac{1}{2}\left(6\frac{2}{3} + \frac{1}{4}\right) - \frac{5}{24}$

27. $\dfrac{3\frac{1}{4}}{6\frac{5}{7}}$

28. $3\frac{6}{7} \times 2\frac{1}{3} \div 4\frac{1}{2} \div \frac{1}{3}$

29. Evaluate: $xyz + x^2 + y^2 - x$ if $x = 12$, $y = 6$, $z = \frac{1}{9}$

30. Write 116213561.82 in words.

LESSON 63 Using both rules to solve equations

63.A using both rules to solve equations

The answer to all four of these equations is 6.

(a) $\frac{x}{3} = 2$ (b) $3x = 18$ (c) $x + 2 = 8$ (d) $x - 3 = 3$

We use the multiplication-division rule to solve (a) and (b) and use the addition-subtraction rule to solve (c) and (d).

(a) $3 \cdot \frac{x}{3} = 2 \cdot 3$

$x = 6$

(b) $\frac{3x}{3} = \frac{18}{3}$

$x = 6$

(c) $x + 2 - 2 = 8 - 2$

$x = 6$

(d) $x - 3 + 3 = 3 + 3$

$x = 6$

Often it is necessary to use both rules to solve the same equation. When it is necessary to use both rules, we always use the addition-subtraction rule first. Then we use the multiplication-division rule.

example 63.A.1 Solve: $3x + 2 = 7$

solution First we subtract 2 from both sides.

$$3x + 2 - 2 = 7 - 2$$
$$3x = 5$$

Now we divide both sides by 3.

$$\frac{3x}{3} = \frac{5}{3} \longrightarrow \boldsymbol{x = \frac{5}{3}}$$

example 63.A.2 Solve: $\frac{2}{3}x - \frac{1}{2} = \frac{10}{3}$

solution First we add $\frac{1}{2}$ to both sides.

$$\frac{2}{3}x - \frac{1}{2} + \frac{1}{2} = \frac{10}{3} + \frac{1}{2}$$

$$\frac{2}{3}x = \frac{23}{6} \qquad \text{simplified}$$

Now we multiply both sides by $\frac{3}{2}$.

$$\frac{3}{2} \cdot \frac{2}{3}x = \frac{23}{6} \cdot \frac{3}{2} \qquad \text{multiplied by } \frac{3}{2}$$

$$\boldsymbol{x = \frac{23}{4}} \qquad \text{simplified}$$

example 63.A.3 Solve: $2\frac{1}{3}x - 1\frac{1}{5} = \frac{23}{10}$

solution We begin by writing the mixed numbers as improper fractions.

$$\frac{7}{3}x - \frac{6}{5} = \frac{23}{10}$$

Next we eliminate the $-\frac{6}{5}$ by adding $\frac{6}{5}$ to both sides.

$$\frac{7}{3}x - \frac{6}{5} + \frac{6}{5} = \frac{23}{10} + \frac{6}{5}$$

$$\frac{7}{3}x = \frac{35}{10} \qquad \text{simplified}$$

$$\frac{7}{3}x = \frac{7}{2} \qquad \text{reduced}$$

Now we multiply both sides by $\frac{3}{7}$.

$$\frac{3}{7} \cdot \frac{7}{3}x = \frac{7}{2} \cdot \frac{3}{7} \qquad \text{multiplied by } \frac{3}{7}$$

$$\boldsymbol{x = \frac{3}{2}} \qquad \text{simplified}$$

problem set 63

1. Roland heard the clarion from afar and increased the speed of the column to 4 miles per hour. How far could he go in 16 hours?

2. Charlemagne's column covered the first 12 miles in 4 hours. He then increased the speed by 2 miles per hour. If the total distance of the trip was 52 miles, how long did it take to finish the trip?

3. Alison traveled at 5 miles per hour for 12 hours. Then she traveled at 8 miles per hour for 8 hours. How far did she go in all?

4. The roses had to be packed 48 to a shipping box. If 4800 boxes were available, how many roses could be shipped?

5. Complete the following table. Begin by inserting the reference numbers.

Fraction	*Decimal*	*Percent*
$\frac{3}{5}$	(a)	(b)

6. Write $.002\frac{3}{17}$ as a fraction of whole numbers.

7. What decimal of 240 is 160?

8. What fraction of 576 is 36?

9. $5\frac{2}{3}$ of what number is $3\frac{1}{4}$?

10. The radius of a circle is 7 feet. What is the (a) circumference and (b) the area?

11. The base of a solid is shown. If the walls of the solid are 1 foot high, what is the volume of the solid in cubic inches? Dimensions are in inches.

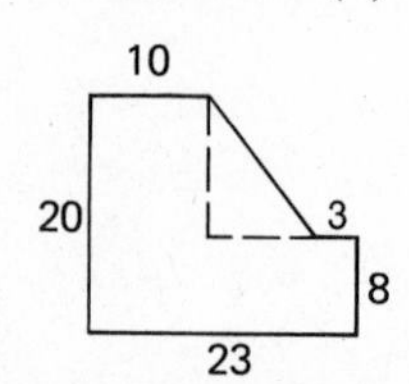

12. Find the least common multiple of 12, 20, and 30.

13. Write $\frac{621}{31}$ as a mixed number.

Solve:

14. $\dfrac{3\frac{1}{5}}{6\frac{3}{10}} = \dfrac{5}{x}$ **15.** $\dfrac{\frac{7}{12}}{\frac{5}{24}} = \dfrac{5}{y}$ ***16.** $2\frac{1}{3}x - 1\frac{1}{5} = \frac{23}{10}$

17. $2\frac{1}{2}x - 2\frac{1}{4} = 3\frac{2}{5}$ **18.** $3x + 7 = 25$

Simplify:

19. $22 + 2[(6 + 1)(3 - 2)3 + 1]$ **20.** $2^2 + 2[2^2(4^2 - 3^2) - 3^3]$

21. $\frac{3}{7} + 3\frac{1}{2} \cdot \frac{2}{3}$ **22.** $6\frac{7}{10} - 4\frac{4}{15} + \frac{7}{30}$ **23.** $17.82 \times .0011$

24. $178.22 - 19.621$ **25.** $\dfrac{192.61}{.005}$ **26.** $\frac{1}{3}\left(2\frac{1}{3} + 6\frac{1}{2}\right) - \frac{7}{24}$

27. $\dfrac{2\frac{1}{3}}{3\frac{1}{5}}$ **28.** $2\frac{1}{4} \times 3\frac{1}{2} \div \frac{3}{4} \times 1\frac{1}{4}$

29. Evaluate: $xy + yz + xyz + x^y$ if $x = 6, y = 2, z = \frac{1}{2}$

30. Write 611,253,612.013 in words.

LESSON 64 Fractional-part word problems

64.A fractional parts

The equation

$$F \times of = is$$

is one of the most important equations in all mathematics. This equation is used to solve many problems that are encountered in everyday life. Also, this equation is almost exactly the same as one form of the percent equation, as we will see in a later lesson.

example 64.A.1 Five-eighths of the gnomes in the magic forest had happy faces. If 840 gnomes had happy faces, how many gnomes lived in the magic forest?

solution We can change the wording of this problem to

$$\frac{5}{8} \text{ of what number is 840?}$$

which we can solve by using the fractional-part-of-a-number equation.

$$F \times of = is$$

We substitute and solve.

$$\frac{5}{8}WN = 840$$

$$\frac{8}{5}\cdot\frac{5}{8}WN = 840\cdot\frac{8}{5} \qquad \text{multiplied by } \frac{8}{5}$$

$$\mathbf{WN = 1344} \qquad \text{simplified}$$

So 1344 gnomes lived in the forest. If 840 had happy faces, then 504 did not have happy faces.

example 64.A.2 On Monday, $2\frac{4}{5}$ times the acceptable number took refuge in the catacombs. If 640 was the acceptable number, how many took refuge in the catacombs on Monday?

solution We can restate the problem as

$$2\frac{4}{5} \text{ of } 640 \text{ is what number?}$$

We will use the fractional part of a number equation, substitute and solve.

$$F \times of = is$$

$$\frac{14}{5}\cdot 640 = WN$$

$$\mathbf{1792 = WN}$$

Thus 640 was an acceptable number but 1152 more took refuge for a total of 1792.

example 64.A.3 When the fog lifted, 400 ghouls were spied skulking near the outskirts. If 240 ghouls were not spied, what fraction of the ghouls were not spied?

solution Since there are 640 ghouls in all, we can restate the problem as

What fraction of 640 is 240?

We use the fractional-part-of-a-number equation, substitute, and solve.

$$F \times of = is$$

$$WF\cdot 640 = 240$$

$$\frac{WF\cdot 640}{640} = \frac{240}{640}$$

$$\mathbf{WF = \frac{3}{8}}$$

Thus $\frac{3}{8}$ of the ghouls were not spied.

problem set 64

*1. Five-eighths of the gnomes in the magic forest had happy faces. If 840 gnomes had happy faces, how many gnomes lived in the magic forest?

*2. On Monday $2\frac{4}{5}$ times the acceptable number took refuge in the catacombs. If 640 was the acceptable number, how many took refuge in the catacombs on Monday?

*3. When the fog lifted, 400 ghouls were spied skulking near the outskirts. If 240 ghouls were not spied, what fraction of the ghouls were not spied?

4. The cutters cut out 400 in 8 hours. If they doubled their rate, how many could they cut out in the next 16 hours?

5. Complete the following table. Begin by inserting the reference numbers.

Fraction	*Decimal*	*Percent*
(a)	.18	(b)

6. Write $.003\frac{5}{16}$ as a fraction of whole numbers.

7. What decimal of 360 is 300?

8. What fraction of 360 is 300?

9. $2\frac{1}{3}$ of $3\frac{1}{4}$ is what number?

10. The radius of a circle is 3 inches. What is the (a) circumference of the circle and (b) what is the area of the circle?

11. The base of a solid is shown. If the solid is 20 centimeters high, what is the volume of the solid in cubic centimeters? Dimensions are in centimeters.

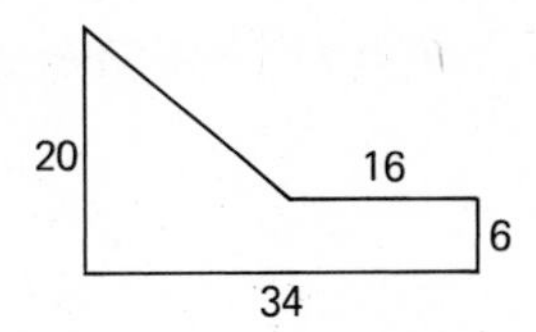

12. Find the least common multiple of 32, 36, and 48.

13. Write $\frac{3211}{7}$ as a mixed number.

Solve:

14. $\dfrac{2\frac{1}{3}}{3\frac{7}{10}} = \dfrac{6}{x}$

15. $\dfrac{\frac{6}{7}}{\frac{12}{21}} = \dfrac{4}{x}$

16. $3\frac{1}{4}x - 2\frac{1}{3} = \frac{17}{12}$

17. $3\frac{1}{6}x - \frac{3}{8} = 1\frac{17}{24}$

18. $17x - 12 = 19$

Simplify:

19. $12 + 2[(6 - 2)(4 - 3)2 - 5]$

20. $3^2 + 2^3[2(3^2 - 4) - 3]$

21. $\frac{2}{7} + 2\frac{1}{3} \cdot \frac{1}{2}$

22. $13\frac{2}{3} - 4\frac{1}{2} + 2\frac{5}{6}$

23. $181.2 \times .013$

24. $1921.61 - 19.897$

25. $\dfrac{175.61}{.7}$

26. $\frac{1}{4}\left(3\frac{1}{5} + 2\frac{1}{2}\right) - \frac{3}{10}$

27. $3\frac{1}{3} \times 2\frac{1}{2} \div \frac{2}{3} \div 2\frac{5}{6}$

28. Evaluate: $xyz + xy + yz$ if $x = \frac{1}{3}$, $y = 36$, $z = 2$

29. Write 621723131.72 in words.

30. Round off 61237899721.2 to the nearest thousand.

LESSON 65 Semicircles

65.A semicircles

Half of a circle is called a **semicircle.** The perimeter of a semicircle is one-half the perimeter of a whole circle. The area of a semicircle is one-half of the area of a whole circle.

$$\text{Perimeter of a semicircle} = \frac{\pi D}{2} \qquad \text{Area of a semicircle} = \frac{\pi R^2}{2}$$

Sometimes it is necessary to find the perimeter or the area of a figure that contains one or more semicircles.

example 65.A.1 Find (a) the perimeter and (b) the area of this figure. Dimensions are in meters.

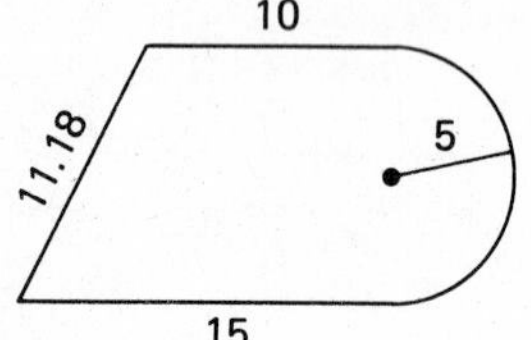

solution (a) The perimeter includes the length of the semicircle.

$$\text{Perimeter} = 11.18 + 15 + 10 + \frac{3.14(10)}{2}$$

$$= 11.18 + 15 + 10 + 15.7 = \mathbf{51.88\ m}$$

(b) The area is the sum of the areas of the triangle, the rectangle, and the semicircle.

$$\text{Area} = \frac{10 \cdot 5}{2} + 10 \cdot 10 + \frac{3.14(5)^2}{2}$$

$$= 25 + 100 + 39.25 = \mathbf{164.25\ m^2}$$

example 65.A.2 Find (a) the perimeter and (b) the area of this figure. Dimensions are in centimeters.

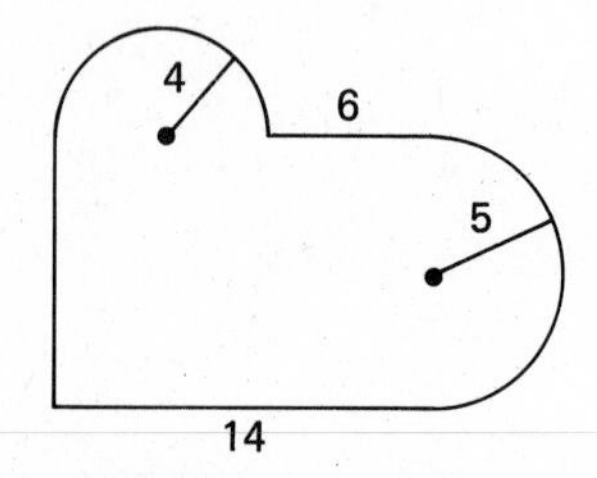

solution (a) The perimeter includes two semicircles.

$$\text{Perimeter} = 10 + 14 + 6 + \frac{3.14(8)}{2} + \frac{3.14(10)}{2}$$

$$= 10 + 14 + 6 + 12.56 + 15.7 = \mathbf{58.26\ cm}$$

(b) The area includes two semicircles.

$$\text{Area} = 10 \cdot 14 + \frac{3.14(4)^2}{2} + \frac{3.14(5)^2}{2}$$

$$= 140 + 25.12 + 39.25 = \mathbf{204.37\ cm^2}$$

problem set 65

1. Three-fourths of the fans did not carry banners. If 1000 fans attended the game, how many fans did carry banners?

2. Seven-sixteenths of the children smirked when they heard the joke. If 210 children smirked, how many children heard the joke?

3. When the score was announced over the PA system, 480 fans were happy. If 3360 fans attended the game, what fraction of the fans did the announcement make happy?

4. The hikers slogged 10 miles through the muck in 5 hours. Then they came to the pavement and were able to triple their rate. How long would it take them to cover 24 miles on the pavement?

5. Complete the following table. Begin by inserting the reference numbers.

Fraction	*Decimal*	*Percent*
$\frac{2}{5}$	(a)	(b)

6. Write $.0003\frac{1}{3}$ as a fraction of whole numbers.

7. .6 of what number is 144?

8. What fraction of 35 is 21?

9. What decimal of 810 is 270?

10. Find (a) the perimeter and (b) the area of this figure. Dimensions are in meters.

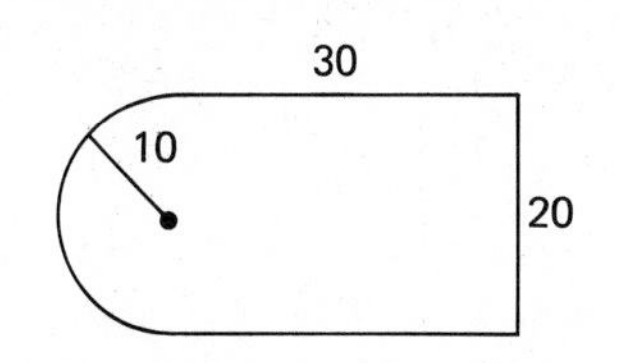

11. The figure shown forms the base of a solid whose sides are 10 centimeters high. What is the volume of the solid in cubic centimeters? Dimensions are in centimeters.

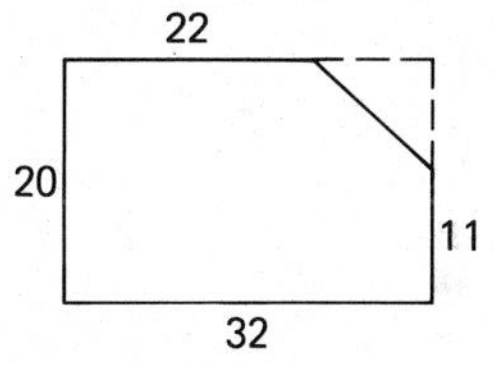

12. Find the least common multiple of 12, 16, and 30.

13. Write $6\frac{12}{17}$ as an improper fraction.

Solve:

14. $\dfrac{3\frac{1}{4}}{2\frac{7}{8}} = \dfrac{6}{x}$

15. $\dfrac{\frac{5}{7}}{\frac{10}{21}} = \dfrac{6}{p}$

16. $4\frac{1}{7}x - 2\frac{2}{3} = 3\frac{1}{4}$

17. $4\frac{1}{3}x - \frac{5}{8} = 1\frac{5}{24}$

18. $18x + 14 = 27$

Simplify:

19. $13 + 3[(3 - 1)(6 - 4)2 - 1]$

20. $2^3 + 3[3(3^2 - 5) + \sqrt{4}]$

21. $1\frac{2}{7} + 3\frac{3}{4} \cdot 2\frac{1}{3}$

22. $12\frac{4}{5} - 4\frac{2}{3} + 11\frac{7}{15}$

23. $191.4 \times .0012$

24. $9218.98 - 178.621$

25. $\dfrac{161.82}{.06}$

26. $\frac{1}{5}\left(2\frac{1}{6} - 1\frac{1}{4}\right) - \frac{1}{30}$

27. $4\frac{3}{4} \times 3\frac{7}{12} \div 2\frac{1}{3} \times \frac{1}{6}$

28. Convert 6 miles to inches.

29. Write 6921321172.1 in words.

30. Evaluate: $xyz + y^2 + x^2 - xy$ if $x = 6, y = 3, z = \frac{1}{9}$

LESSON 66 Ratio word problems

66.A ratio word problems

Often it is desirable to maintain a fixed ratio between two quantities. The conditions are stated in ratio word problems. Three of the four parts are usually given. We use the statement of the problem to help us write a conditional proportion. Then we solve the conditional proportion to find the missing part.

example 66.A.1 The ratio of the number of parrots to the number of macaws was 5 to 7. How many macaws were there when the parrots numbered 750?

solution We were told that the ratio of the number of parrots to the number of macaws was 5 to 7. The first number given (5) goes with the first word (parrots), and the second number given (7) goes with the second word (macaws). Thus we write

$$\frac{P}{M} = \frac{5}{7}$$

Since there are 750 parrots, we replace P with 750. Then we cross multiply and solve.

$$\frac{750}{M} = \frac{5}{7} \quad \text{replaced } P \text{ with 750}$$

$$750 \cdot 7 = 5M \quad \text{cross multiplied}$$

$$\frac{750 \cdot 7}{5} = \frac{5M}{5} \quad \text{divided by 5}$$

$$\mathbf{1050 = M} \quad \text{simplified}$$

Thus, there were 1050 macaws when there were 750 parrots.

example 66.A.2 The ratio of the number of wrigglers to the number of squirmers was 13 to 2. When 26 students had the wriggles, how many were squirming?

solution We begin by writing

$$\frac{W}{S} = \frac{13}{2}$$

Next we replace W with 26. Then we cross multiply and solve.

$$\frac{26}{S} = \frac{13}{2} \quad \text{replaced } W \text{ with 26}$$

$$26 \cdot 2 = 13S \quad \text{cross multiplied}$$

$$\frac{26 \cdot 2}{13} = \frac{13S}{13} \quad \text{divided by 13}$$

$$\mathbf{4 = S} \quad \text{simplified}$$

Thus, when 26 students were wriggling, 4 were squirming.

example 66.A.3 It took $2\frac{1}{2}$ tons of fertilizer to fertilize 75 acres. How many tons would it take to fertilize 450 acres?

solution Some ratio problems do not use the word ratio, but you can tell a constant ratio is implied. In this problem, the ratio is between tons and acres. Either tons or acres can be on top. If tons is on top (or on the bottom) on one side, it must be on top (or on the bottom) on the other side. Thus we can write either

$$\text{(a)}\quad \frac{T}{A} = \frac{T}{A} \qquad \text{or} \qquad \text{(b)}\quad \frac{A}{T} = \frac{A}{T}$$

We decide to put tons on top so we use equation (a).

$$\frac{2\frac{1}{2}}{75} = \frac{T}{A}$$

We were given 450 acres so we replace A with 450. Then we cross multiply and then we solve.

$$\frac{2\frac{1}{2}}{75} = \frac{T}{450} \qquad \text{replaced } A \text{ with } 450$$

$$2\frac{1}{2} \cdot 450 = 75T \qquad \text{cross multiplied}$$

$$\frac{2\frac{1}{2} \cdot 450}{75} = \frac{75T}{75} \qquad \text{divided by } 75$$

$$\mathbf{15 = T} \qquad \text{simplified}$$

Thus it would require 15 tons of fertilizer for 450 acres.

problem set 66

*1. The ratio of the number of parrots to the number of macaws was 5 to 7. How many macaws were there when the parrots numbered 750?

*2. The ratio of the number of wrigglers to the number of squirmers was 13 to 2. When 26 students had the wriggles, how many were squirming?

*3. It took $2\frac{1}{2}$ tons of fertilizer to fertilize 75 acres. How many tons would it take to fertilize 450 acres?

4. Matildabelle traveled the first 100 miles in 4 hours. Then she quadrupled her speed. How long would it take her to travel the next 1400 miles?

5. Complete the following table. Begin by inserting the reference numbers.

Fraction	*Decimal*	*Percent*
(a)	(b)	25

6. Write $.0012\frac{4}{5}$ as a fraction of whole numbers.

7. .3 of what number is 123?

8. What fraction of 36 is 32?

9. What decimal of 360 is 300?

10. $3\frac{1}{2}$ of what number is $4\frac{2}{3}$?

11. Find (a) the perimeter and (b) the area of this figure. Dimensions are in feet.

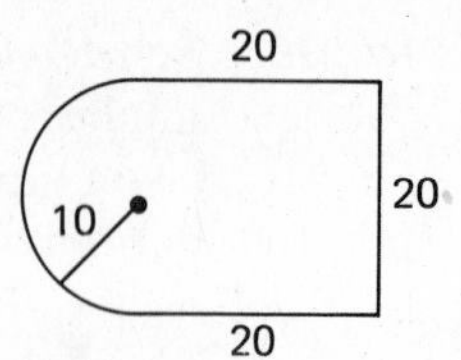

12. The figure shown forms the base of a solid whose sides are 3 meters high. What is the volume of the solid in cubic centimeters? Dimensions are in centimeters.

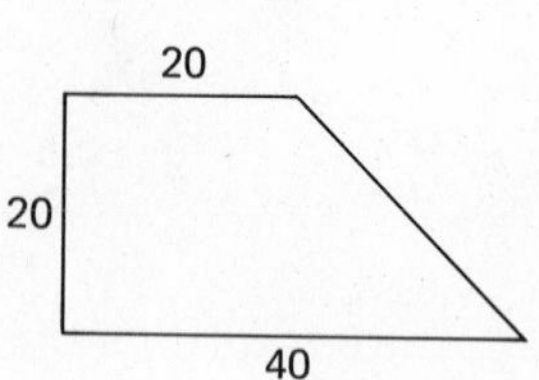

13. Find the least common multiple of 12, 20, and 36.

14. Write $\frac{6131}{5}$ as a mixed number.

Solve:

15. $\frac{2\frac{3}{4}}{3\frac{7}{8}} = \frac{8}{x}$

16. $\frac{\frac{6}{7}}{\frac{10}{21}} = \frac{9}{x}$

17. $5\frac{1}{2}x + 3\frac{1}{3} = 10\frac{1}{7}$

18. $4\frac{6}{7}x - \frac{4}{21} = 11\frac{5}{14}$

19. $18x + 14 = 72$

Simplify:

20. $14 + 2[2^2(2 - 1)(3^2 - 2^3) + 1]$

21. $3^2 + 3[5(3^2 - 2^3) + \sqrt{4}]$

22. $1\frac{3}{7} + 2\frac{1}{3} \cdot 6\frac{1}{2}$

23. $13\frac{4}{5} - 6\frac{1}{8} + 2\frac{3}{40}$

24. $291.8 \times .0013$

25. $10{,}818.17 - 689.891$

26. $\frac{1921.1}{.004}$

27. $\frac{1}{10}\left(2\frac{1}{5} + 7\frac{1}{2}\right) - \frac{3}{10}$

28. $3\frac{1}{4} \times 11\frac{5}{6} \div 2\frac{1}{8} \div \frac{1}{4}$

29. Evaluate: $xyz + x^2 + y^2 + z^z$ if $x = 3, y = 3, z = 2$

30. Round off $182.\overline{157}$ to eight decimal places.

LESSON 67 ***Price per unit***

67.A ratio in pricing

One of the most common uses of ratios is to determine the unit price of things that are for sale.

example 67.A.1 If 20 pounds of beans sold for \$1.20, what was the price of the beans per pound?

solution We write a ratio (fraction) with the total price on top and the number of pounds on the bottom. Note that we write \$1.20 as 120 cents.

$$\frac{120 \text{ cents}}{20 \text{ pounds}}$$

Next we divide 20 into 120 and find that beans are 6 cents per pound.

$$\frac{120 \text{ cents}}{20 \text{ pounds}} \quad \text{means} \quad 6\,\frac{\textbf{cents}}{\textbf{pound}}$$

example 67.A.2 The big can held 16 ounces and cost 80 cents. The small can held 12 ounces and cost 72 cents. Which can was the better buy?

solution We compute the cost per ounce for both cans.

$$\text{Big can} = \frac{80 \text{ cents}}{16 \text{ ounces}} = 5\,\frac{\text{cents}}{\text{ounce}} \qquad \text{Small can} = \frac{72 \text{ cents}}{12 \text{ ounces}} = 6\,\frac{\text{cents}}{\text{ounce}}$$

Thus, the **big can** at 5 cents per ounce is a better buy than the small can at 6 cents per ounce.

problem set 67

*1. If 20 pounds of beans sold for $1.20, what was the price of beans per pound?

*2. The big can held 16 ounces and cost 80 cents. The small can held 12 ounces and cost 72 cents. Which can was the better buy?

3. The ratio of painted faces to natural faces was 3 to 5. If 1800 girls had painted their faces, how many wore no makeup?

4. Seven-eighths of the worker's income was used to pay for necessities. If $5600 was spent on necessities, how much was the worker's income?

5. Complete the following table. Begin by inserting the reference numbers.

Fraction	*Decimal*	*Percent*
(a)	.37	(b)

6. Write $.0015\frac{6}{7}$ as a fraction of whole numbers.

7. .4 of what number is 216?

8. What fraction of 39 is 24?

9. What decimal of 480 is 450?

10. $4\frac{2}{3}$ of what number is $5\frac{4}{5}$?

11. Find (a) the perimeter and (b) the area of this figure. Dimensions are in inches.

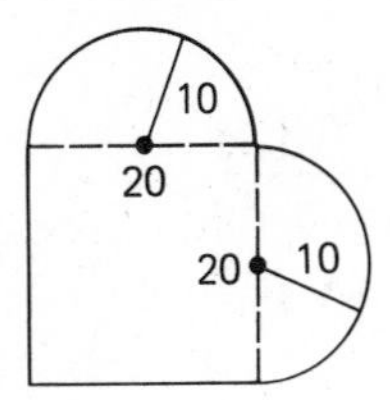

12. What is the volume of a solid whose base is the figure shown in Problem 11 and whose walls are 7 inches tall?

13. Find the least common multiple of 24, 36, and 40.

14. Write $7\frac{31}{32}$ as an improper fraction.

Solve:

15. $\dfrac{3\frac{1}{5}}{2\frac{1}{3}} = \dfrac{4}{x}$

16. $\dfrac{\frac{3}{4}}{\frac{10}{21}} = \dfrac{8}{x}$

17. $6\frac{1}{3}x - 4\frac{7}{8} = 3\frac{3}{4}$

18. $5\frac{6}{7}x - \frac{5}{21} = 11\frac{3}{14}$ **19.** $19x + 6 = 121$

Simplify:

20. $13 + 2[2(3 - 1)(2^2 - 1) + 1]$

21. $2^2 + 3[2^2(2^4 - 3^2) - \sqrt{4}]$ **22.** $2\frac{2}{3} + \frac{1}{6} \cdot 2\frac{1}{4}$

23. $16\frac{11}{12} - 3\frac{5}{6} + 2\frac{7}{18}$ **24.** $12.21 \times .0017$

25. $11{,}716.181 - 891.7891$ **26.** $\frac{136.18}{.03}$

27. $\frac{1}{3}\left(2\frac{1}{3} + 3\frac{1}{4}\right) - \frac{13}{36}$ **28.** $2\frac{1}{3} \times 12\frac{1}{6} \div 1\frac{5}{12} \div \frac{1}{4}$

29. Evaluate: $xy + xyz - x + y^2$ if $x = 6, y = 8, z = \frac{1}{24}$

30. Write 621321611.01 in words.

LESSON 68 *Forms of the percent equation · Percents less than 100*

68.A forms of the percent equation

There are three forms of the percent equation that are commonly used. They are the fractional form (a), the ratio form (b), and the rate form (c).

(a) $\frac{P}{100} \times \text{of} = \text{is}$ (b) $\frac{P}{100} = \frac{\text{is}}{\text{of}}$ (c) Rate $\times$ of = is

We will look closely at forms (a) and (b) now and save form (c) until the next book. In forms (a) and (b), the letter P stands for percent, and the words **of** and **is** are used the same way as in the fractional-part-of-a-number problem. We will use these forms in this lesson to investigate percents less than 100. In a later lesson, we will look at percents greater than 100.

68.B percents less than 100

When the percent is less than 100, the problem is discussing dividing a number into two parts. A two-part diagram that shows this division is helpful. We will use one of these diagrams to discuss the statement

30 percent of 140 is 42

On the left we show the number 140. This is the before diagram and represents 100 percent.

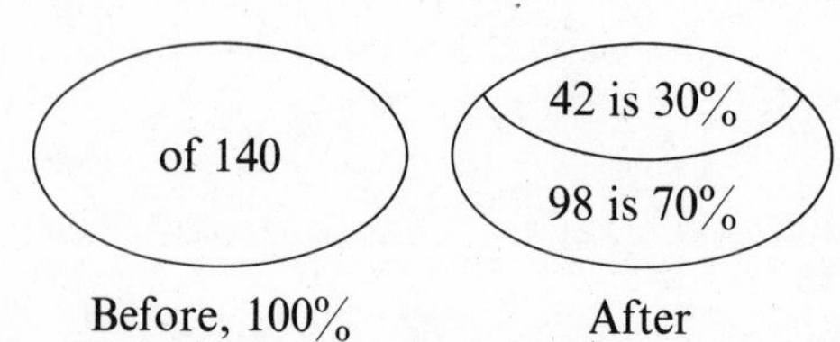

On the right we show 140 divided into two parts. One of the parts is 42, which is 30 percent. The other part is 98, which is 70 percent. The sum of the numbers 42 and 98 is 140. The sum of the 30 percent and 70 percent is 100 percent. **In the diagram, the sum of the two "after" numbers must always equal the "before" number. The sum of the two "after" percents must equal the "before" percent which is always 100 percent.** We will practice by working percent problems and then drawing the two-part diagrams that give us a picture of the solution.

example 68.B.1 Twenty percent of what number is 240? Draw a diagram of the problem.

solution We see that the percent is 20, that *of* goes with "what number" and that *is* goes with 240. We will work the problem twice. We will use the fractional form of the percent equation and then rework the problem using the ratio form of the percent equation to demonstrate that both forms will give the same answer.

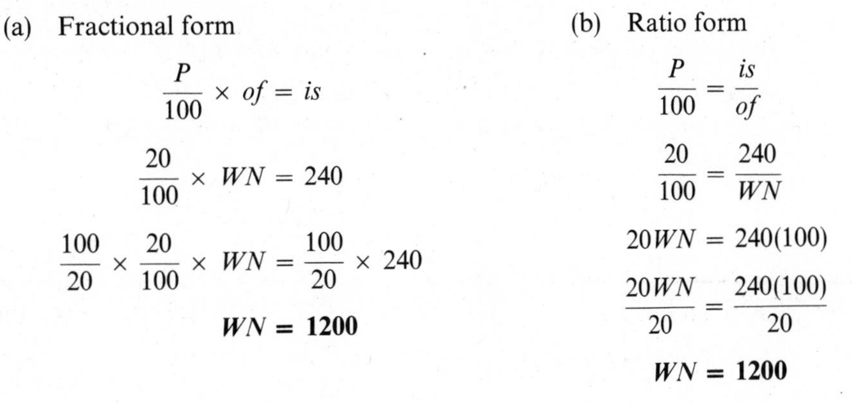

(a) Fractional form

$$\frac{P}{100} \times of = is$$

$$\frac{20}{100} \times WN = 240$$

$$\frac{100}{20} \times \frac{20}{100} \times WN = \frac{100}{20} \times 240$$

$$\mathbf{WN = 1200}$$

(b) Ratio form

$$\frac{P}{100} = \frac{is}{of}$$

$$\frac{20}{100} = \frac{240}{WN}$$

$$20WN = 240(100)$$

$$\frac{20WN}{20} = \frac{240(100)}{20}$$

$$\mathbf{WN = 1200}$$

The diagram shows 1200 as the "before" number. One part is 240, which is 20 percent. Thus the other part must be 960, which is 80 percent.

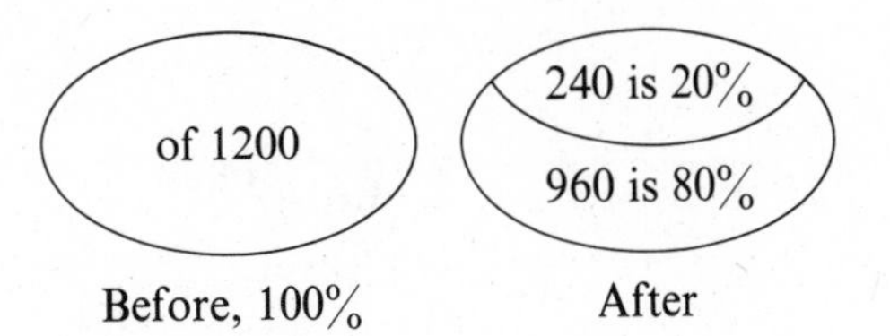

example 68.B.2 Twenty is what percent of 400? Draw a diagram of the problem.

solution This time we will use the fractional form of the percent equation.

$$\frac{P}{100} \times of = is$$

We replace *percent* with *WP*, replace *is* with 20, and replace *of* with 400.

$$\frac{WP}{100} \times 400 = 20$$

To solve we multiply both sides by $\frac{100}{400}$.

$$\frac{100}{400} \cdot \frac{WP}{100} \cdot 400 = 20 \cdot \frac{100}{400}$$

$$\mathbf{WP = 5}$$

Now we draw the diagram (not exactly to scale).

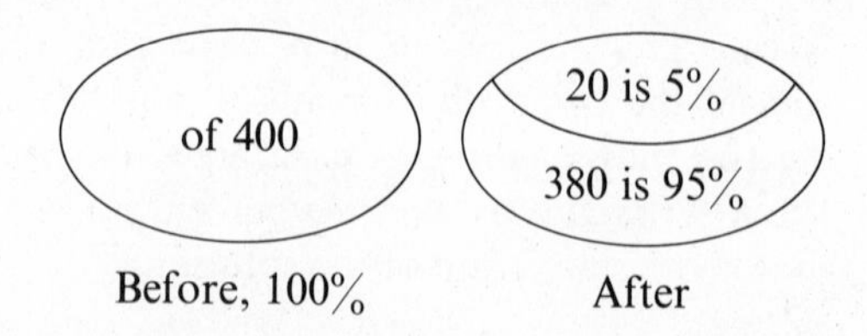

The "after" percents must add to 100 percent so the other percent is 95. The sum of the "after" numbers must equal the "before" number. Thus, the other "after" number must be 380, as shown.

example 68.B.3 Forty percent of 270 is what number? Draw a diagram of the problem.

solution This time we choose to use the ratio form of the percent equation.

$$\frac{P}{100} = \frac{is}{of}$$

We make the replacements, cross multiply, and solve.

$$\frac{40}{100} = \frac{WN}{270} \qquad \text{substituted}$$

$$40 \times 270 = 100WN \qquad \text{cross multiplied}$$

$$\frac{40 \times 270}{100} = \frac{100WN}{100} \qquad \text{divided by 100}$$

$$\mathbf{108} = WN \qquad \text{solved}$$

Now we can draw the diagram.

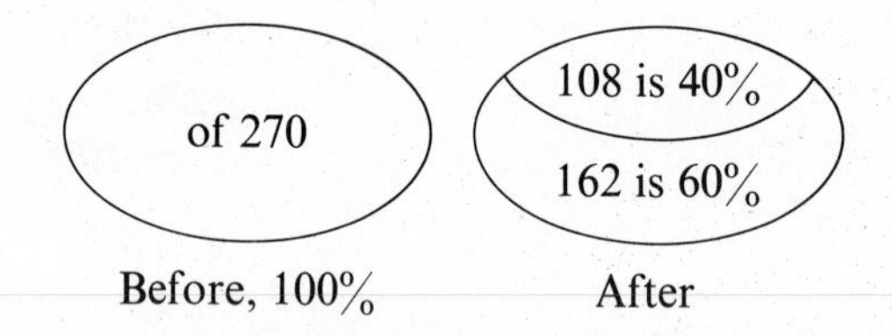

problem set 68

1. The ratio of students who rode horses to school to students who walked to school was 2 to 9. If 18 students walked to school, how many rode their horses to school?

2. Three-and-one-half times as many came as were invited. If 840 came, how many had been invited?

3. Carol and Janet found that three-sixteenths of the horses were defective. If 3200 horses had been purchased, how many were defective?

4. Hyunah managed the first 40 miles in 8 hours. If she then doubled her speed, how long did it take her to manage the next 20 miles?

*5. Twenty percent of what number is 240? Draw a diagram of the problem.

6. Twenty-five is what percent of 250? Draw a diagram of the problem.

7. Thirty percent of 210 is what number? Draw a diagram of the problem.

8. Write $.0016\frac{3}{8}$ as a fraction of whole numbers.

9. Complete the following table. Begin by inserting reference numbers.

Fraction	*Decimal*	*Percent*
(a)	.30	(b)

10. What decimal of 300 is 120?

11. $2\frac{2}{3}$ of what number is $6\frac{1}{7}$?

12. Find (a) the perimeter and (b) the area of this figure. Dimensions are in feet.

13. What is the volume of a solid whose base is the figure shown in Problem 12 and whose walls are 2 feet tall?

14. Write $\frac{6211}{5}$ as a mixed number.

15. Find the least common multiple of 20, 24, and 36.

Solve:

16. $\dfrac{4\frac{1}{3}}{2\frac{6}{7}} = \dfrac{7}{x}$

17. $\dfrac{\frac{4}{5}}{\frac{16}{15}} = \dfrac{x}{3}$

18. $7\frac{1}{3}x + 2\frac{3}{10} = 3\frac{2}{5}$

19. $8\frac{1}{4}x - \frac{5}{11} = 8\frac{13}{22}$

20. $18x + 7 = 122$

Simplify:

21. $14 + 2[2^2(3 - 1)(5 - 2) + 1]$

22. $2^3 + 2^2[3(2^2 - 1) + \sqrt{16}]$

23. $6\frac{1}{3} + \frac{2}{3} \cdot 3\frac{3}{4}$

24. $17\frac{5}{12} - 4\frac{5}{6} + 1\frac{3}{4}$

25. $18.21 \times .0018$

26. $18{,}217.81 - 876.897$

27. $\dfrac{161.17}{.004}$

28. $\frac{1}{4}\left(3\frac{1}{3} + 2\frac{1}{4}\right) - \frac{11}{24}$

29. $3\frac{1}{4} \times 4\frac{5}{12} \div 3\frac{7}{8} \div \frac{1}{4}$

30. Evaluate: $x^y + y^x + xy$ if $x = 2, y = 3$

LESSON 69 Signed numbers · Adding signed numbers

69.A signed numbers

We use a number line to show how numbers are arranged in order. To make a number line, we make evenly spaced marks on a straight line.

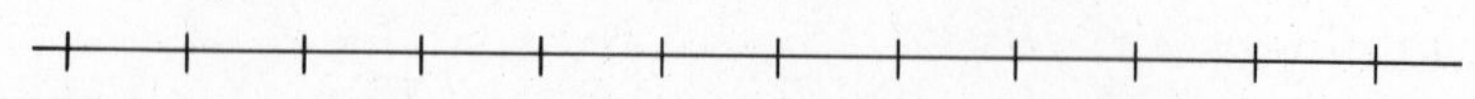

Next we associate the number 0 with one of the marks. We call this point on the line the **origin.**

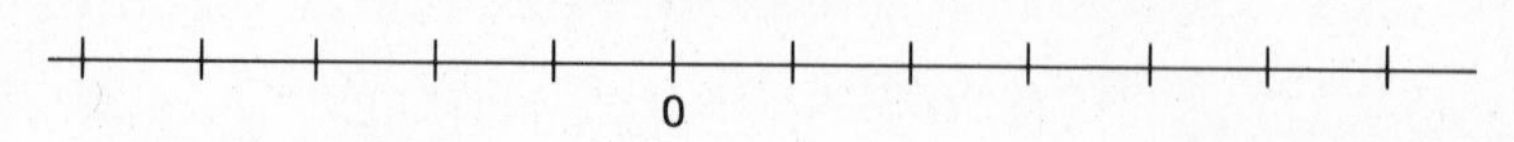

Then we associate the other whole numbers with the points on the right side of the origin.

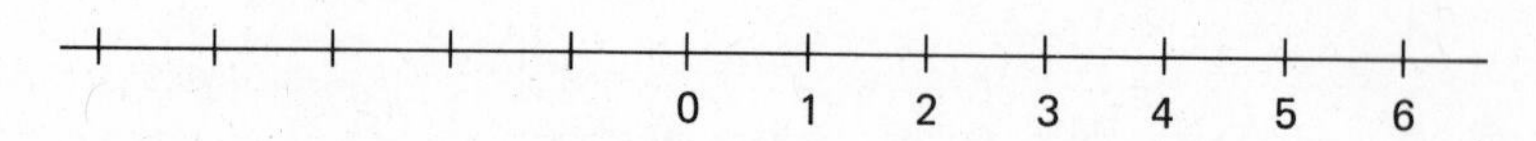

Every other number of arithmetic has a place on the right side of the origin. We call the mark we use to indicate the location of a number the **graph** of the number. When we make a mark to indicate the location of a number, we say that we have **graphed the number.** To demonstrate, we will graph the numbers $2\frac{1}{2}$, $4\frac{3}{5}$, and 5.16.

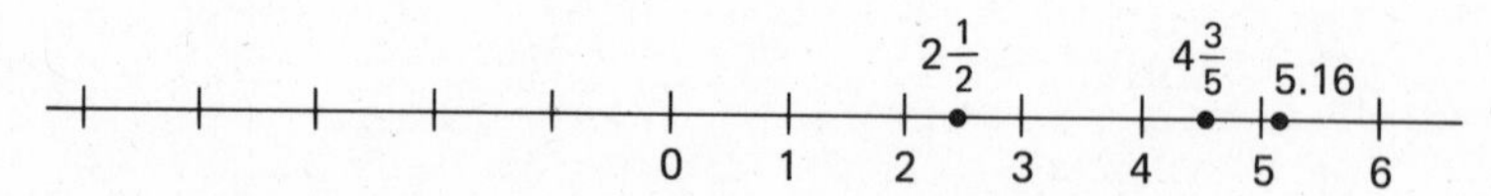

In algebra we call the numbers whose graphs are to the right of the origin the positive numbers. **We can designate a positive number by writing a + sign in front of the numeral or by writing a numeral and not writing a sign.** Thus, both

$$+2 \qquad \text{and} \qquad 2$$

designate the number positive 2.

Every positive number has an opposite whose graph is the same distance on the opposite side of the origin. We call these numbers the **negative numbers.**

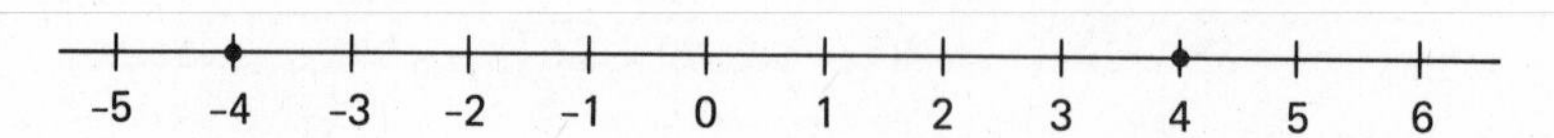

We can see that the graph of $+4$ is 4 units to the right of the origin. The graph of -4 is 4 units to the left of the origin. **We must use a minus sign every time we write a negative number.** It is not necessary to use a plus sign to write a positive number.

-4 means negative 4

$+4$ and 4 both mean positive 4

69.B absolute value

Signed numbers have two qualities. One of the qualities is designated by the sign. The other quality is the "bigness" of the number. The "bigness" is designated by the numerical part of the number. We say that the quality of bigness is the **absolute value** of the number. One of these numbers

$$+4 \qquad -4$$

is positive and one is negative, but both have a bigness or absolute value of 4. We draw vertical lines to indicate the absolute value of a number as we show here.

$$|+4| = 4 \qquad |-4| = 4$$

69.C adding signed numbers

When we are using both positive and negative numbers, we are using numbers that have signs. We call these numbers **signed numbers.** There are rules for adding signed numbers.

Many authors use arrows to help explain the rules for adding signed numbers. Positive numbers are represented by arrows that point to the right. Negative numbers are indicated by arrows that point to the left.

To use these arrows to demonstrate the addition of $+2$ and -3, we begin at 0 on the number line, and draw the $+2$ arrow. From the head of this arrow, we draw the -3 arrow. The head of the -3 arrow is over -1

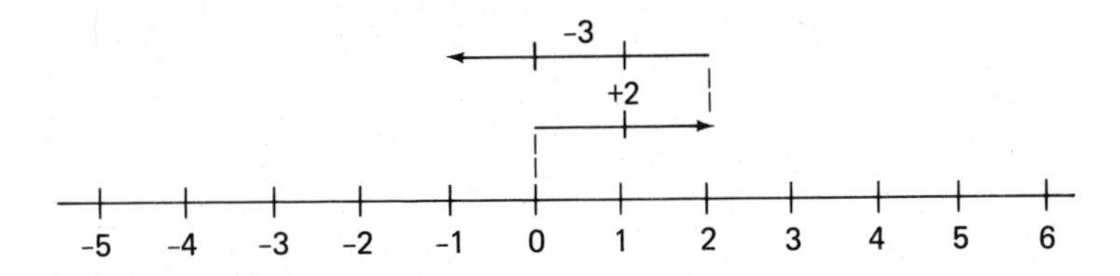

on the number line. Thus,

$$(+2) + (-3) = -1$$

example 69.C.1 Use arrows and a number line to add -3 and $+1$.

solution Either arrow can be drawn first. We will draw the -3 arrow first.

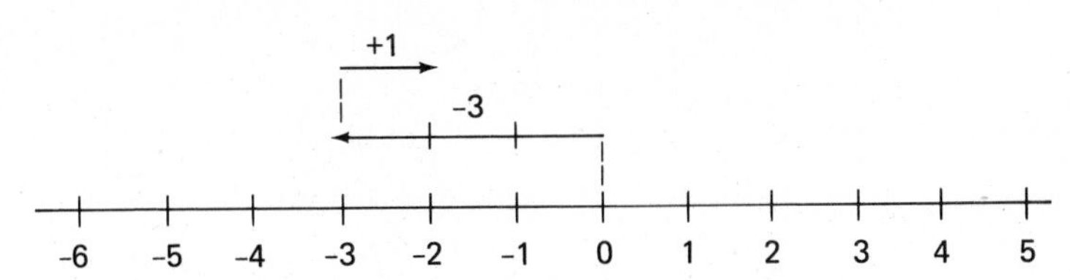

The head of the $+1$ arrow is above -2 on the line, so we say that the answer is -2.

$$(-3) + (+1) = \mathbf{-2}$$

example 69.C.2 Use arrows and a number line to add $(+2) + (+1)$ and $(-2) + (-1)$.

solution We will use a number line for each addition.

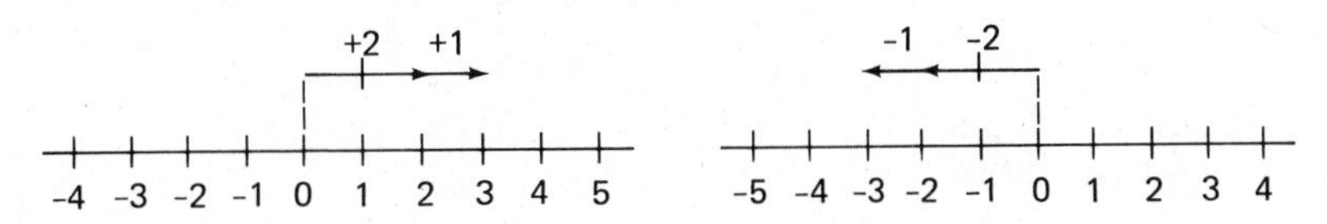

From these diagrams, we see that

$$(+2) + (+1) = \mathbf{+3} \qquad \text{and} \qquad (-2) + (-1) = \mathbf{-3}$$

problem set 69

1. Twenty dozen starters cost a total of \$5. What was the cost of one dozen starters? What would it cost to buy only 10 dozen starters?

2. Three-fifths of the clowns had solid red noses. If there were 150 clowns in all, how many had solid red noses?

3. The ratio of big spenders to the penurious ones was 2 to 5. If 1400 were big spenders, how many were penurious?

4. The cortege wended slowly through the halls of the dead. If it covered 400 yards in the first 20 minutes, how far would it go in the next hour?

*5. Use arrows on a number line to add -3 and $+1$.

*6. Use arrows and number lines to add $(+2) + (+1)$ and $(-2) + (-1)$.

7. Thirty percent of what number is 360? Draw a diagram of the problem.

8. Seventy percent of 420 is what number? Draw a diagram of the problem.

9. Thirty is what percent of 150? Draw a diagram of the problem.

10. Complete the following table. Begin by inserting the reference numbers.

Fraction	*Decimal*	*Percent*
(a)	(b)	40

11. Write $.0015\frac{7}{10}$ as a fraction of whole numbers.

12. What decimal of 420 is 147?

13. Find (a) the perimeter and (b) the area of this figure. Dimensions are in inches.

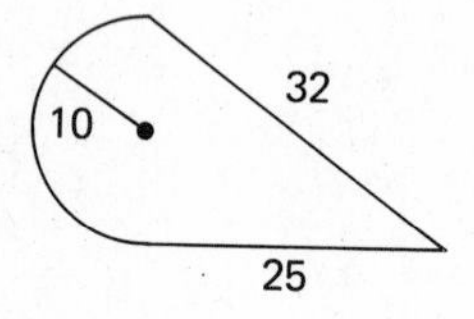

14. What is the volume of a solid whose base is the figure shown in Problem 13 and whose walls are 5 inches high?

15. Write $36\frac{11}{12}$ as an improper fraction.

16. Find the least common multiple of 16, 24, and 36.

Solve:

17. $\dfrac{6\frac{1}{4}}{2\frac{1}{3}} = \dfrac{5}{x}$

18. $\dfrac{\frac{3}{5}}{\frac{9}{25}} = \dfrac{p}{3}$

19. $6\frac{2}{3}x - 4\frac{1}{4} = 3\frac{1}{12}$

20. $20x + 18 = 132$

Simplify:

21. $13 + 2[2^3(2^2 - 1)(3 - 1) + 1]$

22. $2^3 + 2^2[3(2^2 - 1) + \sqrt{16}]$

23. $2\frac{11}{48} - \frac{1}{6} \cdot 3\frac{3}{8}$

24. $18\frac{5}{12} - 12\frac{1}{6} + 1\frac{3}{4}$

25. $19.31 \times .0021$

26. $61,131.812 - 817.819$

27. $\frac{181.31}{.005}$

28. $\frac{1}{5}\left(3\frac{1}{4} - 2\frac{1}{3}\right) + \frac{13}{60}$

29. $6\frac{1}{4} \times 2\frac{1}{3} \div 1\frac{5}{12} \div \frac{1}{6}$

30. Evaluate: $x^x + y^y + x^y + y^x$ if $x = 3, y = 2$

LESSON 70 *Video numbers*

70.A video numbers

The number line is a good way to explain how numbers are arranged in order. The use of the number line and arrows to represent directed numbers helps us understand the rules for adding signed numbers. There are many other visual aids that can be used to understand the addition of signed numbers. Some people use a thought process that could be made into a video game. The players are signed numbers. Players of opposite signs can destroy each other. Thus, -2 could destroy $+2$ parts of a $+5$ and leave $+3$.

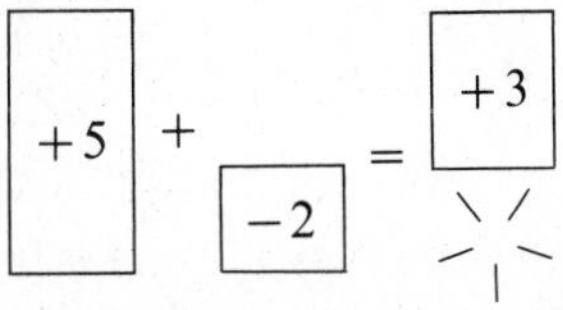

+5 + −2 = +3

$$(+5) + (-2) = \mathbf{+3}$$

If we add -6 and $+4$, the $+4$ will destroy -4 of the -6 and leave -2.

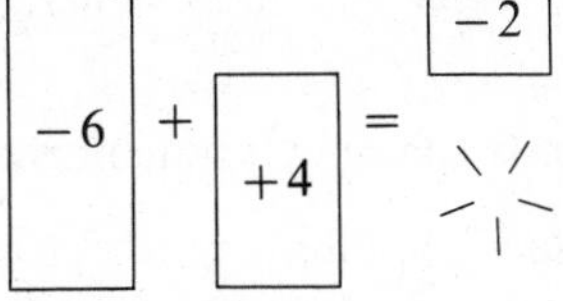

−6 + +4 = −2

$$(-6) + (+4) = \mathbf{-2}$$

Of course, two numbers whose signs are different and whose sizes (absolute values) are equal would destroy all of each other and leave nothing.

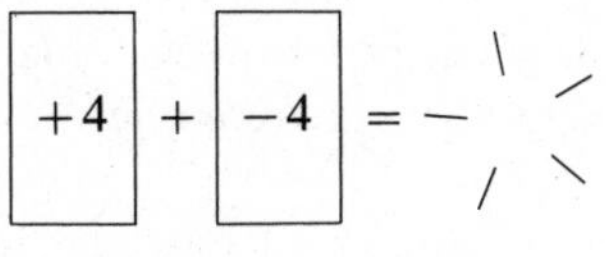

+4 + −4 =

$$(+4) + (-4) = \mathbf{0}$$

example 70.A.1 Use video numbers to explain the addition $(+3) + (-7) + (+1)$.

solution First we draw the boxes.

$$(+3) + (-7) + (+1) = \mathbf{-3}$$

example 70.A.2 Directed numbers (arrows) and video numbers both help explain the rules for adding signed numbers. Use both ways to demonstrate the addition $(+2) + (+1) + (-4)$.

solution On the left we use arrows and on the right we use video numbers.

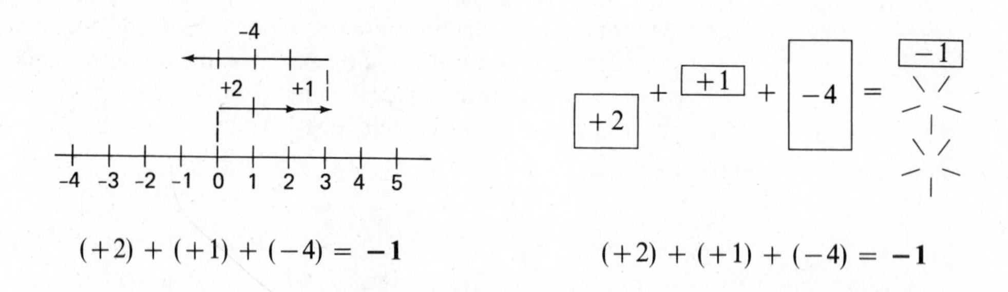

$$(+2) + (+1) + (-4) = \mathbf{-1} \qquad (+2) + (+1) + (-4) = \mathbf{-1}$$

problem set 70

1. The result was 5400. This was $2\frac{1}{2}$ times the expected number. What was the expected number?
2. The ratio of anomalies to normal students was 15 to 2. If 3000 students were anomalies, how many were normal?
3. Seventy-seven hundred made it on time. If the ratio of those on time to those late was 7 to 2, how many were late?
4. The rate of consumption was increased to 1400 bottles per hour. How many bottles could the machine consume if it operated for 8 hours?

*5. Use video numbers to explain the addition $(+3) + (-7) + (+1)$.

*6. Use directed numbers (arrows) and video numbers to demonstrate the addition $(+2) + (+1) + (-4)$.

7. Twenty-five percent of what number is 625? Draw a diagram of the problem.
8. Eighty percent of 280 is what number? Draw a diagram of the problem.
9. Sixty is what percent of 150? Draw a diagram of the problem.
10. Complete the following table. Begin by inserting the reference numbers.

Fraction	*Decimal*	*Percent*
(a)	(b)	23

11. Write $.0016\frac{9}{11}$ as a fraction of whole numbers.

12. What decimal of 600 is 360?

13. Find (a) the perimeter and (b) the area of this figure. Dimensions are in meters.

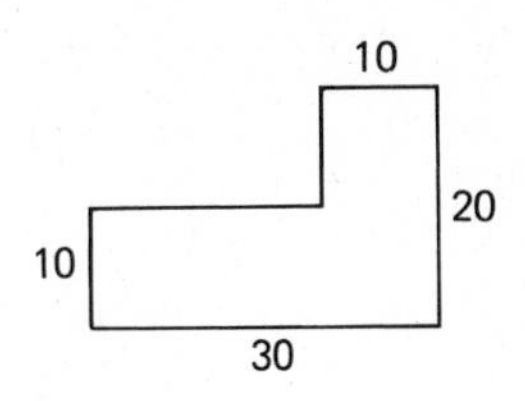

14. What is the volume of a solid whose base is the figure shown in Problem 13 and whose walls are 200 centimeters high?

15. Write $\frac{6213}{30}$ as a mixed number.

16. Find the least common multiple of 18, 24, and 30.

Solve:

17. $\frac{5\frac{1}{3}}{3\frac{1}{6}} = \frac{6}{x}$

18. $\frac{\frac{4}{5}}{\frac{8}{25}} = \frac{x}{5}$

19. $5\frac{1}{3}x - 2\frac{1}{4} = 4\frac{7}{12}$

20. $16x + 19 = 131$

Simplify:

21. $14 + 2[2^2(2^3 - 2^2)(3 - 1) - 1]$

22. $\sqrt[3]{8} + \sqrt{4}[(2^3 - 4)3 + 1]$

23. $3\frac{7}{24} - \frac{1}{8} \cdot 2\frac{5}{6}$

24. $19\frac{6}{7} - \frac{3}{14} + 1\frac{4}{21}$

25. $18.87 \times .0032$

26. $18{,}123.18 - 619.98$

27. $\frac{192.18}{.006}$

28. $\frac{1}{4}\left(3\frac{1}{5} - 2\frac{1}{3}\right) + \frac{7}{60}$

29. $5\frac{1}{3} \times 2\frac{1}{2} \div 2\frac{1}{6} \div 1\frac{1}{12}$

30. Evaluate: $x^y + y^x + xy - x - y$ if $x = 3, y = 2$

LESSON 71 *Rules for addition*

71.A rules for addition

We remember that we call the quality of bigness of a number the absolute value of the number. The two numbers

$$+13 \quad \text{and} \quad -13$$

both have a bigness or absolute value of 13. We use the absolute value in our formal statement of rules for the addition of numbers. First we will state the rule for the addition of numbers whose signs are different. We will use video numbers.

$(+30) + (-14) = \mathbf{+16}$ $\qquad$ $(-30) + (+14) = \mathbf{-16}$

In both cases, we see that the absolute value of the answer is the difference in the absolute values of the numbers. Also the sign of the answer is the same as the sign of the number with the greatest absolute value. We will use these two observations to state formally the rule for adding numbers whose signs are different.

> **The absolute value of the sum of two numbers of opposite sign is the difference of the absolute values of the numbers. The sign of the sum is the same as the sign of the number with the greatest absolute value.**

We will use video numbers to develop the rule for the addition of numbers whose signs are the same.

$(+4) + (+6) = \mathbf{+10}$ $\qquad$ $(-4) + (-6) = \mathbf{-10}$

In both cases the absolute values combine to form the absolute value of the sum. The sign of the sum is the same as the signs of the numbers. We will now state the rule formally.

> **The absolute value of the sum of two numbers of the same sign is the sum of the absolute values of the numbers. The sign of the sum is the same as the sign of the numbers.**

example 71.A.1 Add: $(-4) + (5) + (-10) + (+2)$

solution The numbers can be added in any order, but we will add from left to right.

$(+1) + (-10) + (+2)$ added (-4) and (5)

$(-9) + (+2)$ added $(+1)$ and (-10)

$\mathbf{-7}$ added (-9) and $(+2)$

example 71.A.2 Add: $(-11) + (-3) + (-2) + (7)$

solution Again we will add from left to right.

$(-14) + (-2) + (7)$	added (-11) and (-3)
$(-16) + (7)$	added (-14) and (-2)
$\mathbf{-9}$	added (-16) and (7)

problem set 71

1. The white cell count was only 980. The patient was in trouble as the count should have been at least $5\frac{1}{2}$ times this number. What was the least the white cell count should have been?
2. Doctor Hal and his explorer scouts traveled 32 miles in 8 hours. The next day they had to travel 49 miles, so they increased their speed by 3 miles per hour. How long did it take them to travel the 49 miles?
3. The ratio of winners to losers was 7 to 2. If 4200 of those who came were losers, how many winners were in the throng?
4. Three-fourths of the bats were in the belfry. If 42 bats were in the belfry, how many bats were there in all?
5. Seventy-five percent of 224 is what number? Draw a diagram of the problem.
6. Thirty-five is what percent of 175? Draw a diagram of the problem.
7. Thirty percent of what number is 123? Draw a diagram of the problem.
8. (a) What is 1 percent of 34? (b) What is 115 percent of 34?
9. Complete the following table. Begin by inserting reference numbers.

Fraction	*Decimal*	*Percent*
$\frac{37}{100}$	(a)	(b)

10. Write $.0012\frac{6}{7}$ as a fraction of whole numbers.
11. What decimal of 500 is 125?
12. Find (a) the perimeter and (b) the area of this figure. Dimensions are in meters.

13. What is the volume of a solid whose base is the figure shown in Problem 12 and whose sides are 50 centimeters high?
14. Write $16\frac{11}{12}$ as an improper fraction.
15. Find the least common multiple of 12, 15, and 25.

Solve:

16. $\dfrac{4\frac{2}{3}}{1\frac{5}{6}} = \dfrac{8}{x}$
17. $\dfrac{\frac{3}{5}}{\frac{9}{20}} = \dfrac{10}{x}$
18. $2\frac{1}{4}x + 4\frac{1}{3} = 5\frac{7}{12}$

Simplify:

19. $13 + 2[2^2(3^2 - 2^3)(2^2 + 1) - 15]$

***20.** $(-4) + (5) + (-10) + (+2)$

21. $(-11) + (-2) + (20) + (-7)$

22. $(12) + (-7) + (-10) + (1)$

23. $(-21) + (7) + (-2) + (6)$

24. $16\frac{7}{8} - 2\frac{1}{4} \cdot 1\frac{1}{2} + 1\frac{7}{16}$

25. $166.5 \times .0024$

26. $19{,}621.81 - 698.971$

27. $\frac{213.19}{.005}$

28. $\frac{1}{5}\left(4\frac{2}{5} - 2\frac{1}{3}\right) + \frac{4}{15}$

29. Write 625361811.01 in words.

30. Evaluate: $x^y + y^x + xy + x^2 - y^2$ if $x = 3, y = 2$

LESSON 72 *Graphing inequalities*

72.A graphing inequalities

We have discussed how the number line can be used to help understand the addition of signed numbers. The number line can also be used to help us understand how numbers are arranged in order. Before algebra books began using the number line in the 1960s, many students had difficulty in understanding how -2 could be greater than -5. Now all books use the number line to help define greater than, and we can see on the number line why we say that -2 is greater than -5.

One number is greater than another number if its graph on the number line is to the right of the graph of the other number.

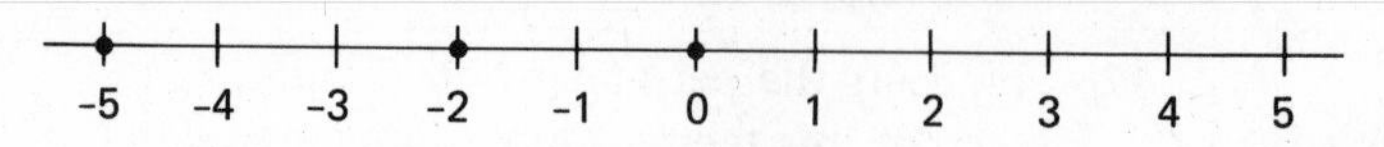

We see that -2 is greater than -5 because the graph of -2 is to the right of the graph of -5. For the same reason, we see that 0 is greater than either -5 or -2.

We call the symbols

$$> \quad \text{or} \quad <$$

the **greater than** and **less than** symbols. The open end is read as "greater than." The closed or pointed end is read as "less than." We only read the end that we come to first. The notation

$$4 > 1 \qquad \text{true}$$

is read from left to right as "4 is greater than 1" and from right to left as "1 is less than 4." If we add half an equals sign, we indicate that **equals to** can also be part of the solution. We read

$$-4 \geq 2 \qquad \text{untrue}$$

from left to right as "-4 is greater than or equal to 2" (which is untrue) or from right to left as "2 is less than or equal to -4" (which is also untrue). The statement

$$4 \geq 2 + 2 \qquad \text{true}$$

is read from left to right as "4 is greater than or equal to 2 + 2." This is true because 4 is equal to 2 + 2.

Inequalities are not always true inequalities. Some statements of inequality are false.

$$4 > 10 \qquad \text{false}$$

This is a false statement because 4 is not greater than 10. Some statements of inequality are conditional inequalities.

$$x > 2 \qquad \text{conditional}$$

The truth or falsity of this inequality depends on the number that is used to replace x. If the replacement number makes the inequality a true inequality, we say the number is a **solution of the inequality.** In the following examples, we show how graphs on the number line can be used to indicate the numbers that satisfy an inequality.

example 72.A.1 Graph: $x > 2$

solution We are asked to graph all numbers that are greater than 2.

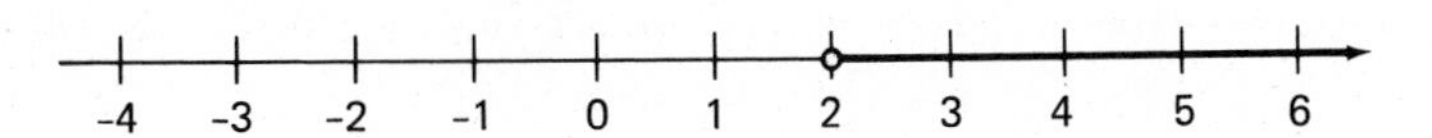

The open circle at 2 tells us that 2 is not a solution because 2 is not greater than 2. The arrow means that all numbers to the right are also solutions.

example 72.A.2 Graph: $x \leq 2$

solution This time we are asked to indicate all numbers that are less than or equal to 2.

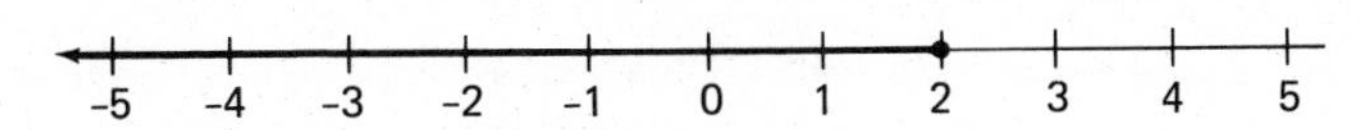

This time the circle at 2 is a solid circle because 2 is a solution to this inequality, as well as all numbers to the left of 2.

problem set 72

1. Rockabilly was old hat, but most of the fans still enjoyed it. If there were 29,000 fans, what fraction enjoyed rockabilly if 3000 fans were in this category?

2. When the trumpet sounded the charge, seven-eights of the malingerers disappeared. If there were 16,000 malingerers in all, how many disappeared?

3. The stagecoach clattered and rattled but still made the 40-mile trip in 5 hours. If the speed for the next leg was increased by 4 miles per hour, how long would it take to travel the 36 miles that remained?

4. The ratio of poltergeists to barghests was 7 to 5. If 112 poltergeists rattled in the closets, how many barghests howled on the hill?

***5.** Graph: $x > 2$

6. Graph: $x \leq 4$

7. Sixty percent of 480 is what number? Draw a diagram of the problem.

8. Forty percent of what number is 220? Draw a diagram of the problem.

9. Forty is what percent of 200? Draw a diagram of the problem.

10. Complete the following table. Begin by inserting the reference numbers.

Fraction	*Decimal*	*Percent*
(a)	.39	(b)

11. Write $.006\frac{4}{11}$ as a fraction of whole numbers.

12. .4 of what number is 148?

13. Find (a) the perimeter and (b) the area of this figure. Dimensions are in feet.

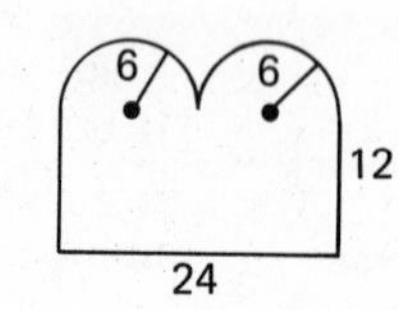

14. What is the volume of a solid whose base is the figure shown in Problem 13 and whose sides are 1 yard high?

15. Write $6\frac{11}{13}$ as an improper fraction.

16. Find the least common multiple of 12, 18, and 30.

Solve:

17. $\dfrac{6\frac{1}{2}}{7\frac{1}{3}} = \dfrac{4}{x}$

18. $\dfrac{\frac{3}{7}}{\frac{9}{21}} = \dfrac{z}{3}$

19. $6\frac{1}{2}x + 3\frac{1}{4} = 5\frac{1}{8}$

Simplify:

20. $4^2 + \sqrt{16}[2^2(3^2 - 2^2)(4^2 - 3^2) - 10]$

21. $(-6) + (4) + (-10) + (+2)$

22. $(-11) + (-2) + (6) + (+3)$

23. $(-36) + (21) + (-6) + (-3)$

24. $13\frac{3}{4} - 1\frac{4}{5} \cdot \frac{3}{2} + \frac{7}{20}$

25. $124.3 \times .0018$

26. $19{,}611.62 - 687.91$

27. $\dfrac{218.31}{.006}$

28. $\frac{1}{6}\left(2\frac{1}{3} \cdot \frac{1}{2} - \frac{3}{4}\right) + \frac{5}{24}$

29. Round off 162189921.618 to the nearest hundred million.

30. Evaluate: $xy + x^2 + y^2 + x^y$ if $x = 4$, $y = 2$

LESSON 73 Inserting parentheses

73.A inserting parentheses

The signed number addition problems thus far have been like this problem.

$$(+4) + (-3) + (-2) + (3)$$

Each number is enclosed in parentheses. Most addition problems do not have parentheses. If they do not have parentheses, the parentheses can be inserted as the first step. If we have this problem

$$-4 + 3 - 2 - 6$$

we consider that the signs tell if the numbers are positive or negative. We enclose each number in parentheses and insert plus signs in between the parentheses.

$$(-4) + (+3) + (-2) + (-6)$$

Now we simplify by adding the signed numbers.

$(-1) + (-2) + (-6)$	added (-4) and $(+3)$
$(-3) + (-6)$	added (-1) and (-2)
$\mathbf{-9}$	added (-3) and (-6)

example 73.A.1 Add: $-2 + 3 - 5 + 7$

solution First we insert parentheses with plus signs between the parentheses.

$$(-2) + (+3) + (-5) + (+7)$$

Now we add.

$(+1) + (-5) + (+7)$	added (-2) and $(+3)$
$(-4) + (+7)$	added $(+1)$ and (-5)
$\mathbf{+3}$	added (-4) and $(+7)$

example 73.A.2 Add: $-3 - 2 - 6 + 2 - 7$

solution We insert parentheses and plus signs as necessary.

$$(-3) + (-2) + (-6) + (+2) + (-7)$$

Now we add.

$(-5) + (-6) + (+2) + (-7)$	added (-3) and (-2)
$(-11) + (+2) + (-7)$	added (-5) and (-6)
$(-9) + (-7)$	added (-11) and $(+2)$
$\mathbf{-16}$	added (-9) and (-7)

73.B order of addition

Numbers that are enclosed in parentheses can be added in any order. Some people begin by adding the positive numbers together and by adding the negative numbers together. Then they add these two sums.

example 73.B.1 Add: $-3 - 2 + 3 - 4 + 5$

solution We begin by inserting the necessary parentheses.

$$(-3) + (-2) + (3) + (-4) + (+5)$$

Now we add numbers whose signs are the same and get

$(-9) + (8)$	combined positives and negatives
$\mathbf{-1}$	added (-9) and (8)

problem set 73

1. Three-sevenths of the cherubs had angelic faces. If there were 420,000 cherubs in the firmament, how many did not have angelic faces?

2. The ratio of chimeras to gargoyles was 2 to 11. If there was a total of 380 chimeras, what was the total number of gargoyles?

3. Joe traveled the first 100 miles in 5 hours. Then he doubled his speed for the last part of the trip. If the total trip was 300 miles, how many hours did it take Joe to travel the last part?

4. Four dozen roses cost $24. Homeratho could only afford six roses. How much did he pay for them?

5. Graph: $x \leq 3$

6. Graph: $x > -3$

7. Eighty percent of 620 is what number? Draw a diagram of the problem.

8. Thirty percent of what number is 120? Draw a diagram of the problem.

9. Sixty is what percent of 300? Draw a diagram of the problem.

10. Complete the following table. Begin by inserting the reference numbers.

Fraction	*Decimal*	*Percent*
(a)	(b)	47

11. Write $.003\frac{6}{11}$ as a fraction of whole numbers.

12. $2\frac{1}{4}$ of what number is $5\frac{3}{8}$?

13. Find the volume in square inches of a solid whose base is the figure shown and whose sides are 1 foot high. Dimensions are in inches.

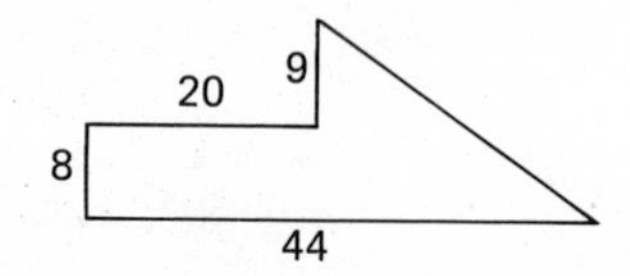

14. What is the surface area of a rectangular solid whose length is 10 feet, whose width is 5 feet, and whose height is 2 feet?

15. Write $\frac{1621}{7}$ as a mixed number.

16. Find the least common multiple of 10, 12, and 18.

Solve:

17. $\dfrac{1\frac{2}{3}}{6\frac{3}{4}} = \dfrac{\frac{3}{4}}{x}$

18. $\dfrac{\frac{4}{7}}{\frac{8}{28}} = \dfrac{p}{6}$

19. $5\frac{1}{2}x + 2\frac{1}{3} = 4\frac{5}{6}$

Simplify:

20. $3^2 + \sqrt{25}[2^2(2^2 - 3)(5^2 - 4^2) - 20]$

*21. $-2 + 3 - 5 + 7$

22. $-4 - 6 + (-3) + 1 - 6$

23. $-6 + (-3) + 4 - 5 + 10$

24. $-2 + (-4) + 5 - 6 + (-7)$

25. $16\frac{9}{20} - 2\frac{1}{2} \cdot 3\frac{2}{5} + \frac{9}{10}$

26. $19.81 \times .0061$

27. $3689.87 - 98.176$

28. $\dfrac{198.71}{.005}$

29. $\dfrac{1}{8}\left(3\dfrac{1}{2} \cdot \dfrac{1}{2} - \dfrac{4}{5}\right) + \dfrac{9}{40}$

30. Evaluate: $\sqrt{x} + x^y + xy$ if $x = 4, y = 3$

LESSON 74 Percents greater than 100

74.A percents greater than 100

In percent problems we always let the initial amount equal 100 percent. If the amount increases, then the final percent will be greater than 100 percent. In these problems, the "after" diagram will always be larger than the "before" diagram.

example 74.A.1 What number is 140 percent of 80? Find the number and then draw a diagram that depicts the problem.

solution This time we will work the problem twice. On the left we use the fractional format and on the right we use the ratio format.

FRACTIONAL FORMAT	RATIO FORMAT
$\dfrac{P}{100} \times of = is$	$\dfrac{P}{100} = \dfrac{is}{of}$

Now we make the necessary substitutions and solve.

$\dfrac{140}{100} \cdot 80 = WN$	$\dfrac{140}{100} = \dfrac{WN}{80}$
$\dfrac{11{,}200}{100} = WN$	$140 \cdot 80 = 100WN$
$\mathbf{112 = WN}$	$\dfrac{11{,}200}{100} = \dfrac{100WN}{100}$
	$\mathbf{112 = WN}$

The "after" diagram must be larger than the "before" diagram because 112 is greater than 80.

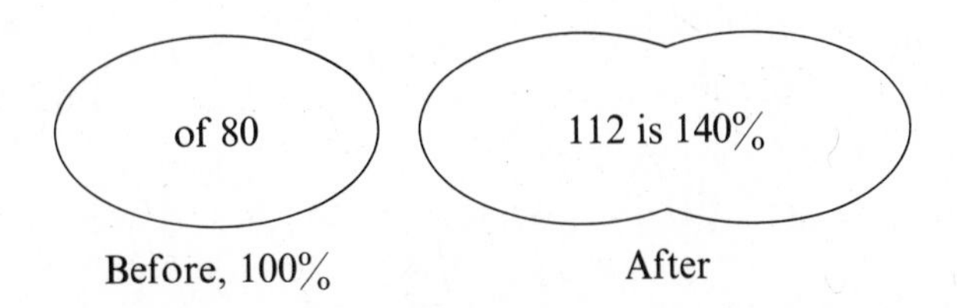

example 74.A.2 If 160 is increased by 160 percent, what is the result? Work the problem and then draw a diagram that depicts the problem.

solution **We must be careful here. The original number is 160 and is 100 percent. If we increase the percent by 160, the final percent will be 260 percent.** We will use the fractional format this time.

$$\frac{260}{100} \times 160 = WN \longrightarrow 2.6 \times 160 = WN \longrightarrow \boldsymbol{WN = 416}$$

Our diagrams show a "before" of 160 and an "after" of 416, which is 260 percent.

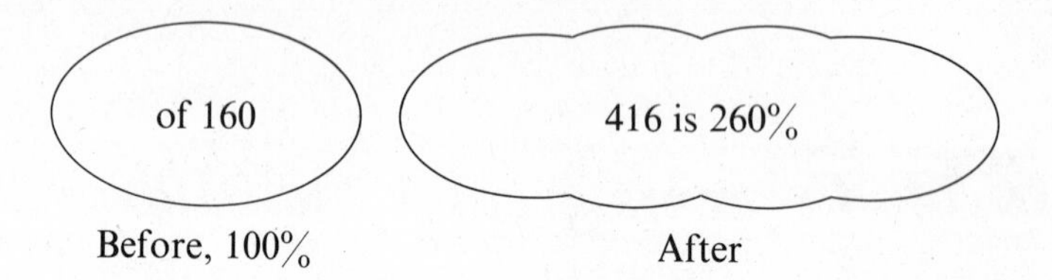

example 74.A.3 What percent of 90 is 306? Work the problem and then draw a diagram that depicts the problem.

solution This time we will use the ratio format.

$$\frac{P}{100} = \frac{is}{of} \longrightarrow \frac{WP}{100} = \frac{306}{90} \longrightarrow 90WP = 30{,}600$$

$$\longrightarrow \frac{90WP}{90} = \frac{30{,}600}{90} \longrightarrow \boldsymbol{WP = 340}$$

of 90 | 306 is 340%

Before, 100% | After

problem set 74

1. From her turret the chatelaine could see 440 sheep in the meadow. If she could only see four-ninths of the sheep, how many sheep were there in all?

2. The ratio of brigands to highway robbers was 5 to 7. If 450 brigands were skulking in the shadows, how many highway robbers also skulked there?

3. Falstaff ate 160 ounces of comestibles and bragged that he could have eaten $3\frac{1}{2}$ times that much. How many ounces did Falstaff think he could have eaten?

4. Jasper had to travel 480 kilometers before nightfall. His speed would be 20 kilometers per hour. How long would his trip take?

5. Graph: $x \geq -2$

6. Graph: $x < 5$

*7. What number is 140 percent of 80? Find the number and then draw a diagram that depicts the problem.

8. If 160 is increased by 140 percent, what is the result? Work the problem and then draw a diagram that depicts the problem.

9. What percent of 90 is 405? Work the problem and then draw a diagram that depicts the problem.

10. Complete the following table. Begin by inserting the reference numbers.

Fraction	*Decimal*	*Percent*
(a)	.12	(b)

11. Write $.0013\frac{3}{7}$ as a fraction of whole numbers.

12. What fraction of 40 is 24?

13. Find the volume of a solid whose base is the figure shown and whose sides are 2 yards high. Dimensions are in feet.

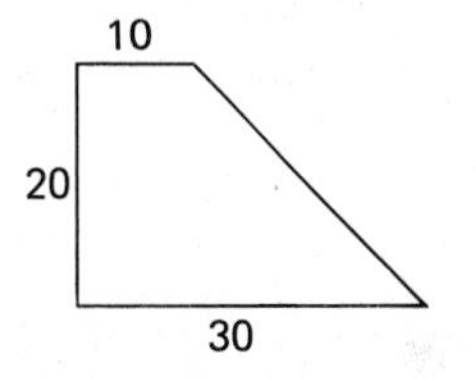

14. What is the surface area of this prism? Dimensions are in feet.

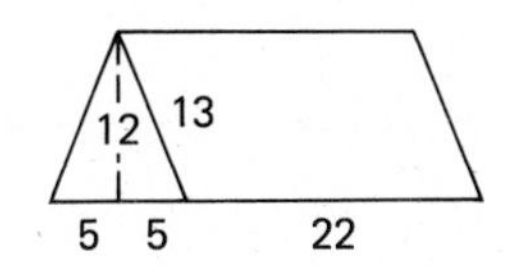

15. Write $6\frac{12}{29}$ as an improper fraction.

16. Find the least common multiple of 12, 18, and 27.

Solve:

17. $\dfrac{1\frac{3}{4}}{3\frac{3}{8}} = \dfrac{\frac{1}{3}}{x}$

18. $\dfrac{\frac{5}{6}}{\frac{25}{36}} = \dfrac{p}{12}$

19. $6\frac{1}{3}x + 1\frac{5}{6} = 7\frac{2}{3}$

Simplify:

20. $2^3 + \sqrt{36}[3^2(3^2 - 2^3)(4^2 - 2^4) + 3]$

21. $-5 + 6 + (-5) + 7$

22. $-4 - 6 + 10 + (+6)$

23. $-2 - 2 + (-3) - 6 + (+3)$

24. $-6 + (-3) - 6 - 10 + 1$

25. $17\frac{3}{5} - 1\frac{3}{5} \cdot \frac{1}{2} + 3\frac{1}{5}$

26. $.1315 \times .0012$

27. $1328.613 - 12.3621$

28. $\frac{1}{6}\left(2\frac{1}{2} \cdot \frac{1}{3} - \frac{2}{3}\right) + \frac{25}{36}$

29. Evaluate: $xyz + yz + y^z$ if $x = \frac{1}{3}$, $y = 6$, $z = 2$

30. Write 6289116131.2 in words.

LESSON 75 *Multiplication and division of signed numbers*

75.A rules for multiplication and division

It is easy to memorize the rules for multiplying and dividing signed numbers. These rules are difficult to explain and to understand but not difficult to use. In this book, we will concentrate on learning how to use the rules.

The rules for the signs of the answers are the same for multiplication and division.

> RULES FOR BOTH MULTIPLICATION AND DIVISION
>
> 1. **The answer is a positive number if the signs of the numbers are alike.**
> 2. **The answer is a negative number if the signs of the numbers are different.**

example 75.A.1 Simplify: (a) $(4)(2)$ (b) $(-4)(-2)$ (c) $(-4)(2)$ (d) $(4)(-2)$

solution The answer to (a) and (b) is $+8$ because both numbers have the same sign. **Note that (-4) times (-2) equals $+8$.**

(a) $(4)(2) = \mathbf{8}$ (b) $(-4)(-2) = \mathbf{8}$

The answers to (c) and (d) are -8 because the numbers have different signs.

(c) $(-4)(2) = \mathbf{-8}$ (d) $(4)(-2) = \mathbf{-8}$

example 75.A.2 Simplify: (a) $\frac{8}{2}$ (b) $\frac{-8}{-2}$ (c) $\frac{8}{-2}$ (d) $\frac{-8}{2}$

solution The answers to (a) and (b) are both positive because the numbers have the same signs. **Note that a negative number divided by a negative number equals a positive number.**

(a) $\frac{8}{2} = \mathbf{4}$ (b) $\frac{-8}{-2} = \mathbf{4}$

The answers to (c) and (d) are both negative because the signs of the numbers are different.

(c) $\frac{8}{-2} = \mathbf{-4}$ (d) $\frac{-8}{2} = \mathbf{-4}$

problem set 75

1. Four-fifths of the artichokes had disfiguring growths. If 500 did not have these growths, how many artichokes were there in all?
2. Chicanery was rampant at the medicine show. If seven-eighths of the customers were chicaned and there were 3200 customers, how many were not chicaned?
3. The demagogue pandered to five-sixths of those who were prejudiced. If he pandered to 1200 who were prejudiced, how many were prejudiced in all?
4. After traveling 120 miles in 3 hours, Rudolph reduced the speed of the sled by 10 miles per hour. How long did it take him to travel the next 450 miles?
5. Graph: $x \geq -3$
6. Graph: $x < 4$

7. What number is 180 percent of 60? Draw a diagram that depicts the problem.

8. If 140 is increased by 160 percent, what is the result? Draw a diagram that depicts the problem.

9. What percent of 80 is 200? Draw a diagram that depicts the problem.

10. Complete the following table. Begin by inserting the reference numbers.

Fraction	*Decimal*	*Percent*
$\frac{29}{100}$	(a)	(b)

11. Write $.0011\frac{8}{25}$ as a fraction of whole numbers.

12. What decimal of 480 is 360?

13. Express in cubic centimeters the volume of a solid whose base is the figure shown and whose sides are 1 meter high. Dimensions are in centimeters.

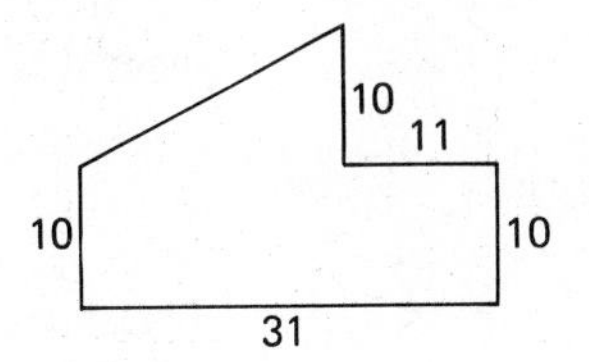

14. Find the perimeter of the following figure. Dimensions are in feet.

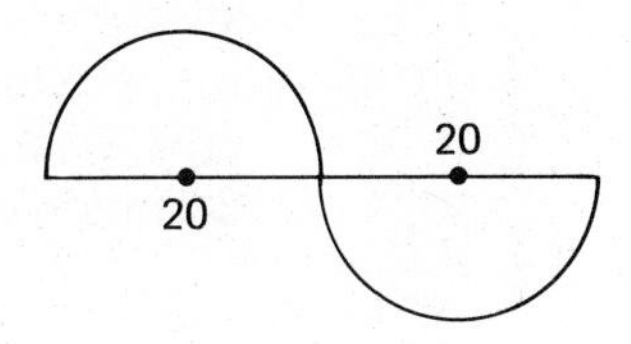

15. Write $\frac{12{,}311}{5}$ as a mixed number.

16. Find the least common multiple of 12, 20, and 30.

Solve:

17. $\dfrac{2\frac{6}{7}}{1\frac{1}{14}} = \dfrac{\frac{1}{2}}{y}$

18. $\dfrac{\frac{6}{5}}{\frac{12}{25}} = \dfrac{5}{p}$

19. $2\frac{1}{5}x + 2\frac{1}{10} = 6\frac{3}{20}$

Simplify:

20. $2^4 + \sqrt{25}\,[2^2(3^2 - 2^3)(2^2 + 1) - 1]$

*21. (a) $(4)(2)$ (b) $(-4)(-2)$ (c) $(-4)(2)$ (d) $(4)(-2)$

22. (a) $(6)(2)$ (b) $(-6)(-2)$ (c) $(-6)(2)$ (d) $(6)(-2)$

23. (a) $\frac{6}{2}$ (b) $\frac{-6}{-2}$ (c) $\frac{6}{-2}$ (d) $\frac{-6}{2}$

24. $-6 + (-2) - 5 - 4 + 3$

25. $-7 + 3 - 31 + (+3)$

26. $.1319 \times .0014$

27. $17\frac{1}{3} - 3\frac{1}{2} \cdot 1\frac{1}{3} + \frac{5}{6}$

28. $\frac{1}{6}\left(2\frac{1}{3} \cdot \frac{1}{2} - \frac{1}{6}\right) + \frac{13}{36}$

29. Express 2540 as the product of primes.

30. Evaluate: $\sqrt[3]{x} + \sqrt{y} + xy$ if $x = 8, y = 16$

LESSON 76 Algebraic addition · Using mental parentheses

76.A algebraic addition

If we have to simplify the expression

$$8 - 5$$

there are two ways to do it. On the left (below) we let the minus sign indicate subtraction. We simplify by subtracting 5 from 8.

$8 - 5$	$(8) + (-5)$
3	3
subtracted	added

On the right we let the minus sign indicate that -5 is a negative number. We inserted parentheses and then we added. We call this procedure **algebraic addition.** It might seem that we are making a hard problem out of an easy problem. This is not so because in a later lesson we will discuss the simplification of complicated expressions such as

$$-(-4) + \{-[-(-2)]\}$$

These problems are easy to simplify if we use algebraic addition but are much more difficult if we use subtraction. In problems that involve signed numbers, we should try to avoid the use of subtraction because using subtraction often leads to trouble. Algebraic subtraction is defined, however, and some people use algebraic subtraction. The definition of algebraic subtraction is given here.

> ALGEBRAIC SUBTRACTION
>
> To subtract algebraically, we change the sign of the subtrahend and add.

To use this rule to simplify

$$(8) - (+5)$$

we change the sign of the 5 from + to − and add instead of subtracting.

$$(8) + (-5) = 3$$

This is exactly the same process we use for algebraic addition except that this time we used the definition of subtraction to explain what we did.

76.B using mental parentheses

Writing the parentheses is not necessary. When we see an expression such as

$$-4 + 2 - 5 - 7$$

we consider that the signs go with the numbers. We can look at this problem and think

$$(-4) + (+2) + (-5) + (-7)$$

but not write down the parentheses.

example 76.B.1 Simplify: $-3 - 2 + 6 - 4$

solution We insert the parentheses mentally and do not write them down. Then we add.

$$-5 + 6 - 4 \qquad \text{added } -3 \text{ and } -2$$

$$1 - 4 \qquad \text{added } -5 \text{ and } +6$$

$$\mathbf{-3} \qquad \text{added 1 and } -4$$

problem set 76

1. The ratio of the bumptious to the gracious was 7 to 3. If 1400 were bumptious, how many were gracious?

2. The Visigoths executed only $\frac{3}{11}$ of the number the Vandals executed. If the Visigoths executed 2202, how many did the Vandals execute?

3. Five hours was allotted for the trip. If Apocrypha's top speed was 77 miles per hour, how far could she go in the time allotted?

4. There was not enough food to go around because gate crashers had swelled the total to $2\frac{1}{2}$ times the number expected. If 1200 came, how many had been expected?

5. Graph: $x > -2$

6. Graph: $x \leq 4$

7. What number is 175 percent of 220? Draw a diagram that depicts the problem.

8. If 240 is increased by 140 percent, what is the resulting number? Draw a diagram that depicts the problem.

9. What percent of 60 is 105? Draw a diagram that depicts the problem.

10. Complete the following table. Begin by inserting the reference numbers.

Fraction	*Decimal*	*Percent*
(a)	.37	(b)

11. Write $.0010\frac{3}{5}$ as a fraction of whole numbers.

12. $3\frac{1}{2}$ of what number is $4\frac{1}{4}$?

13. Find the volume in cubic centimeters of a solid whose base is shown and sides are 2 meters high. Dimensions are in centimeters.

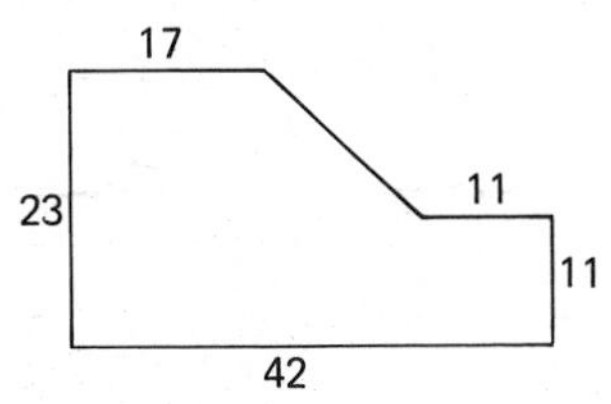

14. What is the surface area of a rectangular solid whose length, width, and height are 5 meters, 10 meters, and 3 meters, respectively?

15. Write $14\frac{11}{12}$ as an improper fraction.

16. Find the least common multiple of 16, 24, and 32.

Solve:

17. $\dfrac{6\frac{1}{4}}{2\frac{1}{3}} = \dfrac{\frac{5}{12}}{x}$

18. $\dfrac{\frac{7}{12}}{\frac{21}{24}} = \dfrac{p}{12}$

19. $2\frac{1}{2}x + 2\frac{1}{2} = 6\frac{3}{8}$

Simplify:

20. $2^3 + \sqrt{36}[2^3(2^2 - 1)(5^2 - 22) - 1]$

21. (a) $(3)(-2)$ (b) $(-3)(2)$ (c) $(-3)(-2)$

22. (a) $\frac{12}{-4}$ (b) $\frac{-12}{4}$ (c) $\frac{-12}{-4}$

***23.** $-3 - 2 + 6 - 4$

24. $-5 - 6 + 3 - 2$

25. $-2 + 3 - 6 + (-2) - 5$

26. $1.218 \times .0016$

27. $16\frac{2}{7} - 2\frac{1}{2} \cdot 1\frac{1}{7} + \frac{9}{14}$

28. $\frac{1}{3}\left(1\frac{1}{4} \cdot \frac{7}{8} - \frac{1}{4}\right) + \frac{17}{32}$

29. Round off $.00\overline{516}$ to the nearest ten-thousandth.

30. Evaluate: $xy + x^y + y^x + \sqrt{x}$ if $x = 4, y = 2$

LESSON 77 ***Opposites***

77.A opposites

The graph of the number 2 is 2 units to the right of the origin. The graph of the number -2 is 2 units to the left of the origin. Both graphs are the same distance from the origin, but the graphs are on opposite sides of the origin.

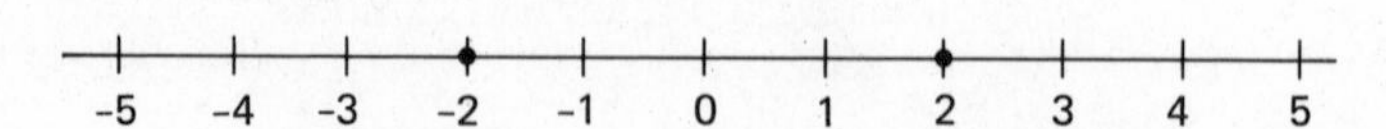

We say that these numbers are a pair of **opposites.** The number -2 is the opposite of 2 and the number 2 is the opposite of -2.

Sometimes it is helpful if we say the words **the opposite of** instead of saying **negative.** If we do this we read

$$-2$$

as the opposite of 2. And we would read

$$-(-2)$$

as the opposite of the opposite of 2.

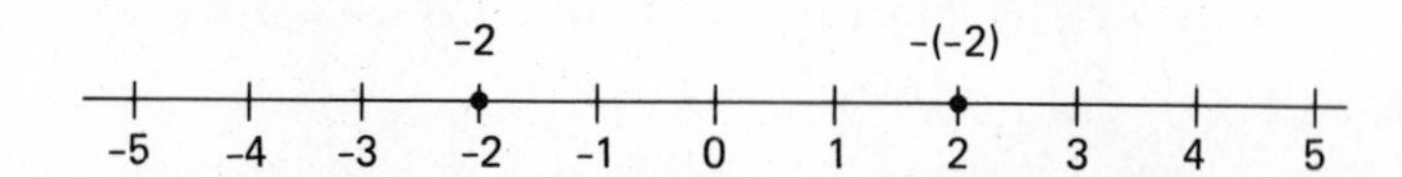

From this diagram, we see that -2 is to the left of the origin and thus the opposite of this is on the right of the origin. Thus,

$-(-2)$ = the opposite of
the opposite of 2 which is 2

If we record another minus sign and write $-[-(-2)]$, we move back to the left side.

$-[-(-2)]$ = the opposite of
the opposite of
the opposite of 2 which is -2

Every time we record another minus sign, we switch sides again. We can see this in the pattern developed here.

	READ AS	WHICH IS
2	2	2
-2	the opposite of 2	-2
$-(-2)$	the opposite of the opposite of 2	2
$-[-(-2)]$	the opposite of the opposite of the opposite of 2	-2
$-\{-[-(-2)]\}$	the opposite of the opposite of the opposite of the opposite of 2	2

The expressions in the left-hand column are all equivalent expressions for 2 or for -2. If we look at the right-hand column, we see that every time an additional $(-)$ is included in the left-hand expression, the right-hand expression changes sign.

A similar alternation in sign occurs whenever a particular number is multiplied by a negative number. For instance,

$$-2 = -2$$

$$(-2)(-2) = +4$$

$$(-2)(-2)(-2) = -8$$

$$(-2)(-2)(-2)(-2) = +16$$

$$(-2)(-2)(-2)(-2)(-2) = -32$$

From the patterns above, we can deduce that when there is an even number of negative factors, the answer is positive and when there is an odd number of negative factors, the answer is negative.

example 77.A.1 Simplify: $-(-\{-[-(-7)]\})$

solution There are five $-$ signs. Five is an odd number so

$$-(-\{-[-(-7)]\}) = \mathbf{-7}$$

example 77.A.2 Simplify: $-\{-[-(-6)]\}$

solution There are four $-$ signs. Four is an even number so

$$-\{-[-(-6)]\} = \mathbf{6}$$

example 77.A.3 Simplify: $(-4)(-2)(-5)$

solution The product of three negative numbers is a negative number.

$$(-4)(-2)(-5) = \mathbf{-40}$$

example 77.A.4 Simplify: $(-2)(-2)(-3)(-3)(-1)(-1)$

solution There are six factors that are negative numbers. Six is an even number, and the product of an even number of negative factors is a positive number. So

$$(-2)(-2)(-3)(-3)(-1)(-1) = \mathbf{+36}$$

problem set 77

1. The Braves started the season by hitting 1.875 times their last season's average. If their last season's average was .368, what was their beginning average this season?

2. Fortunately, the ratio of lovers to misanthropes was 13 to 2. If 3900 were lovers, how many were misanthropes?

3. For 4 hours Sam traveled at 40 miles per hour. Then for the next 3 hours, he increased his speed to 60 miles per hour. How far did he go in the 7 hours he traveled?

4. Only a vestige of the original remained. The vestige weighed 14 ounces and represented one-thirty-fifth ($\frac{1}{35}$) of the original. What did the original weigh?

5. Graph: $x > -3$

6. Graph: $x \leq 2$

7. What number is 160 percent of 120? Draw a diagram that depicts the problem.

8. If 220 is increased by 120 percent, what is the resulting number? Draw a diagram that depicts the problem.

9. What percent of 50 is 125? Draw a diagram that depicts the problem.

10. Complete the following table. Begin by inserting the reference numbers.

Fraction	*Decimal*	*Percent*
(a)	(b)	43

11. Write $.0011\frac{3}{4}$ as a fraction of whole numbers.

12. What decimal of 800 is 280?

13. Find the volume in cubic centimeters of a solid whose base is shown and whose sides are 150 centimeters high. Dimensions are in centimeters.

10
10
10
20
30

14. What is the perimeter of the following figure? Dimensions are in yards.

6
12

15. Write $\frac{1811}{15}$ as a mixed fraction.

16. Find the least common multiple of 15, 18, and 27.

Solve:

17. $\dfrac{7\frac{1}{4}}{2\frac{1}{2}} = \dfrac{\frac{1}{6}}{x}$

18. $\dfrac{\frac{3}{4}}{\frac{9}{16}} = \dfrac{z}{6}$

19. $3\frac{1}{5}x + 2\frac{1}{3} = 5\frac{2}{5}$

Simplify:

20. $3^2 + 2^3[\sqrt{36}(2^2 - 3)(2^2 + 1) - 20]$

21. (a) $(6)(-3)$ (b) $(-6)(-3)$ (c) $(-6)(3)$

22. (a) $\dfrac{24}{-6}$ (b) $\dfrac{-24}{6}$ (c) $\dfrac{-24}{-6}$

23. $-\{-[-(-3)]\}$

***24.** $-(-\{-[-(-7)]\})$

25. $-5 - 4 - (-2) + (6) + (-3)$

26. $-2 - (2) + (-4) - (-2)$

27. $3\frac{3}{7} - 1\frac{1}{7} \cdot 3\frac{1}{3} + \frac{8}{21}$

28. $\frac{1}{3}\left(2\frac{1}{4} \cdot \frac{1}{2} - \frac{3}{4}\right) + \frac{11}{24}$

29. Write 136121134.5 in words.

30. Evaluate: $tv + t^v + v^t + v^2$ if $t = 3$, $v = 3$

LESSON 78 *Evaluation with signed numbers*

78.A evaluation with signed numbers

The order of operations for simplifying expressions with signed numbers is the same as the order of operations for the numbers of arithmetic. We always do the multiplications before we do the algebraic additions.

example 78.A.1 Simplify: $2(-4) - 3 - (-6)(-3)$

solution First we will do the multiplications and get

$$-8 - 3 - 18$$

We finish by adding algebraically and get -29 for our answer.

$$-8 - 3 - 18 = \mathbf{-29}$$

example 78.A.2 Evaluate: $-x - xy$ if $x = -2$ and $y = -4$

solution Some people find that parentheses and brackets help prevent making mistakes in signs. We will write parentheses for each variable.

$$-(\quad) - (\quad)(\quad)$$

Now we insert the proper numbers in the parentheses.

$$-(-2) - (-2)(-4)$$

We remember to multiply before we add, so the final result is -6,

$$2 - 8 = \mathbf{-6}$$

example 78.A.3 Evaluate: $-a - xa$ if $x = -3$ and $a = 2$

solution We will use parentheses and write

$$-(2) - (-3)(2)$$

Now we simplify, remembering that we always multiply first.

$-2 - (-6)$	multiplied (-3) and (2)
$-2 + 6$	simplified $-(-6)$
4	added

problem set 78

1. Harry traveled 200 miles before he saw that it would be necessary to increase his speed to 60 miles per hour. If the entire trip was 440 miles, how long did he travel at 60 miles per hour?

2. Forty tons cost \$20,000. What would be the cost for only 12 tons?

3. The ratio of fans seated in season box seats to fans in bleacher seats was 2 to 17. If 68,000 screamed in the bleachers, how many were ensconced in box seats?

4. Fourteen hundred remained standing after the crash. If this was one-seventh of the total, how many were no longer standing?

5. Graph: $x > -1$

6. Graph: $x \le -2$

7. What number is 140 percent of 140? Draw a diagram that depicts the problem.

8. If 230 is increased by 160 percent, what is the resulting number? Draw a diagram that depicts the problem.

9. What percent of 70 is 98? Draw a diagram that depicts the problem.

10. Complete the following table. Begin by inserting the reference numbers.

Fraction	*Decimal*	*Percent*
$\frac{4}{5}$	(a)	(b)

11. Write $.0012\frac{2}{3}$ as a fraction of whole numbers.

12. Six tenths of what number is 750?

13. Find the volume of a solid whose base is shown and whose sides are 3 inches high. Dimensions are in inches.

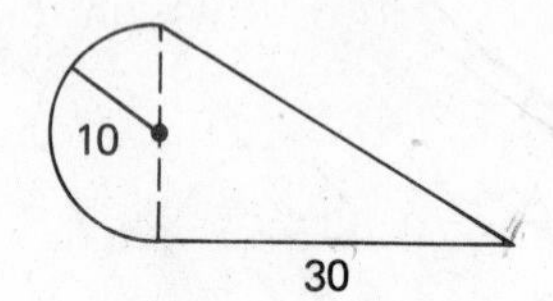

14. Convert 4 miles to inches.

15. Write $\frac{1357}{12}$ as a mixed number.

16. Find the least common multiple of 10, 15, and 25.

Solve:

17. $\dfrac{3\frac{3}{4}}{6\frac{3}{8}} = \dfrac{\frac{1}{4}}{x}$

18. $\dfrac{\frac{3}{5}}{\frac{27}{25}} = \dfrac{z}{5}$

19. $2\frac{1}{3}x + 3\frac{1}{2} = 6\frac{1}{6}$

Simplify:

20. $3^2 + 2^2[2^3(\sqrt{36} - 2^2)(1 + \sqrt{16}) - 75]$

21. (a) $3(-6)$ (b) $(-2)(-5)$ (c) $(-1)(+7)$

22. (a) $\dfrac{15}{-3}$ (b) $\dfrac{-20}{4}$ (c) $\dfrac{-36}{-12}$

23. $-(-\{-[-(-1)]\})$

24. $-(-\{-[-(+2)]\})$

25. $-6 \div 3 + (-2) \div (+3)$

26. $2(-4) - (-6)(-3)$

Evaluate:

***27.** $-x - xy$ if $x = -2$ and $y = -4$

28. $-a + ya$ if $a = -1$ and $y = -2$

29. $ay - a - y$ if $a = -7$ and $y = -3$

30. $x^y + y^x + 2xy$ if $x = 4$, $y = 3$

LESSON 79 Formats for equations · Negative coefficients

79.A formats for the addition rule

Thus far, we have been adding and subtracting positive numbers to solve equations. To solve

$$x + 2 = 4$$

we have subtracted 2 from both sides of the equation.

$$x + 2 - 2 = 4 - 2 \qquad \text{subtract 2}$$

$$x = 2 \qquad \text{simplified}$$

From now on we will consider that we are adding a negative number to both sides rather than subtracting. We can also use a different format.

example 79.A.1 Use algebraic addition to solve: $x + 5 = 7$

solution We will solve by adding -5 to both sides.

$$\begin{array}{rcrl} x + 5 & = & 7 & \text{equation} \\ -5 & = & -5 & \text{add } -5 \text{ to both sides} \\ \hline \boldsymbol{x} & \boldsymbol{=} & \boldsymbol{2} & \text{solution} \end{array}$$

example 79.A.2 Solve: $2x + 3 = 7$

solution First we will add -3 to both sides. Then we will divide by 2.

$$\begin{array}{rcrl} 2x + 3 & = & 7 & \text{equation} \\ -3 & & -3 & \text{add } -3 \text{ to both sides} \\ \hline 2x & = & 4 & \end{array}$$

$$\frac{2x}{2} = \frac{4}{2} \qquad \text{divide by 2}$$

$$\boldsymbol{x = 2} \qquad \text{simplified}$$

79.B negative coefficients

Often we encounter variables with negative coefficients. In both the following equations, the coefficient of x is -2:

$$-2x = 4 \qquad -2x = -6$$

These equations can be solved by dividing both sides by -2.

$$\frac{-2x}{-2} = \frac{4}{-2} \qquad \frac{-2x}{-2} = \frac{-6}{-2}$$

$$\boldsymbol{x = -2} \qquad \boldsymbol{x = 3}$$

Beginners often make mistakes when they divide by negative numbers. This kind of mistake can be avoided by using an extra step. Before we divide, we multiply both sides of the equation by -1. This changes the signs on both sides. Then we can solve by dividing both sides by a positive number.

example 79.B.1 Solve: $-3x = 12$

solution **As the first step, we mentally multiply both sides by -1. This changes both signs,** and we get

$$3x = -12$$

Now we solve by dividing by positive 3.

$$\frac{3x}{3} = \frac{-12}{3} \quad \text{divided by 3}$$

$$\mathbf{x = -4} \quad \text{simplified}$$

example 79.B.2 Solve: $-3x + 4 = 10$

solution We always use the addition rule first. Thus, we begin by adding -4 to both sides.

$$\begin{array}{rcrl} -3x + 4 & = & 10 & \text{equation} \\ -4 & & -4 & \text{add } -4 \text{ to both sides} \\ \hline -3x & = & 6 & \end{array}$$

Next we multiply mentally by -1 and get

$$3x = -6$$

Now we divide by 3 to solve.

$$\frac{3x}{3} = \frac{-6}{3} \quad \text{divided by 3}$$

$$\mathbf{x = -2} \quad \text{simplified}$$

problem set 79

1. Wanataxa peered into the gloom and could positively identify only 2000. If there was a total of 16,000 in the gloom, what fraction of the total did he identify?
2. The ratio of defective components to non-defective components was 2 to 19. If 380 parts were non-defective, how many were defective?
3. The delegates crowded into the hall until they numbered $1\frac{1}{4}$ times the limit set by the fire marshall. If the fire marshall's limit was 800, how many delegates were in the hall?
4. For the first 50 miles, Harold traveled at 25 miles per hour. Then he doubled his speed. If the total trip was to be 450 miles, how long did the total trip take?

Graph:

5. $x \geq -2$
6. $x < 2$
7. What number is 240 percent of 120? Draw a diagram that depicts the problem.
8. If 160 is increased by 185 percent, what is the resulting number? Draw a diagram that depicts the problem.
9. What percent of 60 is 96? Draw a diagram that depicts the problem.

10. Complete the following table. Begin by inserting the reference numbers.

Fraction	*Decimal*	*Percent*
(a)	.41	(b)

11. Write $.0013\frac{1}{3}$ as a fraction of whole numbers.

12. What decimal of 620 is 217?

13. Express in cubic meters the volume of a solid whose base is shown and whose sides are 10 centimeters high. Dimensions are in meters.

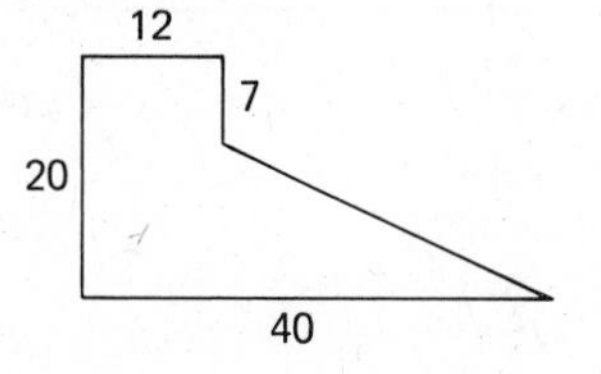

14. Convert 1265.81 centimeters to kilometers.

15. Convert 1385.31 kilometers to centimeters.

16. Find the least common multiple of 60, 120, and 180.

Solve:

***17.** $-3x + 4 = 10$

18. $-5x + 7 = 12$

19. $\dfrac{2\frac{2}{3}}{\frac{4}{5}} = \dfrac{\frac{3}{4}}{x}$

20. $-2\frac{1}{4}x + \frac{1}{2} = 3\frac{3}{4}$

Simplify:

21. $2^3 + 2^2[2(\sqrt{49} - \sqrt{36})(1 + \sqrt{9}) - 3]$

22. $1\frac{3}{4} \cdot 2\frac{1}{3} - \frac{5}{12}$

23. (a) $(4)(-2)$ (b) $\dfrac{16}{-2}$ (c) $(-6)(-2)$

24. $-(-\{-[-(5)]\})$

25. $4 - 3(5) - 7(-6) - 4(-5)$

26. $\frac{1}{2}\left(1\frac{2}{3} \cdot 1\frac{4}{5} - \frac{1}{2}\right) + \frac{5}{6}$

Evaluate:

27. $-x + xy$ if $x = -1, y = -3$

28. $xyz + xy$ if $x = -1, y = -2, z = -3$

29. $xy + y^x + x^x$ if $x = 2, y = 3$

30. Write 6256183122 in words.

LESSON 80 Algebraic phrases

80.A algebraic phrases

In algebra we learn to write conditional algebraic equations that have the same meanings as statements made by using words. Then we find the solutions to the equations. These solutions give us the answers to the questions asked. When we write the conditional equations, we use algebraic phrases. These algebraic phrases have the same meanings as the word phrases. There are several keys to writing algebraic phrases. The word **sum** means things are added, as do the words **greater than** or **increased by.**

WORD PHRASE	ALGEBRAIC PHRASE
The sum of a number and 7	$N + 7$
7 greater than a number	$N + 7$
A number increased by 7	$N + 7$

The words **less than** or **decreased by** mean to subtract (add the opposite of).

7 less than a number	$N - 7$
A number decreased by 7	$N - 7$

The word **product** means things are multiplied.

The product of a number and 7	$7N$

If we use N to represent an unknown number, then we will use $-N$ to represent the opposite of the unknown number. Thus twice a number and 9 times the opposite of a number could be written as follows:

Twice a number	$2N$
9 times the opposite of a number	$9(-N)$

Sometimes two or more operations are designated by a single phrase.

The sum of twice a number and -8	$2N - 8$
5 times the sum of twice a number and -8	$5(2N - 8)$

Cover the answers in the right-hand column and see if you can write the algebraic phrase that is indicated.

The sum of a number and 9	$N + 9$
A number decreased by 9	$N - 9$
The opposite of a number, decreased by 6	$-N - 6$
The sum of the opposite of a number and -3	$-N - 3$
The product of 5 times a number and 6	$6(5N)$
The sum of twice a number and -3	$2N - 3$
4 times the sum of twice a number and -3	$4(2N - 3)$
The product of -2, and a number increased by 5	$-2(N + 5)$

problem set 80

1. Francis got the first 40 pictures framed for a total of $120. How much would it cost her to get 800 pictures framed?

2. The ratio of roses to snapdragons was 4 to 5. If there were 26,000 roses on the float, how many snapdragons were there?

3. If $7\frac{4}{5}$ of the total came to 109,200, what was the total?

4. The last leg of the trip was traveled at 680 miles per hour. If the last leg took 10 hours to travel, what was the length of the last leg?

5. Write the algebraic phrase that is described.
*(a) The sum of twice a number and -8.
(b) The product of -3, and a number increased by 5.
(c) The opposite of a number, decreased by 4.
(d) 5 times the sum of twice a number and 6.

6. Graph: $x \geq -1$

7. What number is 190 percent of 340? Draw a diagram that depicts the problem.

8. If 200 is increased by 160 percent, what is the resulting number? Draw a diagram that depicts the problem.

9. What percent of 80 is 128? Draw a diagram that depicts the problem.

10. Complete the following table. Begin by inserting the reference numbers.

Fraction	*Decimal*	*Percent*
$\frac{3}{4}$	(a)	(b)

11. Write $.0015\frac{2}{5}$ as a fraction of whole numbers.

12. Express in cubic yards the volume of a solid whose base is shown and whose sides are 2 yards high. Dimensions are in yards.

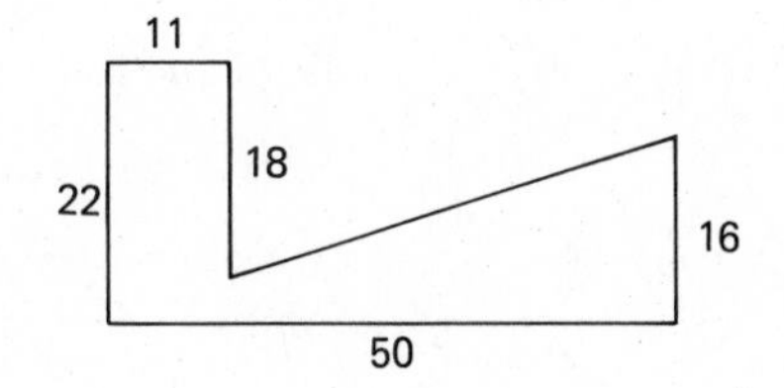

13. Convert 612 inches to yards.

14. Find the least common multiple of 25, 36, and 45.

15. Express 720 as a product of primes.

Solve:

16. $-6x - 6 = -3$

17. $-7x - 3 = 12$

18. $\dfrac{3\frac{1}{3}}{6\frac{1}{2}} = \dfrac{\frac{4}{13}}{x}$

19. $-3\frac{1}{3}x + \frac{3}{4} = 2\frac{5}{12}$

Simplify:

20. $2^2 + 2^3[3(\sqrt{49} - \sqrt{36})(\sqrt{16} + \sqrt{4}) - \sqrt{9}]$

21. $2\frac{2}{3} \cdot 3\frac{1}{4} - \frac{5}{12}$

22. (a) $(3)(-2)$ (b) $\dfrac{-6}{-3}$ (c) $(-1)(-1)$

23. $-(-\{-[-(+6)]\})$

24. $-4(-3) + (-2)(-5)$

25. $\dfrac{621.21}{.004}$

26. $\dfrac{1}{3}\left(2\dfrac{1}{4} \cdot \dfrac{3}{4} - \dfrac{15}{16}\right) + \dfrac{11}{48}$

27. $118.21 \times .0014$

Evaluate:

28. $yz - xz$ if $x = -1$, $y = 1$, $z = 2$

29. $z + y + yz$ if $x = -10$, $y = -1$, $z = -2$

30. $y^y + x^x + y^x + x^y$ if $x = 2$, $y = 3$

LESSON 81 Trichotomy · Symbols of negation

81.A trichotomy

We have discussed how we compare numbers by using the words **less than, equal to,** or **greater than.** We recall that we use the number line to define what we mean by these words. On this number line, we have graphed the numbers -4, -2, 0, and 2.

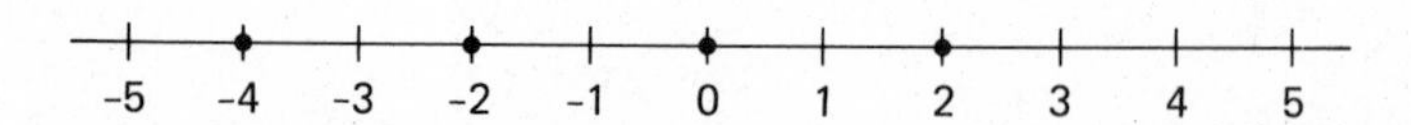

We remember that a number is greater than another number if its graph is to the right of the graph of the other number. By looking at the graphs above, we can verify that all of the following inequalities are true inequalities.

$$0 > -4 \qquad -2 < 0 \qquad 2 > -2 \qquad -4 < 2$$

It is interesting to note that if we have two numbers, then one and only one of the following statements is true.

(a) The first number is greater than the second number.
(b) The first number is equal to the second number.
(c) The first number is less than the second number.

Because there are only three possibilities, this is often called the **trichotomy axiom** from the Greek word *tricha,* which means "in three parts."

81.B symbols of negation

The symbols shown here

$$< \qquad > \qquad = \qquad \geq \qquad \leq$$

are used to denote that things are equal or are not equal. We can negate each of these symbols by drawing a slash through the symbol. Thus, if we are asked to read the following inequalities:

(a) $4 \not< 1$ (b) $4 \neq 2 + 6$ (c) $7 \not\geq 10$

we can read them from left to right as

(a) $4 \not< 1$ 4 is not less than 1 true

(b) $4 \neq 2 + 6$ — 4 is not equal to 2 + 6 — true

(c) $7 \ngeq 10$ — 7 is not greater than or equal to 10 — true

Negated inequalities can also be false inequalities.

$-4 \nless 10$ — -4 is not less than 10 — false

$7 \ngeq 3$ — 7 is not greater than or equal to 3 — false

When we graph conditional inequalities, we indicate the numbers that will make them true inequalities.

example 81.B.1 Graph: $x \nless 2$

solution If x is not less than 2, then x must be greater than or equal to 2.

$$x \geq 2$$

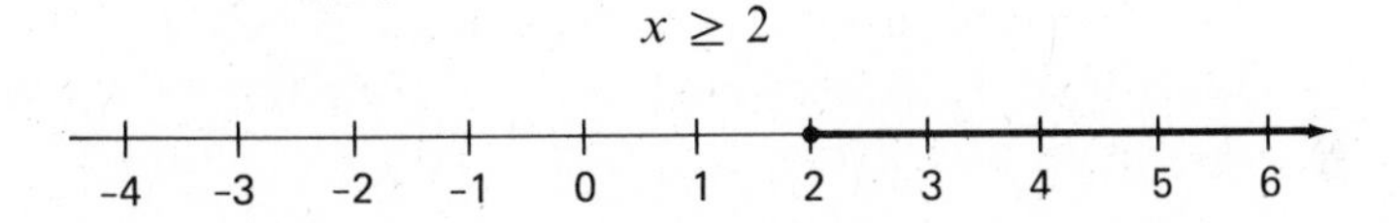

example 81.B.2 Graph: $x \ngeq -1$

solution If x is not greater than or equal to -1, then x must be less than -1.

$$x < -1$$

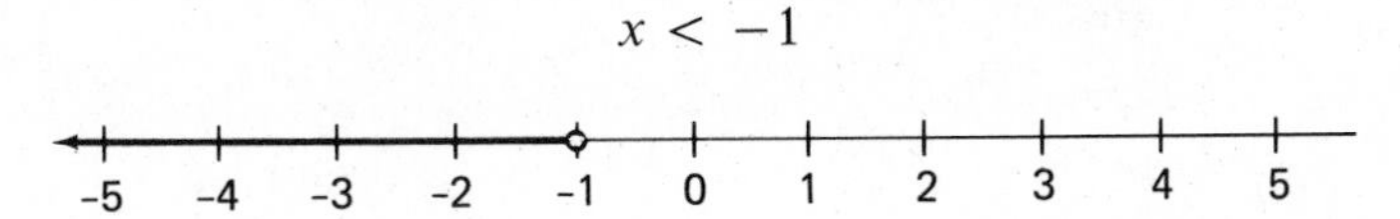

example 81.B.3 Graph: $x \nless -2$

solution If x is not less than -2, then x must be greater than or equal to -2.

$$x \geq -2$$

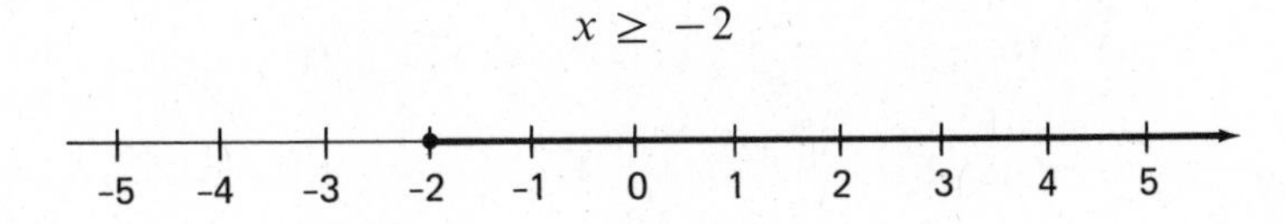

Note that in this example, we used a solid circle at -2 because of the $\geq$ sign. In the preceding example, the circle at -1 was not solid because the inequality $<$ did not contain part of an equals sign.

problem set 81

1. The rain came down in torrents. When three-eighths of the rain had fallen, the stream was 21 inches deep. How deep was the stream when all the rain had fallen?

2. The ratio of fisherman to fish caught was 2 to 23. How many fish were caught by 460 fishermen?

3. Forty of the expensive ones could be purchased for \$28,000. What would 72 of the expensive ones cost?

4. After traveling the first 480 miles in 12 hours, the old man increased his speed by 20 miles per hour. How long did it take him to travel the last 300 miles?

5. Write the algebraic phrase that is described.
   (a) Four times the sum of three times a number and -6
   (b) The product of -2, and a number increased by 5
   (c) The opposite of a number, decreased by 25

Graph:

*6. $x \nless -2$ *7. $x \ngtr -1$

8. What number is 210 percent of 250? Draw a diagram that depicts the problem.

9. If 250 is increased by 170 percent, what is the resulting number? Draw a diagram that depicts the problem.

10. What percent of 90 is 126? Draw a diagram that depicts the problem.

11. Write $.0013\frac{2}{7}$ as a fraction of whole numbers.

12. Express in cubic centimeters the volume of a solid whose base is shown and whose sides are 1 meter high. Dimensions are in centimeters.

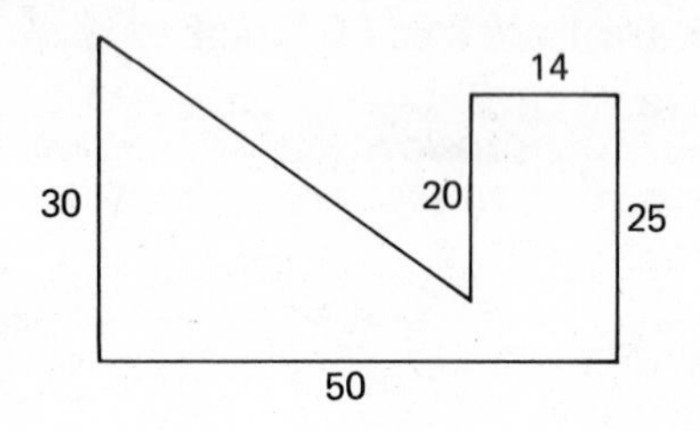

13. Convert 2 miles to inches.

14. Find the least common multiple of 30, 360, and 45.

Solve:

15. $-7x - 7 = -2$ 16. $-2x + 3 = 7$

17. $\dfrac{2\frac{1}{4}}{1\frac{1}{2}} = \dfrac{1\frac{5}{6}}{x}$ 18. $-2\frac{1}{3}x - \frac{3}{4} = 3\frac{7}{12}$

Simplify:

19. $2^3 + 2^2[3(\sqrt{64} - \sqrt{49})(\sqrt[3]{8} + 1) - 1]$

20. $1\frac{2}{3} \cdot 3\frac{1}{4} - \frac{7}{12}$

21. (a) $2(-3)$ (b) $\dfrac{-12}{-3}$ (c) $(-1)(-2)(3)$

22. $-2(-3) + (-2)(-4) - (-2)$ 23. $(-2)(3) - (-2)(-4) - 3(-2)$

24. $\frac{1}{2}\left(3\frac{2}{5} \cdot \frac{3}{4} - 1\frac{1}{4}\right)$ 25. $3\frac{1}{2} \times 2\frac{1}{3} \div \frac{5}{6} \times \frac{1}{3}$

26. $\dfrac{118.02}{.004}$ 27. $672.8913 - 19.8761$

Evaluate:

28. $-xy + y$ if $x = -2$, $y = -3$

29. $xy + yz + xz$ if $x = -1$, $y = -2$, $z = -3$

30. $x^3 + y^3 + 3x^2y + 3xy^2$ if $x = 2$, $y = 3$

LESSON 82 Number word problems

82.A number word problems

To solve word problems, we look for statements that describe equal quantities. Then we use an equals sign and algebraic phrases to write equations that make the same statement of equality. The problems in this lesson can be solved with one equation.

We will avoid the use of the meaningless variables x and y when we write these equations. We will try to use variables whose meaning is easy to remember. The problems in this lesson discuss some unknown number. We will use the letter N to represent the unknown number.

example 82.A.1 The sum of twice a number and 42 is 128. Find the number.

solution The word **is** means **equals.** Thus the sum of twice a number and 42 equals 128.

$$\begin{array}{rcll} 2N + 42 &=& 128 & \text{equation} \\ -42 && -42 & \text{add } -42 \text{ to both sides} \\ \hline 2N &=& 86 & \\ N &=& \mathbf{43} & \text{divided by 2} \end{array}$$

We will use 43 for N in the original equation to check.

$$2(43) + 42 = 128 \longrightarrow 86 + 42 = 128 \longrightarrow 128 = 128 \quad \text{check}$$

example 82.A.2 Twenty-seven less than 3 times a number is 144. What is the number?

solution If we use N for the number, then 3 times the number is $3N$.

$$\begin{array}{rcll} 3N - 27 &=& 144 & \text{equation} \\ +27 && +27 & \text{add 27 to both sides} \\ \hline 3N &=& 171 & \\ \frac{3N}{3} &=& \frac{171}{3} & \text{divided by 3} \\ N &=& \mathbf{57} & \text{simplified} \end{array}$$

We will use 57 for N in the original equation to check.

$$3(57) - 27 = 144 \longrightarrow 171 - 27 = 144 \longrightarrow 144 = 144 \quad \text{check}$$

example 82.A.3 Four more than 33 times a number equals -95. What is the number?

solution First we write the equation. Then we solve.

$$\begin{array}{rcll} 33N + 4 &=& -95 & \text{equation} \\ -4 && -4 & \text{add } -4 \text{ to both sides} \\ \hline 33N &=& -99 & \\ N &=& \mathbf{-3} & \text{divided by 33} \end{array}$$

Now we use -3 for N in the original equation to check.

$$33(-3) + 4 = -95 \longrightarrow -99 + 4 = -95 \longrightarrow -95 = -95 \quad \text{check}$$

problem set 82

1. The sum of twice a number and 42 is 126. Find the number.

*2. Twenty-seven less than 3 times a number is 144. What is the number?

*3. Four more than 33 times a number equals -95. What is the number?

4. For the first 20 miles, the stage crawled along at 4 miles per hour. When the robbers were sighted, the speed was quadrupled. How long did it take to travel the last 80 miles at the new speed?

Graph:

5. $x \not\leq 3$ 6. $x \not> -2$ 7. $x \geq -2$

8. What number is 225 percent of 140? Draw a diagram that depicts the problem.

9. If 300 is increased by 125 percent, what is the resulting number? Draw a diagram that depicts the problem.

10. What percent of 640 is 896? Draw a diagram that depicts the problem.

11. Complete the following table. Begin by inserting reference numbers.

Fraction	*Decimal*	*Percent*
(a)	(b)	40

12. Express in cubic meters the volume of a solid whose base is shown and whose sides are 50 centimeters high. Dimensions are in meters.

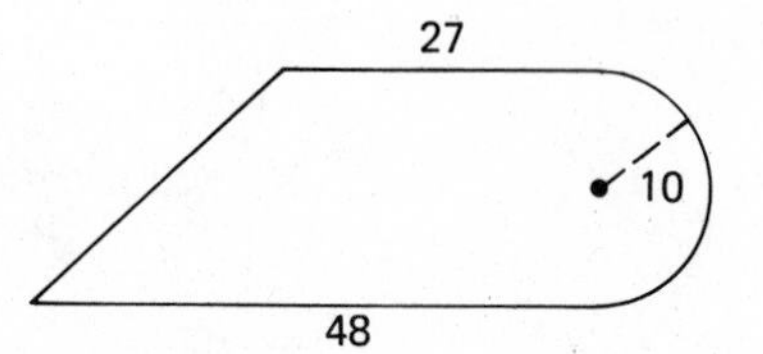

13. Find the volume of a rectangular solid whose length, width, and height measure 12 inches, 10 inches, and 3 inches, respectively.

14. Find the least common multiple of 12, 32, and 36.

Solve:

15. $-6x - 5 = -1$

16. $-2x + 6 = 1$

17. $\dfrac{3\frac{1}{6}}{2\frac{1}{3}} = \dfrac{\frac{1}{2}}{x}$

18. $-3\frac{1}{3}x - \frac{1}{4} = 6\frac{1}{2}$

Simplify:

19. $2^2 + 2[\sqrt{9}(2^3 - 2 \cdot 3)(3^2 - 4 \cdot 2)]$

20. $2\frac{1}{3} \cdot 1\frac{2}{3} - \frac{7}{9}$

21. (a) $3(-2)$ (b) $\dfrac{-18}{3}$ (c) $(-2)(-2)(5)$

22. $-(-3) - (-4) + 6 - (-5)$

23. $-[-(-2)] - (-1) - 5 + (-10)$

24. $\frac{1}{3}\left(2\frac{1}{4} \cdot 1\frac{1}{5} - \frac{17}{20}\right)$

25. $4\frac{1}{3} \div 1\frac{2}{3} \times 2\frac{3}{4} \div \frac{1}{4}$

26. $\dfrac{117.03}{.005}$

27. $1182.6218 - 13.6211$

Evaluate:

28. $xy + yz$ if $x = -1, y = -2, z = -3$

29. $zy - zx$ if $x = -2, y = -3, z = -1$

30. $x^2 + y^2 + 2xy$ if $x = 2, y = 3$

LESSON 83 Operations with signed numbers · Roots by cut and try

83.A operations with signed numbers

We always multiply before we add algebraically. If an expression contains symbols of inclusion, we simplify these first, always beginning with the innermost symbols of inclusion.

example 83.A.1 Simplify: $-2(-2 - 3 + 1) - (-3 - 2)$

solution We first simplify within the parentheses.

$$-2(-4) - (-5)$$

Now we multiply.

$$8 - (-5) \qquad \text{multiplied } (-2) \text{ and } (-4)$$

Then we add

$$\mathbf{13} \qquad \text{added}$$

example 83.A.2 Simplify: $-2(-2 - 3 \cdot 5) - 2(3 - 5)$

solution First we simplify within the parentheses.

$$-2(-17) - 2(-2) \qquad \text{simplified}$$
$$34 - (-4) \qquad \text{multiplied}$$
$$\mathbf{38} \qquad \text{added}$$

example 83.A.3 Simplify: $-2(-2) - 2 - (-3 - 5) - 2$

solution First we simplify within parentheses. Then we multiply, and then we add.

$$-2(-2) - 2 - (-8) - 2 \qquad \text{simplified}$$
$$4 - 2 + 8 - 2 \qquad \text{multiplied}$$
$$\mathbf{8} \qquad \text{added}$$

83.B roots by cut and try

We have restricted our investigation of roots of numbers to roots that are whole numbers. The square root of 16 is a whole number. The fourth root of 16 is also a whole number.

$$\sqrt{16} = 4 \qquad \sqrt[4]{16} = 2$$

The square root of 5 is not a whole number. Neither is the third root (cube root) of 5.

$$\sqrt{5} = ? \qquad \sqrt[3]{5} = ?$$

The easiest way to simplify these expressions is to use a pocket calculator. There are rote ways to find the roots of numbers but we will not use them. Instead, we will use a method that is sometimes called cut and try. It could be called "make a guess and then guess again." In this lesson we will look at square roots, and in Lesson 92 we will look at third roots.

example 83.B.1 Use the method of cut and try to estimate $\sqrt{5}$ to one decimal place.

solution We are searching for the number that multiplied by itself equals 5.

$$(?)(?) = 5$$

Since 2 times 2 equals 4 and 3 times 3 equals 9,

$$(2)(2) = 4$$
$$(?)(?) = 5$$
$$(3)(3) = 9$$

the number we seek is between 2 and 3 and is closer to 2. Let's try 2.1 as our first guess.

$$(2.1)(2.1) = ? \qquad \begin{array}{r} 2.1 \\ 2.1 \\ \hline 2\,1 \\ 42 \\ \hline 4.41 \end{array}$$

Since 4.41 is less than 5, we need to try a larger number. Let's try 2.2.

$$(2.2)(2.2) = ? \qquad \begin{array}{r} 2.2 \\ 2.2 \\ \hline 4\,4 \\ 44 \\ \hline 4.84 \end{array}$$

Since 4.84 is also smaller than 5, we will guess again. We will guess 2.3 this time.

$$(2.3)(2.3) = ? \qquad \begin{array}{r} 2.3 \\ 2.3 \\ \hline 6\,9 \\ 46 \\ \hline 5.29 \end{array}$$

This product is greater than 5. Now we have a guess that is too large and one that is too small.

$$(2.2)(2.2) = 4.84$$
$$(\quad)(\quad) = 5$$
$$(2.3)(2.3) = 5.29$$

Since 4.84 is closer to 5 than 5.29, we will use 2.2 as our estimate for $\sqrt{5}$.

$$\mathbf{\sqrt{5} \approx 2.2}$$

The symbol $\approx$ means "is approximately equal to."

example 83.B.2 Use the method of cut and try to estimate $\sqrt{40}$ to one decimal place.

solution To get a good first guess, we note that 6 times 6 equals 36 and 7 times 7 equals 49.

$$\sqrt{40} = ? \qquad (6)(6) = 36$$
$$(?)(?) = 40$$
$$(7)(7) = 49$$

It looks as though our number is between 6 and 7 and is closer to 6. Let's guess 6.2.

$$(6.2)(6.2) = ? \qquad \begin{array}{r} 6.2 \\ 6.2 \\ \hline 12\,4 \\ 372 \\ \hline 38.44 \end{array}$$

This is smaller than 40 so we increase our guess.

$$(6.3)(6.3) = ? \qquad \begin{array}{r} 6.3 \\ 6.3 \\ \hline 18\,9 \\ 378 \\ \hline 39.69 \end{array}$$

Still not large enough so let's try 6.4.

$$(6.4)(6.4) = ? \qquad \begin{array}{r} 6.4 \\ 6.4 \\ \hline 25\,6 \\ 384 \\ \hline 40.96 \end{array}$$

Now we have 40 bracketed.

$$(6.3)(6.3) = 39.69$$

$$(?)(?) = 40$$

$$(6.4)(6.4) = 40.96$$

Since 39.69 is closer to 40 than 40.96, we say that

$$\boldsymbol{\sqrt{40} \approx 6.3}$$

problem set 83

1. The product of a number and 15 is increased by 4. The result is 49. What is the number?
2. Fifteen less than 10 times a number is -105. What is the number?
3. Hartzler multiplied his magic number by 14. Then he added 8. The result was 50. What was Hartzler's magic number?
4. The ratio of straights to bents was $2\frac{1}{2}$ to 3. If 300 were bent, how many were straight?

Graph:

5. $x \not\leq 1$
6. $x \not> -1$
7. What number is 230 percent of 350? Draw a diagram that depicts the problem.
8. If 200 is increased by 130 percent, what is the resulting number? Draw a diagram that depicts the problem.
9. What percent of 80 is 108? Draw a diagram that depicts the problem.
10. Thirty percent of what number is 240? Draw a diagram of the problem.

11. Complete the following table. Begin by inserting the reference numbers.

Fraction	*Decimal*	*Percent*
(a)	.36	(b)

12. Write $.0016\frac{1}{6}$ as a fraction of whole numbers.

13. Express in cubic yards the volume of a solid whose base is shown and whose sides are 2 yards high. Dimensions are in yards.

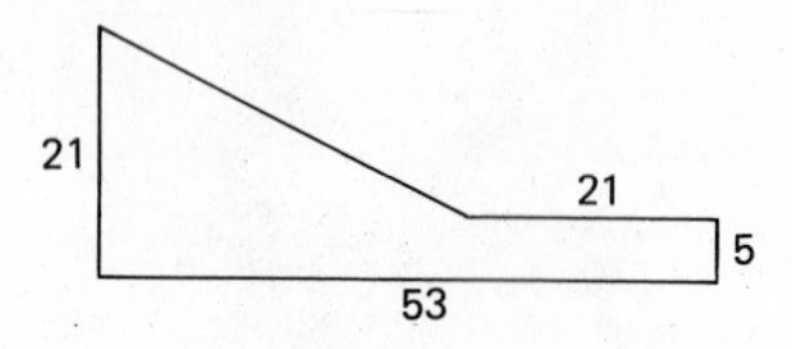

14. Convert 1234.8792 meters to kilometers.

***15.** Use the method of cut and try to find $\sqrt{40}$ to one decimal place.

Solve:

16. $-3x - 10 = -1$

17. $-6x + 4 = 6$

18. $\dfrac{3\frac{1}{3}}{2\frac{1}{4}} = \dfrac{1\frac{1}{2}}{x}$

19. $-3\frac{1}{4}x - \frac{3}{8} = 4\frac{1}{2}$

Simplify:

20. $2^3 + 2^3[2^2(\sqrt{16} - \sqrt{9})(\sqrt{9} + 2^2)]$

***21.** $-2(-2 - 3 \cdot 5) - 2(3 - 5)$

22. $-3(-2 - 6 \cdot 2) - 4(2 - 4)$

23. $-(-2)(-3) + (-6)(-1) - 2(-1 - 3 \cdot 5)$

24. $\frac{1}{3}\left(2\frac{1}{4} \cdot \frac{2}{3} - \frac{1}{6} \cdot \frac{1}{2}\right)$

25. $2\frac{1}{3} \times 1\frac{1}{4} \div \frac{3}{8} \div \frac{1}{2}$

26. $21.62 \times .0012$

27. $\dfrac{615.01}{.003}$

Evaluate:

28. $yy - xy \quad$ if $x = -1, y = -2$

29. $abc + ab + bc \quad$ if $a = -2, b = -1, c = 3$

30. $a^2 + b^2 + c^2 + 2ab + 2bc + 2ac \quad$ if $a = 1, b = 2, c = 3$

LESSON 84 Order of division

84.A order of division

The procedure for simplifying expressions that require multiplication, division, and algebraic addition is as follows:

1. Do the multiplications and divisions from left to right in the order that they are encountered.
2. Do the algebraic additions.

To use these steps to simplify

$$32 \div 2 \cdot 4 - 3 \cdot 5 + 2$$

we move from left to right, doing only multiplication and division.

$16 \cdot 4 - 3 \cdot 5 + 2$ divided 32 by 2

$64 - 15 + 2$ multiplied 16 by 4 and 3 by 5

Now we do the algebraic additions.

$49 + 2$ added 64 and -15

51 added 49 and 2

The notation used in this problem is unduly complicated, and arithmetic teachers are often the only people who can correctly simplify problems like this one. If symbols of inclusion are used and if a fraction line is used to designate division, the same expression can be written in a way that clearly indicates the desired division.

$\left(\frac{32}{2} \cdot 4\right) - (3 \cdot 5) + 2$ expression

$64 - 15 + 2$ simplified parentheses

51 added algebraically

For this reason we will avoid the use of the symbol $\div$, and instead we will use a fraction line to indicate division. We will also use parentheses, brackets, and braces to clarify our expressions. We will begin by simplifying within these symbols of inclusion and will remember to multiply before performing algebraic addition.

example 84.A.1 Simplify: $\frac{4 - (3 - 7) + 3 \cdot 2 + 6}{2(-4 - 1)}$

solution We will simplify above and below and will divide as the last step.

$\frac{4 - (-4) + 3 \cdot 2 + 6}{2(-5)}$ simplified parentheses

$\frac{4 - (-4) + 6 + 6}{-10}$ multiplied

$\frac{20}{-10}$ added

$\mathbf{-2}$ divided

example 84.A.2 Simplify: $\frac{6(-4 + 3) - (-2 - 6)2}{3(2 - 4)}$

solution Again we simplify above and below.

$\frac{6(-1) - (-8)2}{3(-2)}$ simplified parentheses

$\frac{-6 + 16}{-6}$ multiplied

$\frac{10}{-6}$ added

$\mathbf{-\frac{5}{3}}$ simplified

We leave the answer as an improper fraction.

problem set 84

1. A number was multiplied by 4. Then this product was increased by 4. If the result was 24, what was the number?

2. A number was multiplied by -4. Then this product was increased by -4. If the result was -24, what was the number?

3. Seventy-five less than 30 times a number is -225. What is the number?

4. The container overflowed because the kids tried to pour in $3\frac{3}{5}$ times the amount it would hold. If they tried to pour in 288 gallons, how much would the container hold?

5. Graph: $x \not< 2$

6. Graph: $x \not\geq -2$

7. What number is 130 percent of 220? Draw a diagram that depicts the problem.

8. If 230 is increased by 60 percent, what is the resulting number? Draw a diagram that depicts the problem.

9. What percent of 70 is 112? Draw a diagram that depicts the problem.

10. Seventy-six is what percent of 95? Draw a diagram that depicts the problem.

11. Complete the following table. Begin by inserting reference numbers.

Fraction	*Decimal*	*Percent*
(a)	.53	(b)

12. Write $.0014\frac{2}{3}$ as a fraction of whole numbers.

13. Express in cubic inches the volume of a solid whose base is shown and whose sides are 2 feet high. Dimensions are in inches.

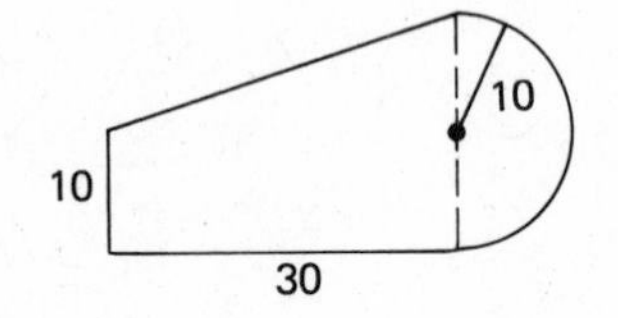

14. Find the perimeter of the following figure. Dimensions are in yards.

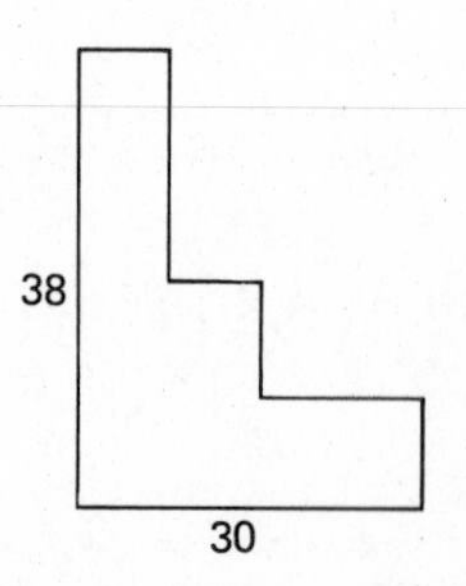

15. Convert 12.687 kilometers to centimeters.

Solve:

16. $-2x - 12 = -2$

17. $-3x + 6 = 4$

18. $\dfrac{4\frac{1}{5}}{2\frac{1}{3}} = \dfrac{1\frac{1}{2}}{x}$

19. $-2\frac{1}{3}x - \frac{3}{4} = 1\frac{3}{8}$

Simplify:

20. $3^2 + 2^3[2^2(\sqrt{16} - \sqrt{9}) - (\sqrt{25} - \sqrt{4})]$

21. $-2(-6 - 2 \cdot 5) + 2(-2 - 1)$

22. $-1(6 - 2 \cdot 2) - 3(2 \cdot 1 - 3)$

***23.** $\dfrac{4 - (3 - 7) + 3 \cdot 2 + 6}{2(-4 - 1)}$

24. $\dfrac{5 - (4 - 6)2 + 4 \cdot 2}{3(-2 - 1)}$

25. $\dfrac{3 - (6 - 3)2 + 3 \cdot 2}{4(-1 + 3)}$

26. $\dfrac{1}{4}\left(2\dfrac{1}{3} \cdot \dfrac{1}{4} - \dfrac{5}{6} \cdot \dfrac{1}{2}\right)$

27. $3\dfrac{1}{3} \times 2\dfrac{1}{4} \div \dfrac{1}{8} \times \dfrac{1}{2}$

Evaluate:

28. $-bc - ac$ if $a = -1$, $b = -3$, $c = -5$

29. $a^2 + b^2 + c^2 + 2abc + ab$ if $a = 3$, $b = 2$, $c = 4$

30. Use the method of cut and try to find $\sqrt{20}$ to one decimal place.

LESSON 85 Variables on both sides

85.A variables on both sides

Some equations have a variable on both sides of the equation. To solve these equations, we first add as necessary to eliminate the variables on one side or the other side.

example 85.A.1 Solve: $3x + 3 = x - 5$

solution We can eliminate either the $3x$ or the x. We decide to eliminate the x, so we add $-x$ to both sides.

$$\begin{array}{rcll} 3x + 3 &=& x - 5 & \text{equation} \\ -x & & -x & \text{add } -x \text{ to both sides} \\ \hline 2x + 3 &=& -5 & \end{array}$$

Now we eliminate the 3 by adding -3 to both sides. Then we divide both sides by 2.

$$\begin{array}{rcll} 2x + 3 &=& -5 & \\ -3 & & -3 & \text{add } -3 \text{ to both sides} \\ \hline 2x &=& -8 & \\ \mathbf{x} &=& \mathbf{-4} & \text{divided by 2} \end{array}$$

Check:

$3(-4) + 3 = (-4) - 5 \longrightarrow -12 + 3 = -9 \longrightarrow -9 = -9$ check

example 85.A.2 Solve: $4x + 2 = 2x + 5$

solution We can either eliminate the $4x$ or the $2x$. We decide to eliminate the $4x$, so we add $-4x$ to both sides.

$$\begin{array}{rcll} 4x + 2 &=& 2x + 5 & \text{equation} \\ -4x & & -4x & \text{add } -4x \text{ to both sides} \\ \hline 2 &=& -2x + 5 & \end{array}$$

Now we add -5 to both sides. Then we change signs and divide by 2.

$$\begin{array}{rcll} 2 &=& -2x + 5 & \\ -5 & & \quad -5 & \text{add } -5 \text{ to both sides} \\ \hline -3 &=& -2x & \\ 3 &=& 2x & \text{changed signs} \\ \frac{3}{2} &=& x & \text{divided by 2} \end{array}$$

Check:

$$4\left(\frac{3}{2}\right) + 2 = 2\left(\frac{3}{2}\right) + 5 \longrightarrow 6 + 2 = 3 + 5 \longrightarrow 8 = 8 \qquad \text{check}$$

example 85.A.3 Solve: $4x - 2 = -x$

solution This time we will eliminate the $-x$ on the right by adding $+x$ to both sides.

$$\begin{array}{rcll} 4x - 2 &=& -x & \text{equation} \\ +x & & +x & \text{add } +x \text{ to both sides} \\ \hline 5x - 2 &=& 0 & \end{array}$$

Now we add $+2$ to both sides and then divide by 5.

$$\begin{array}{rcl} 5x - 2 &=& 0 \\ +2 & & +2 \\ \hline 5x &=& 2 \\ x &=& \frac{2}{5} \end{array}$$

Check:

$$4\left(\frac{2}{5}\right) - 2 = -\left(\frac{2}{5}\right) \longrightarrow \frac{8}{5} - \frac{10}{5} = -\frac{2}{5} \longrightarrow -\frac{2}{5} = -\frac{2}{5} \qquad \text{check}$$

problem set 85

1. The product of a number and 5 was decreased by 4 and the result was 96. What was the number?
2. The sum of twice a number and 7 equals 27. What is the number?
3. In 4 hours the big horses pulled the wagon 16 miles. Then they tired and slowed down to half speed. How long did it take them to pull the wagon the last 10 miles?
4. The ratio of red hats to polka-dotted hats was 5 to 13. If 260 hats were polka-dotted, how many hats were red?
5. Graph: $x \ngtr -1$
6. Graph: $x \nleq 3$
7. What number is 160 percent of 240? Draw a diagram that depicts the problem.
8. If 270 is increased by 80 percent, what is the resulting number? Draw a diagram that depicts the problem.
9. What percent of 60 is 96? Draw a diagram that depicts the problem.
10. Twenty percent of what number is 60? Draw a diagram of the problem.
11. Write $.0011\frac{5}{6}$ as a fraction of whole numbers.

12. Express in cubic meters the volume of a solid whose base is shown and whose sides are 2 meters high. Dimensions are in meters.

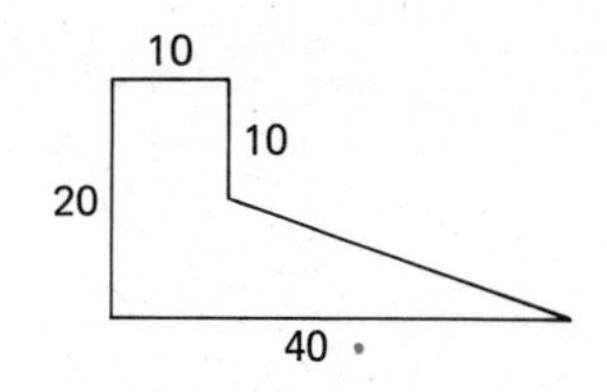

13. Use the method of cut and try to find $\sqrt{11}$ to one decimal place.

Solve:

14. $-3x - 11 = -3$ **15.** $-2x + 3 = 2$ ***16.** $3x + 3 = x - 5$

17. $2x - 5 = -x$ **18.** $6x + 6 = 2x - 4$ **19.** $-3\frac{1}{4}x + \frac{2}{3} = 1\frac{5}{6}$

Simplify:

20. $-2(6 - 3 \cdot 2) - 2(3 \cdot 1 - 2)$ **21.** $-3(6 - 2 \cdot 3) - 3(2 \cdot 7 - 6)$

22. $-4(-2 - 3 \cdot 4) - 2(3 \cdot 4 - 7)$ **23.** $\dfrac{6 - (2 - 3)3 + 4 \cdot 1}{3(-1 + 2)}$

24. $\dfrac{6 - (3 - 5)2 + 1 \cdot 2}{6(-2 + 3)}$ **25.** $\frac{1}{3}\left(2\frac{1}{2} \cdot \frac{1}{3} - \frac{3}{4} \cdot \frac{1}{2}\right)$

26. $2^2 + 2^3[3^2(\sqrt{36} - 5)(2^4 - 3 \cdot 5)]$ **27.** $3\frac{1}{2} \times 2\frac{1}{3} \div 6\frac{1}{3} \div \frac{1}{2}$

Evaluate:

28. $bc - ab$ if $a = -3$, $b = 2$, $c = -1$

29. $xy^2 + yx^2 + x^3y - xy$ if $x = 2$, $y = 3$

30. Write 67821132.3 in words.

LESSON 86 Two-step problems

86.A two-step problems

Some of the problems that have appeared in the problem sets thus far have required two steps for their solutions. In this lesson we will look at two-step problems that require the solution of an equation as the first step. These problems are easy to recognize because they almost always contain the word **if**.

example 86.A.1 If $2x + 4 = 6$, what is the value of $3x - 7$?

solution The word **if** tells us that the problem is a two-part problem. We will solve the first part to find the value of x.

$$\begin{array}{rcll} 2x + 4 &=& 6 & \text{equation} \\ -4 && -4 & \text{add } -4 \text{ to both sides} \\ \hline 2x &=& 2 & \\ x &=& 1 & \text{divided} \end{array}$$

Now we use 1 for x to find the value of $3x - 7$.

$3(1) - 7$	substituted
$3 - 7$	multiplied
$\mathbf{-4}$	added

example 86.A.2 If $4x - 2 = 3$, what is the value of $\frac{2}{5}x - \frac{1}{4}$?

solution We solve the equation to find the value of x.

$$\begin{aligned} 4x - 2 &= 3 &&\text{equation} \\ +2 \quad & \quad +2 &&\text{add } +2 \text{ to both sides} \\ \hline 4x &= 5 \\ x &= \frac{5}{4} \end{aligned}$$

Now we use $\frac{5}{4}$ for x to find the value of $\frac{2}{5}x - \frac{1}{4}$.

$\frac{2}{5}\left(\frac{5}{4}\right) - \frac{1}{4}$	substituted
$\frac{1}{2} - \frac{1}{4}$	multiplied
$\mathbf{\frac{1}{4}}$	added

example 86.A.3 If $\frac{4}{3}x - 2 = 4$, what is the value of $6x - \frac{2}{5}$?

solution We will solve the equation for x.

$$\begin{aligned} \frac{4}{3}x - 2 &= 4 &&\text{equation} \\ +2 \quad & \quad +2 &&\text{add } +2 \text{ to both sides} \\ \hline \frac{4}{3}x &= 6 \end{aligned}$$

$$\frac{3}{4} \cdot \frac{4}{3}x = 6 \cdot \frac{3}{4} \longrightarrow \mathbf{x = \frac{9}{2}} \qquad \text{solved}$$

Now we use $\frac{9}{2}$ for x to find the value of $6x - \frac{2}{5}$.

$6\left(\frac{9}{2}\right) - \frac{2}{5}$	substituted
$27 - \frac{2}{5}$	multiplied
$\mathbf{26\frac{3}{5}}$	added

problem set 86

1. The sum of 7 times a number and 42 was -98. What was the number?
2. A number was multiplied by -4. Then -6 was added. If the final result was -34, what was the number?
3. If 140 of the new ones cost \$980, what would 200 of the new ones cost?
4. In an attempt to cut them off at the pass, the posse rode 14 miles in only 2 hours. The horses were tired so they reduced their speed by 2 miles per hour. If it was still 15 miles to the pass, how much longer did it take to get there?
5. Graph: $x \ngtr 3$
6. Graph: $x \nleq 2$
7. What number is 230 percent of 270? Draw a diagram that depicts the problem.
8. If 260 is increased by 170 percent, what is the resulting number? Draw a diagram that depicts the problem.
9. What percent of 80 is 116? Draw a diagram that depicts the problem.
10. Thirty percent of what number is 180? Draw a diagram that depicts the problem.
11. Complete the following table. Begin by inserting the reference numbers.

Fraction	*Decimal*	*Percent*
$\frac{9}{10}$	(a)	(b)

12. Write $.0012\frac{6}{7}$ as a fraction of whole numbers.
13. Express in cubic centimeters the volume of a solid whose base is shown and whose sides are 12 centimeters high. Dimensions are in centimeters.

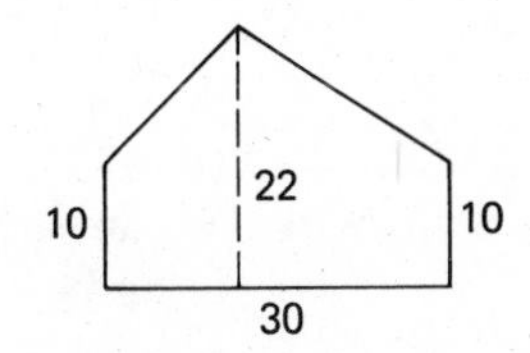

14. Find the perimeter of the following figure. Dimensions are in feet.

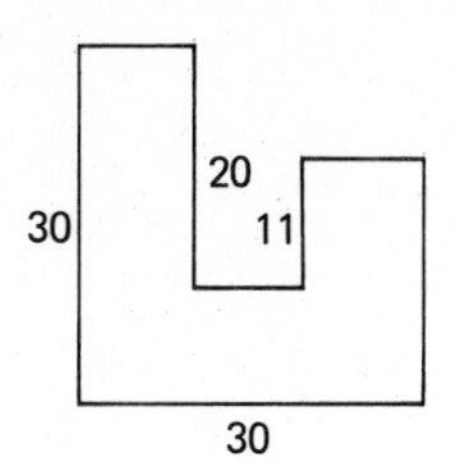

15. Convert 1287.321 centimeters to kilometers.
*16. If $2x + 4 = 6$, what is the value of $3x - 7$?
17. If $6x - 2 = 10$, what is $\frac{1}{2}x + 3$?
18. If $\frac{3}{4}x - 2 = 1$, what is the value of $3x - 4$?

Solve:

19. $-4x - 12 = -1$
20. $4x + 2 = 2x - 4$
21. $-2\frac{1}{3}x + \frac{2}{3} = 1\frac{4}{6}$
22. $\dfrac{3\frac{1}{2}}{2\frac{1}{3}} = \dfrac{\frac{1}{2}}{x}$

Simplify:

23. $-3(6 - 1 \cdot 4) - 3(2 \cdot 6 - 4)$

24. $-2(8 - 2 \cdot 3) - 4(3 \cdot 1 - 2)$

25. $\dfrac{7 - (2 - 4)2 + 5 \cdot 1}{4(-2 + 3)}$

26. $\dfrac{8 - (4 - 7)3 + 4 \cdot 2}{5(-3 + 5)}$

27. $3^2 + 3[2^3(\sqrt{49} - 2^2)(3^2 - 2^3) - 2^2]$

28. $3\frac{1}{3} \times 2\frac{1}{5} \div 1\frac{2}{3} \div \frac{1}{4}$

Evaluate:

29. $bcb - ac$ if $a = -1, b = -2, c = -5$

30. Use the method of cut and try to find $\sqrt{31}$ to one decimal place.

LESSON 87 *Unequal quantities*

87.A unequal quantities

Some word problems tell us that one quantity is a certain amount greater than another quantity. Other word problems tell us that a quantity is a certain amount less than another quantity. To solve these problems, we must use equations. We add to both sides of the equations as necessary to make the quantities equal.

example 87.A.1 Twice a number is 56 less than -72. What is the number?

solution **We must be careful because the problem tells us about things that are not equal. We begin by writing an equation that we know is incorrect.**

$$2N = -72 \qquad \text{incorrect}$$

The problem said that $2N$ is 56 less than -72. To make the equation correct, we must add 56 to $2N$ or add -56 to -72.

ADDING 56 TO $2N$

$$\begin{aligned} 2N + 56 &= -72 \qquad \text{correct} \\ -56 &\quad\ -56 \\ 2N &= -128 \\ N &= \mathbf{-64} \end{aligned}$$

ADDING -56 TO -72

$$\begin{aligned} 2N &= -72 - 56 \qquad \text{correct} \\ 2N &= -128 \\ N &= \mathbf{-64} \end{aligned}$$

Check:

$$2(-64) + 56 = -72 \longrightarrow -128 + 56 = -72 \longrightarrow -72 = -72 \qquad \text{check}$$

example 87.A.2 Six times a number is 14 less than the opposite of the number. What is the number?

solution Again we begin by writing an equation that we know is incorrect.

$$6N = -N \qquad \text{incorrect}$$

We know that $6N$ is 14 less than $-N$. We can make the equation a correct equation by adding 14 to $6N$ or by adding -14 to $-N$.

ADDING 14 TO $6N$	ADDING -14 TO $-N$
$\begin{array}{r} 6N + 14 = -N \\ +N \qquad +N \\ \hline 7N + 14 = 0 \end{array}$	$\begin{array}{r} 6N = -N - 14 \\ +N \quad +N \\ \hline 7N = -14 \end{array}$

On the left we add -14 to both sides before dividing. On the right we divide.

$$\begin{array}{rr} 7N + 14 = & 0 \\ -14 & -14 \\ \hline 7N \quad = & -14 \\ N = & \mathbf{-2} \end{array} \qquad \begin{array}{r} 7N = -14 \\ N = \mathbf{-2} \end{array}$$

problem set 87

*1. Twice a number is 56 less than -72. What is the number?

*2. Six times a number is 14 less than the opposite of the number. What is the number?

3. Five times a number is 49 greater than the product of 2 and the opposite of the number. What is the number?

4. Seven times a number is 30 greater than the sum of twice the number and 5. What is the number?

5. Graph $x \nleq -1$.

6. Graph $x \ngeq -3$.

7. What number is 320 percent of 250? Draw a diagram that depicts the problem.

8. If 250 is increased by 190 percent, what is the resulting number? Draw a diagram that depicts the problem.

9. What percent of 90 is 162? Draw a diagram that depicts the problem.

10. Forty percent of what number is 62? Draw a diagram that depicts the problem.

11. Write $.003\frac{4}{5}$ as a fraction of whole numbers.

12. Use the method of cut and try to find $\sqrt{43}$ to one decimal place.

13. Find the least common multiple of 20, 24, and 30.

14. Find the volume of a solid whose base is shown and whose sides are 10 centimeters high. Dimensions are in centimeters.

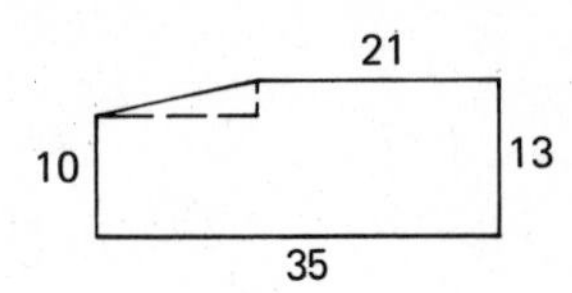

15. Convert 3 miles to yards.

16. If $6x - 2 = 16$, what is the value of $x - 3$?

17. If $3x + 2 = 8$, what is the value of $\frac{1}{3}x + \frac{1}{2}$?

18. If $\frac{3}{2}x - 3 = 1$, what is the value of $2x + 3$?

Solve:

19. $-3x + 12 = 1$

20. $6x + 2 = 3x - 5$

21. $-6x + 2x = 3x - 6$

22. $-3\frac{1}{4}x + \frac{3}{4} = 1\frac{3}{8}$

Simplify:

23. $-6(2 \cdot 3 - 2 \cdot 4) - 2(6 \cdot 1 - 2)$

24. $-2(3 \cdot 2 - 7) - 3(2 \cdot 1 - 5)$

25. $\dfrac{6 - 3(2 - 1) + 4 \cdot 3}{5(-3 + 6)}$

26. $\dfrac{8 - 2(3 - 2) + 4 \cdot 2}{4(2 - 3)}$

27. $2^3 + 3[2^2(\sqrt{36} + \sqrt{4})(\sqrt[3]{8} - 1) + 1]$

28. $2\frac{1}{4} \times 3\frac{1}{3} \div 2\frac{1}{3} \times 1\frac{1}{4}$

Evaluate:

29. $d + c - ac$ if $a = -1, c = -1, d = 1$

30. $abc - a$ if $a = -2, b = -1, c = -3$

LESSON 88 Exponents and signed numbers

88.A exponents and signed numbers

We must be careful when we simplify exponentials whose bases are signed numbers. The exponential

$$4^2$$

indicates that 4 is to be used as a factor twice.

$$(4)(4) = 16$$

The notation

$$-4^2$$

indicates the opposite of "4 squared," which is

$$-(4)(4) = -16$$

If we wish to indicate that -4 is to be used as a factor twice, we must enclose the -4 in parentheses and write

$$(-4)^2$$

This means -4 times -4 so

$$(-4)(-4) = 16$$

example 88.A.1 Simplify: $-4^2 - (-3)^2 - 3^3$

solution First we simplify the exponentials

$$-16 - 9 - 27$$

and then we add and get

$$\mathbf{-52}$$

example 88.A.2 Simplify: $(-3)^3 - (-2)^3 - 2^2$

solution First we simplify the exponentials.

$$-27 - (-8) - 4$$

Then we add

$$-27 + 8 - 4 \quad \text{simplified parentheses}$$

$$\mathbf{-23} \quad \text{added}$$

problem set 88

1. Five times a number is 35 greater than 2 times the opposite of the number. What is the number?
2. Seven times the opposite of a number is 50 less than 3 times the number. What is the number?
3. Only $\frac{3}{5}$ of the students in the school came to hear the distinguished scholar. If there were 1200 students in the school, how many came to hear the distinguished scholar?
4. The ratio of those who stayed awake to those who dozed was $2\frac{1}{2}$ to 7. If 1400 stayed awake, how many dozed?
5. Graph $x \nless -2$.
6. Graph $x \ngtr -1$.
7. What number is 160 percent of 230? Draw a diagram that depicts the problem.
8. If 180 is increased by 180 percent, what is the resulting number? Draw a diagram that depicts the problem.
9. What percent of 60 is 111? Draw a diagram that depicts the problem.
10. Twenty-five percent of what number is 61? Draw a diagram that depicts the problem.
11. Write $.012\frac{1}{2}$ as a fraction of whole numbers.
12. Write $\frac{3121}{7}$ as a mixed number.
13. Use the method of cut and try to find $\sqrt{8}$ to one decimal place.
14. Find the volume of a solid whose base is shown and whose sides are 6 feet high. Dimensions are in feet.

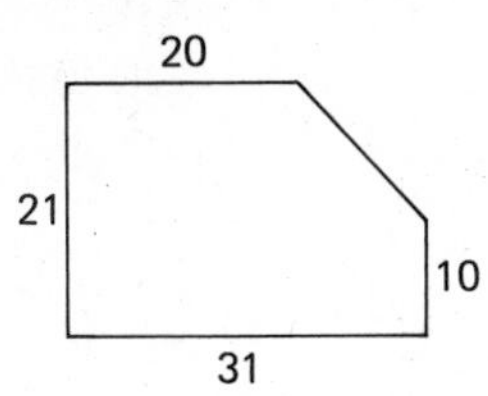

15. Convert 7892.321 centimeters to kilometers.
16. If $3x - 5 = 7$, what is the value of $3x - 2$?
17. If $4x + 2 = 11$, what is the value of $2x - 1$?
18. If $\frac{1}{2}x - 2 = 3$, what is the value of $\frac{1}{2}x + 3$?

Solve:

19. $-4x + 3 = x + 4$
20. $3x - 4 = -x + 6$
21. $-3\frac{1}{3}x - \frac{1}{6} = 1\frac{1}{3}$
22. $\dfrac{6\frac{1}{2}}{2\frac{1}{3}} = \dfrac{\frac{1}{2}}{x}$

Simplify:

*23. $-4^2 - (-3)^2 - 3^3$

24. $-3^2 + (-2)^2$

25. $-2^2(2 \cdot 2 - 3 \cdot 2) - 3(2 \cdot 3 - 2^2)$

26. $\dfrac{3 - 2(3 - 2) + 3 \cdot 2}{4(-2 + 5)}$

27. $\dfrac{-3 - 2(3 \cdot 2 - 1) + 2^2}{2^2(-3 - 2^2)}$

28. $3\frac{1}{2} \times 2\frac{1}{3} \div \frac{1}{4} \div \frac{1}{6}$

Evaluate:

29. $abc + ab$ if $a = -2, b = -3, c = -4$

30. $-ab + bc$ if $a = 2, b = -3, c = -4$

LESSON 89 Advanced ratio problems

89.A advanced ratio problems

When a ratio problem states the ratio between two things, the ratio between each of the things and the total is implied. If the ratio of rye to barley is 7 to 5, then a total of 12 is implied.

$$\text{Rye} = 7$$
$$\text{Barley} = 5$$
$$\text{Total} = 12$$

By looking at what we have written, we can write three proportions. The first two lines give us

$$\text{(a)} \quad \frac{R}{B} = \frac{7}{5}$$

The first line and the last line give us

$$\text{(b)} \quad \frac{R}{T} = \frac{7}{12}$$

and the last two lines give us

$$\text{(c)} \quad \frac{B}{T} = \frac{5}{12}$$

In ratio problems that discuss the total, it is helpful to begin by writing all three equations. Then the problem is reread to see which of the equations should be used.

example 89.A.1 Farmer Dunncowski wanted to plant his farm in wheat and corn in the ratio of 7 to 9. If he had 640 acres, how many should be planted in wheat?

solution First we write down the figures given and also write the total.

$$W = 7$$
$$C = 9$$
$$T = 16$$

With this information, we can write 3 proportions.

(a) $\frac{W}{C} = \frac{7}{9}$ (b) $\frac{W}{T} = \frac{7}{16}$ (c) $\frac{C}{T} = \frac{9}{16}$

We were given the total and asked for wheat so we will use proportion (b). In (b) we replace total with 640.

$$\frac{W}{T} = \frac{7}{16} \quad \text{proportion (b)}$$

$$\frac{W}{640} = \frac{7}{16} \quad \text{replaced } T \text{ with 640}$$

$$16W = 7 \cdot 640 \quad \text{cross multiplied}$$

$$\frac{16W}{16} = \frac{7 \cdot 640}{16} \quad \text{divided both sides by 16}$$

$$\mathbf{W = 280} \quad \text{simplified}$$

If he plants 280 acres in wheat, then 640 − 280 = 360 acres will be planted in corn.

example 89.A.2 When the race began, the ratio of fats to leans was 2 to 17. If 3800 racers were in the race, how many were lean?

solution First we write down the numbers for fats and leans. We add these two numbers to get the total.

$$F = 2$$

$$L = 17$$

$$T = 19$$

Now we can write the three proportions.

(a) $\frac{F}{L} = \frac{2}{17}$ (b) $\frac{F}{T} = \frac{2}{19}$ (c) $\frac{L}{T} = \frac{17}{19}$

We were given a total of 3800 and asked for the number of leans. Thus, we will use proportion (c) and replace T with 3800.

$$\frac{L}{T} = \frac{17}{19} \quad \text{proportion (c)}$$

$$\frac{L}{3800} = \frac{17}{19} \quad \text{replaced } T \text{ with 3800}$$

$$19L = 17 \cdot 3800 \quad \text{cross multiplied}$$

$$\frac{19L}{19} = \frac{17 \cdot 3800}{19} \quad \text{divided both sides by 19}$$

$$\mathbf{L = 3400} \quad \text{simplified}$$

If 3400 out of 3800 were lean, then 400 must have been fat.

problem set 89

***1.** Farmer Dunncowski wanted to plant his farm in wheat and corn in the ratio of 7 to 9. If he had 640 acres, how many should be planted in wheat?

***2.** When the race began, the ratio of fats to leans was 2 to 17. If 3800 racers were in the race, how many were lean?

3. Four times a number was 22 greater than twice the same number. What was the number?

4. Five-sixteenths of the teenies swooned as the star approached the microphone. If 2200 did not swoon, how many teenies attended the concert?

5. Graph $x \nleq -1$.

6. What number is 140 percent of 75? Draw a diagram that depicts the problem.

7. If 150 is increased by 30 percent, what is the resulting number? Draw a diagram that depicts the problem.

8. What percent of 70 is 91? Draw a diagram that depicts the problem.

9. Thirty-five is what percent of 140? Draw a diagram that depicts the problem.

10. Write $.011\frac{3}{5}$ as a fraction of whole numbers.

11. Use the method of cut and try to find $\sqrt{18}$ to one decimal place.

12. Find the least common multiple of 16, 18, and 20.

13. Complete the following table. Begin by inserting the reference numbers.

Fraction	*Decimal*	*Percent*
(a)	(b)	85

14. Find the volume in cubic feet of a solid whose base is shown and whose sides are 2 yards high. Dimensions are in feet.

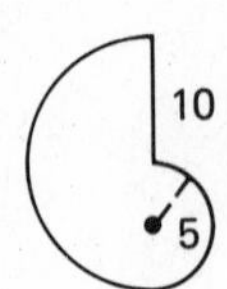

15. Convert 62.131121 kilometers to centimeters.

16. If $2x - 5 = 7$, what is the value of $6x - 1$?

17. If $5x - 3 = 22$, what is the value of $\frac{1}{5}x + 2$?

18. If $\frac{3}{4}x + 2 = 23$, what is the value of $\frac{1}{2}x + 2$?

Solve:

19. $-3x + 6 = 2x + 1$

20. $2x - 4 = -x + 5$

21. $-2\frac{1}{2}x - \frac{1}{3} = 2\frac{2}{3}$

22. $\dfrac{\frac{3}{5}}{\frac{6}{25}} = \dfrac{\frac{1}{3}}{x}$

Simplify:

23. $-2^2 - (-2)^3$

24. $-3^2 - (-3)^2$

25. $-8(2 \cdot 3 - 4) - 3(8 - 9)$

26. $\dfrac{4 - 2(6 - 4) + 2^2 \cdot 3}{3(-3 + 5)}$

27. $\dfrac{-2 + 3(2 \cdot 4 - 3) + 2^2}{3(2^2 - 1)}$

28. $2\frac{1}{2} \div 3\frac{1}{3} \times \frac{1}{6} \div \frac{1}{3}$

Evaluate:

29. $-bc - ab$ if $a = -2, b = -2, c = -5$

30. $ab - ac$ if $a = -1, b = -2, c = -1$

LESSON 90 Multiplication of exponentials

90.A multiplication of exponentials

We use exponential notation to indicate how many times a number is to be used as a factor. Consider the notation

$$3^5$$

The base of this exponential is 3 and the exponent is 5. This means that 3 is to be used as a factor 5 times.

$$3^5 = 3 \cdot 3 \cdot 3 \cdot 3 \cdot 3$$

Now we will consider the notation

$$3^5 \cdot 3^2$$

The first part tells us that 3 is a factor 5 times, and the second part tells us that 3 is a factor twice.

$$(3 \cdot 3 \cdot 3 \cdot 3 \cdot 3)(3 \cdot 3)$$

This means that 3 is a factor 7 times, which is written

$$3^7$$

We see from this that the number of times 3 is to be used as a factor can be found by adding the exponents.

$$3^5 \cdot 3^2 = 3^{5+2} = 3^7$$

example 90.A.1 Simplify: $5^2 \cdot 5^3 \cdot 5^4$

solution This means

$$(5 \cdot 5)(5 \cdot 5 \cdot 5)(5 \cdot 5 \cdot 5 \cdot 5)$$

and 5 is used as a factor 9 times. Thus we can write

$$5^2 \cdot 5^3 \cdot 5^4 = \mathbf{5^9}$$

example 90.A.2 Simplify: $10^{15} \cdot 10^{17}$

solution The first part has 10 as a factor 15 times, and the second part has 10 as a factor 17 times. Thus,

$$10^{15} \cdot 10^{17} = 10^{15+17} = 10^{32}$$

If we use 10 as a factor 32 times, we get

100,000,000,000,000,000,000,000,000,000,000

This numeral is cumbersome and difficult to work with as it is easy to miscount the number of zeros. Thus the exponential

$$10^{32}$$

is a better way to write this number and is preferred for most applications.

example 90.A.3 Simplify: $2^3 \cdot 5^2$

solution If we write the exponentials in expanded form, we get

$$(2 \cdot 2 \cdot 2)(5 \cdot 5)$$

We have 2 as a factor 3 times and 5 as a factor twice. This shows us why the expression

$$2^3 \cdot 5^2$$

cannot be simplified by adding the exponents. The bases are not the same.

90.B variable bases

Since letters stand for numbers, all the rules for numbers also apply to letters. Thus if we write

$$x^2 \cdot x^3$$

we mean

$$(x \cdot x)(x \cdot x \cdot x)$$

Here we have x used as a factor 5 times, and we can write this as

$$x^5$$

We call the rule demonstrated by this example the product theorem for exponents.

PRODUCT THEOREM FOR EXPONENTS

If m and n are real numbers and $x \neq 0$,

$$x^m \cdot x^n = x^{m+n}$$

This is a very formal way of saying that

We multiply exponentials that have the same bases by adding the exponents.

example 90.B.1 Simplify: $x \cdot x^2 \cdot y^{17} \cdot x^4 \cdot y^6$

solution **When we write x, we mean x^1.** We rearrange the expression and get

$$xx^2x^4y^{17}y^6$$

Now we add exponents of like bases and get

$$\mathbf{x^7y^{23}}$$

example 90.B.2 Simplify: $aa^2mm^5a^4m^6$

solution We will rearrange mentally. Then we add the exponents of like bases and get

$$\mathbf{a^7m^{12}}$$

problem set 90

1. In the hallway the ratio of reds to greens was 2 to 5. If 4900 were standing in the hallway, how many were red?
2. Five times the opposite of a number was 56 less than twice the number. What was the number?
3. Four hundred umbrellas cost $2000. What would 140 umbrellas cost?
4. Wilson bought 60 items for $40 each and 40 items for $150 each. What was the average cost of the items?
5. Graph $x \nleq -2$.
6. What number is 130 percent of 80? Draw a diagram that depicts the problem.
7. If 160 is increased by 40 percent, what is the resulting number? Draw a diagram that depicts the problem.
8. What percent of 60 is 108? Draw a diagram that depicts the problem.
9. Forty percent of what number is 240? Draw a diagram that depicts the problem.
10. Write $.001\frac{3}{4}$ as a fraction of whole numbers.
11. Find the perimeter of the following figure. Dimensions are in inches.

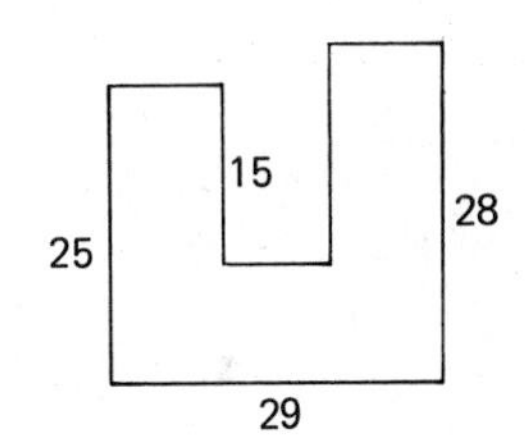

12. Use the method of cut and try to find $\sqrt{55}$ to one decimal place.
13. If $3x - 6 = 9$, what is the value of $3x - 4$?
14. If $4x - 4 = 28$, what is the value of $2x + 2$?
15. If $3x + 4 = 31$, what is the value of $\frac{1}{4}x + 1$?

Solve:

16. $-2x + 4 = 3x + 2$
17. $3x - 2 = -2x + 5$
18. $-3\frac{1}{2}x - \frac{1}{2} = 2\frac{1}{4}$
19. $\dfrac{-\frac{2}{5}}{\frac{4}{25}} = \dfrac{-\frac{1}{2}}{x}$

Simplify:

20. $-2^3 - (-2)^3$
21. $-(-3)^2$
22. $(-3)^2 + (-2)^3$
*23. $x \cdot x^2 \cdot y^{17} \cdot x^4 \cdot y^6$
24. $5^{15} \cdot 5^{17}$
25. $aa^3mm^3am^4a^2$
26. $xx^3x^2yy^5y^6xy^2$
27. $\dfrac{2^3 - 2(3 - 1)(2^3 - 3)}{2^2(2^2 - 1)}$
28. $\dfrac{3 - 2^2(2^2 - 1)(2^3 - 1)}{3^2(3^2 - 2^3)}$
29. $3\frac{1}{2} \div \frac{1}{4} \times 2\frac{1}{3} \div \frac{1}{6}$
30. Evaluate: $-ab + bc$ if $a = -2$, $b = -1$, $c = 1$

LESSON 91 Terms · Adding like terms

91.A terms

We use the word **term** in algebra to describe a part of a larger expression. The following expression has five distinct parts. We say that each of the parts is a term.

$$4 + \frac{6}{2} - 4x + 3(x + 2) + \frac{x + 5}{2}$$

The first term is 4; the second term is $\frac{6}{2}$; the third term is $4x$; the fourth term is $3(x + 2)$; and the fifth term is $\frac{x + 5}{2}$. From this we see that terms can be made up of a single symbol, symbols multiplied, or symbols divided. If a term has a number in front, we call the number the **coefficient** of the term. The coefficient of all three terms in the following expression is 5.

$$5xym + 5y^2 + \frac{5xm}{y}$$

91.B like terms

To see if

$$4xy \qquad \text{and} \qquad 6yx$$

are **like terms,** we consider only the letters so we discard the numbers and have left

$$xy \qquad \text{and} \qquad yx$$

If these both represent the same number regardless of the numbers that are used for x and y, we say that the expressions are **like terms.** To investigate, we let $x = -2$ and $y = 3$. Then

$$xy = (-2)(3) = -6 \qquad \text{and} \qquad yx = (3)(-2) = -6$$

Since the order of factors does not affect a product, we see that this will hold true for any numbers that we use for x and y. **So we say that $4xy$ and $6yx$ are like terms.**

example 91.B.1 Are $2xy^2$ and $3y^2x$ and $7yxy$ like terms?

solution We first discard the coefficients and have left

$$xy^2 \qquad \text{and} \qquad y^2x \qquad \text{and} \qquad yxy$$

If we expand these expressions, we get

$$xyy \qquad \text{and} \qquad yyx \qquad \text{and} \qquad yxy$$

Each of these has as factors two y's and one x. Since the order of multiplication does not affect a product, these expressions will represent the same number. To demonstrate we will use -2 for x and 3 for y.

xyy	yyx	yxy
$(-2)(3)(3)$	$(3)(3)(-2)$	$(3)(-2)(3)$
-18	-18	-18

Thus the terms $2xy^2$ and $3y^2x$ and $7yxy$ are like terms.

91.C adding like terms

Multiplication is a shorthand for repeated addition. If we have

$$3 + 3 + 3 + 3 \quad + \quad 3 + 3 + 3 + 3 + 3 \quad + \quad 3 + 3 + 3 + 3 + 3 + 3$$

we have four 3s plus five 3s plus six 3s for a total of fifteen 3s.

$$4(3) + 5(3) + 6(3) = 15(3)$$

If we were adding xy's instead of 3s, we would have

$$xy + xy + xy + xy + xy + xy + xy + xy + xy + xy + xy + xy + xy + xy + xy$$

$$4xy \quad + \quad 5xy \quad + \quad 6xy$$

$$= 15xy$$

Since xy's all stand for the same number, we can add them in the same way we added the 3s. This is the reason for the rule for adding like terms.

To add like terms we add the coefficients of the terms.

example 91.C.1 Simplify by adding like terms: $4x + 3 + 6x + 4 + 2xy + 3yx + 2$

solution We add the numbers for a sum of 9.

$$3 + 4 + 2 = 9$$

We add the x terms and get $10x$.

$$4x + 6x = 10x$$

And we add the xy terms and get $5xy$.

$$2xy + 3yx = 5xy \qquad \text{(or } 5yx \text{ if you prefer)}$$

So the sum is

$$\mathbf{9 + 10x + 5xy}$$

example 91.C.2 Add like terms: $4 + 3x^2 + 2xx + 4x + 7 + 3x$

solution We add like terms and get

$$\mathbf{11 + 5x^2 + 7x}$$

problem set 91

1. The ratio of red marbles to blue marbles was 2 to 5. If the bowl contained 350 marbles, how many were blue?
2. The ratio of flooded land to dry land was 2 to 9. If the plantation consisted of 1210 acres, how many acres were flooded?
3. The product of a number and 2 is 20 less than the product of the number and -3. What is the number?
4. The tree harvest was followed by seed planting. The forester was amazed because the number of seeds that sprouted was $2\frac{3}{4}$ times the number that had been planned on. If 140,000 sprouts had been planned on, how many sprouts did the forester get?
5. Graph: $x \geq 1$
6. What number is 75 percent of 220? Draw a diagram that depicts the problem.
7. If 150 is increased by 60 percent, what is the resulting number? Draw a diagram that depicts the problem.

8. What percent of 50 is 80? Draw a diagram that depicts the problem.

9. Use the method of cut and try to find $\sqrt{27.5}$ to one decimal place.

10. Find the area of the following figure. Dimensions are in feet.

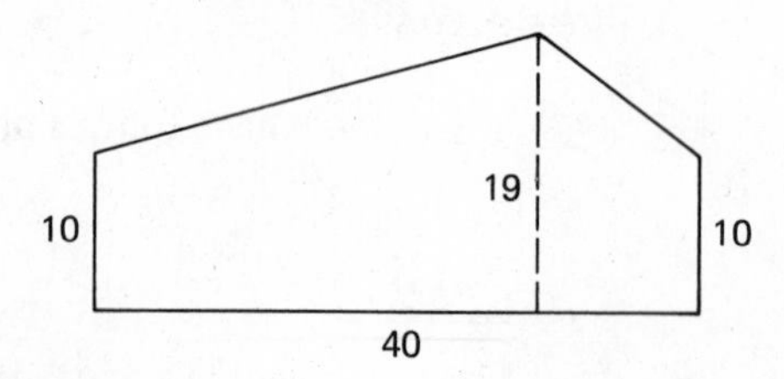

11. Find the least common multiple of 12, 30, and 36.

12. If $6x - 5 = 31$, what is the value of $2x - 3$?

13. If $5x - 3 = 22$, what is the value of $6x - 2$?

Solve:

14. $2x - 5 = 5x - 6$

15. $6x - 2 = 2x + 4$

16. $-4\frac{1}{3}x - \frac{1}{4} = 2\frac{1}{3}$

17. $\dfrac{-\frac{1}{4}}{\frac{1}{2}} = \dfrac{\frac{1}{3}}{x}$

Simplify by adding like terms.

*18. $4x + 3 + 6x + 4 + 2xy + 3yx + 2$

19. $3 + 2ab^2 + 3a + 2abb - 4a$

20. $6x + 2y + 3xy + 2 + 3x + 4y - xy$

Simplify:

21. $-2^3 - (-3)^2$

22. $-(-3)^2 - [-(-1)]$

23. $2^3 - 2[3(2 - 3) + 2(2 - 4)]$

24. $aa^3ba^2a^4b^3b^2$

25. $xxyyx^2y^3x^3y^4$

26. $ababa^2b^2a^2b^2a^3b^3$

27. $\dfrac{2^4 - 3(2^2 - 1)(3^2 - 2^3)}{2^3(2^2 - 1)}$

28. $\dfrac{2^3 - 3^2(2^2 - 1)(3 - 2^2)}{2(3^2 - 2^3)}$

29. $3\frac{1}{3} \div \frac{1}{4} \times 2\frac{1}{3} \div \frac{1}{3}$

30. Evaluate: $a + bc + ca$ if $a = -1, b = -2, c = 3$

LESSON 92 Distributive property · Estimating higher-order roots

92.A distributive property

The notation

$$4(2 + 3)$$

tells us to multiply 4 by what is inside the parentheses. There are two ways we can do this.

$4(2 + 3)$	$4(2 + 3)$
$4(5)$	$4 \cdot 2 + 4 \cdot 3$
20	$8 + 12$
	20

On the left we added 2 and 3 to get 5. Then we multiplied 4 by 5 to get 20. On the right we multiplied 4 by 2 and multiplied 4 by 3 and then added 8 and 12. We get an answer of 20 both times. We call this peculiarity or property of numbers the **distributive property.** This property is especially helpful when we simplify expressions that have variables. If we look at

$$4(x + y)$$

we cannot add x and y and then multiply. But we can multiply x by 4 and then multiply y by 4 and then add the two products.

$$4(x + y) = 4x + 4y$$

example 92.A.1 Use the distributive property to multiply $4(x + 2y - 8)$.

solution The distributive property also applies to the algebraic sum of three or more terms. We multiply 4 by each of the terms.

$$4(x + 2y - 8) = \mathbf{4x + 8y - 32}$$

example 92.A.2 Use the distributive property to multiply: $3x(x^2 - 2 + 2x)$

solution We multiply $3x$ by each of the terms in the parentheses.

$$3x(x^2 - 2 + 2x) = \mathbf{3x^3 - 6x + 6x^2}$$

92.B estimating higher-order roots

We can use the cut-and-try method to find roots to any accuracy we wish. However, the multiplication required is often time-consuming and untidy. Therefore, we will just practice finding whole number approximations of higher-order roots.

example 92.B.1 Estimate $\sqrt[3]{250}$ to the nearest whole number.

solution We want the cube root, so there are three factors.

$$(\ )(\ )(\ ) = 250$$

Let's try 3 as our first guess.

$$(3)(3)(3) = 27$$

That was a poor guess. Let's try 6.

$$(6)(6)(6) = 216$$

This is a little small so let's try 7.

$$(7)(7)(7) = 343$$

Now we have a guess that is too large and one that is too small.

$$(6)(6)(6) = 216$$
$$(\ )(\ )(\ ) = 250$$
$$(7)(7)(7) = 343$$

Since 216 is closer to 250 than 343, we say that the whole number approximation of $\sqrt[3]{250}$ is 6.

$$\sqrt[3]{250} \approx 6$$

example 92.B.2 Estimate $\sqrt[4]{390}$ to the nearest whole number.

solution The fourth root is used as a factor four times so the problem can be stated as

$$(\ )(\ )(\ )(\ ) = 390$$

Let's try 5.

$$(5)(5)(5)(5) = 625$$

That was too large so we will try 4.

$$(4)(4)(4)(4) = 256$$

We now have the answer bracketed.

$$(4)(4)(4)(4) = 256$$
$$(\ )(\ )(\ )(\ ) = 390$$
$$(5)(5)(5)(5) = 625$$

The number we are looking for is between 4 and 5 and is closer to 4. So to the nearest whole number the 4th root of 390 is 4.

$$\sqrt[4]{390} \approx 4$$

problem set 92

1. The football coach shuddered when he realized that the ratio of the dextrous to the inept was 3 to 7. If 140 players were trying to make the team, how many were dextrous?

2. When the score was announced, 420 students were elated and the rest were dejected. If the ratio of elated to dejected was 2 to 9, how many students were there in all?

3. Since Lafayette led the column, the troops marched 24 miles in 6 hours. Then he left and the troops reduced their speed by 1 mile per hour. How long did it take them to march the last 18 miles?

4. Four times a number is 36 greater than the product of the number and -2. What is the number?

5. What number is 45 percent of 200? Draw a diagram that depicts the problem.

6. If 80 is increased by 80 percent, what is the resulting number? Draw a diagram that depicts the problem.

7. What percent of 60 is 93? Draw a diagram that depicts the problem.

8. Find the least common multiple of 20, 25, and 30.

9. Find the surface area of the given prism. Dimensions are in centimeters.

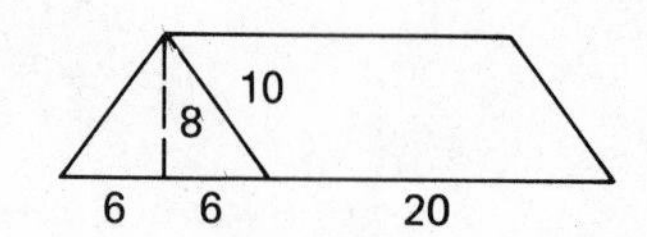

10. If $6x - 3 = 12$, what is the value of $\frac{2}{3}x + 5$?

11. If $3x + 5 = 22$, what is the value of $2x + 1$?

Use the distributive property to multiply.

***12.** $3x(x^2 - 2 + 2x)$ **13.** $2ab(a + b)$ **14.** $5a(6a + 2b + ab)$

Simplify by adding like terms.

15. $3x + 5 + 3y + 2xy + 6x + 5xy$ **16.** $a^2 + 5a^2 + 3b^2 + 2a^2 - 5b^2$

Solve:

17. $2x - 6 = 3x - 4$ **18.** $6x - 10 = -3 + 2x$

19. $-2x + 6 = -4x + 6$ **20.** $\dfrac{-\frac{2}{3}}{\frac{1}{4}} = \dfrac{\frac{1}{6}}{x}$

Simplify:

21. $-3^2 - (-2)^3$ **22.** $-[-(-3)] + (-2)^3$

23. $2^3 - 2^4[2(3 - 5) + 2^2(3^2 - 2^2)]$ **24.** $aa^2ba^3b^2a^2b^3$

25. $xy^2x^2x^3y^3y^2$ **26.** $m^2nn^3m^3n^2n^3$

27. $\dfrac{2^3 - 2(1 - 2^2)(3 - 2^2)}{2(2^3 - 3^2)}$ **28.** $\dfrac{1 - 3(2^2 - 3) + (3^3 - 2 \cdot 3)}{6(2 \cdot 3 - 2)}$

***29.** Use the method of cut and try to estimate $\sqrt[3]{250}$ to the nearest whole number.

***30.** Use the method of cut and try to estimate $\sqrt[4]{390}$ to the nearest whole number.

LESSON 93 *Base 2*

93.A bases, digits, and place value

We remember that the system of numbers that we use is called the Hindu-Arabic system. This system is a base ten system and uses the 10 digits

0, 1, 2, 3, 4, 5, 6, 7, 8, and 9

Whole numbers in this system have a value equal to the sums of the products of the digits and the place values. We remember that the place values in this system are as shown here

BASE 10 PLACE VALUES

etc.	10,000	1000	100	10	1

Thus, the numeral 2001 has the value of

etc.	10,000	1000	100	10	1
		2	0	0	1

2 times 1000 plus 0 times 100 plus 0 times 10 plus 1 times 1

and the numeral 4327 has the value of

etc.	10,000	1000	100	10	1
		4	3	2	7

4 times 1000, plus 3 times 100, plus 2 times 10, plus 7 times 1

All systems do not use 10 digits. Some systems use fewer than 10 digits, and some use more than 10 digits. Each system uses the same number of digits as the base of the system. The base 9 system uses nine digits; the base 6 system uses six digits; the base 4 system uses four digits; etc.

Any whole number greater than 1 can be used as a base of a number system. In this lesson, we will investigate the base 2 number system, which uses the two digits

0 and 1

93.B base 2 numerals

The values of the places in the base 2 system are as shown here. We use base 10 numbers to write the place values.

BASE 2 PLACE VALUES

etc.	128	64	32	16	8	4	2	1

The first place to the left of the decimal point has a value of 1. To get the next value, we multiply 1 by 2 and get 2. To get the next value, we multiply 2 by 2 and get 4. To get the next value, we multiply 4 by 2 and get 8, etc. Each place has a value twice the value of the place to its right.

example 93.B.1 What is the base 10 value of 10101 (base 2)?

solution We begin by making a table of base 2 place values. We use base 10 numerals to write the place values.

BASE 2 PLACE VALUES

64	32	16	8	4	2	1
		1	0	1	0	1

We see that we have one 16, one 4, and one 1.

$$16 + 4 + 1 = 21$$

Since the sum of these numbers is 21, the base 10 value of 10101 (base 2) is 21 (base 10).

10101 (base 2) equals **21 (base 10)**

example 93.B.2 What base 10 number does 110101 (base 2) represent?

solution We make a table of place values and write each digit in its proper place.

BASE 2 PLACE VALUES

64	32	16	8	4	2	1
	1	1	0	1	0	1

We see that we have one 32, one 16, no 8s, one 4, no 2s, and one 1.

$$32 + 16 + 4 + 1 = 53$$

so

110101 (base 2) equals **53 (base 10)**

example 93.B.3 Write 53 (base 10), using base 2 numerals.

solution We always make a table of place values.

BASE 2 PLACE VALUES

64	32	16	8	4	2	1
	1	1		1		1

We can get 53 with one 32 + one 16 + one 4 + one 1. Then we have to write zeros in the other places. So

110101 (base 2) equals 53 (base 10)

example 93.B.4 Write 102 (base 10), using base 2 numerals.

solution We always begin by writing a table of place values.

BASE 2 PLACE VALUES

64	32	16	8	4	2	1

The number 102 can be made from one 64, one 32, one 4, and one 2.

BASE 2 PLACE VALUES

64	32	16	8	4	2	1
1	1			1	1	

Thus we see that we can write

102 (base 10) as **1100110 (base 2)**

problem set 93

1. The racers surged across the bridge at the start of the marathon. Mary estimated that the ratio of male racers to female racers was 17 to 2. If there were 4750 racers in all, how many were males?

2. When the brouhaha commenced, the ratio of natives to intruders was 9 to 5. If 700 were involved, how many were natives?

3. Only one-fourth of those present really believed what the speaker was saying. If 1744 people were present, how many did not believe what the speaker was saying?

4. Five times a certain number was 6 less than 3 times the same number. What was the number?

*5. What base 10 number does 110101 (base 2) represent?

6. What base 10 number does 11111 (base 2) represent?

*7. Write 53 (base 10) using base 2 numerals.

8. What number is 160 percent of 300? Draw a diagram that depicts the problem.

9. If 60 is increased by 75 percent, what is the resulting number? Draw a diagram that depicts the problem.

10. What percent of 80 is 152? Draw a diagram that depicts the problem.

11. Find the volume in cubic inches of a solid whose base is shown and whose height is 1 yard. Dimensions are in inches.

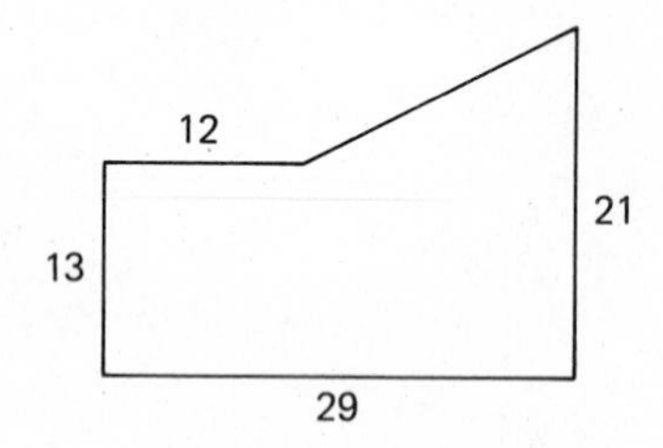

12. If $3x - 7 = 11$, what is the value of $\frac{1}{3}x + 3$?

13. If $2x - 6 = 5$, what is the value of $3x - \frac{1}{2}$?

Use the distributive property to multiply.

14. $2b(3a + b + ab)$

15. $3ab(a + b - 4)$

Simplify by adding like terms.

16. $a^2 + b^2 + 2aa + 3bb - 3ab + 2ab$

17. $a^3 + 3a^3b + b^3 + 2a^3 + 3aa^2b$

Solve:

18. $2x - 5 = 3x + 10$

19. $-6x - 3 = 2x + 6$

20. $-3x - 2 = 5x + 6$

21. $\dfrac{-\frac{4}{5}}{2\frac{1}{2}} = \dfrac{\frac{1}{3}}{x}$

Simplify:

22. $(-2)^2 + 2^2$

23. $-[-(-2)] + (-2)^3$

24. $2^3 - 2[2^2(2^3 - 3^2) + 3^2(2 - 1)]$

25. aba^2b^2ab

26. $xy^3x^2yx^2$

27. $m^3m^2n^2n^3m^3$

28. $\dfrac{2^4 - 3(2 - 2^2) + (3^2 - 2 \cdot 4)}{5(2 \cdot 3 - 3)}$

29. Use the method of cut and try to estimate $\sqrt[4]{562}$ to the nearest whole number.

30. Evaluate: $ab + bc$ if $a = -2, b = 3, c = 1$

LESSON 94 Percent word problems

94.A percent word problems

Many people have difficulty with percent word problems because they don't want to draw diagrams. They feel that somehow, somewhere, there exists a shortcut so that percent problems can be worked by rote and without understanding. No such shortcut exists, and that is the reason for drawing the diagrams. To work a percent word problem, we will first draw the diagrams. Then we will put the given numbers into the diagrams. Then we will use the percent equation to find the missing numbers.

example 94.A.1 Twenty percent of the clowns had red hair. If 40 clowns had red hair, how many did not have red hair?

solution We draw diagrams and insert the numbers that we have.

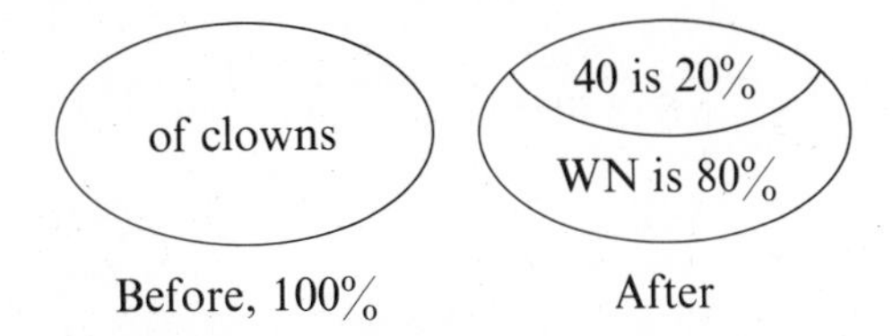

Now we can state the problem as

Forty is 20 percent of the clowns.

Now we use the percent equation to solve.

$$\frac{P}{100} \cdot \text{of} = \text{is} \qquad \text{equation}$$

$$\frac{20}{100} \cdot \text{clowns} = 40 \qquad \text{substituted}$$

$$\frac{100}{20} \cdot \frac{20}{100} \cdot \text{clowns} = 40 \cdot \frac{100}{20} \qquad \text{multiplied by } \frac{100}{20}$$

$$\text{clowns} = 200$$

If there were 200 clowns in all and 40 had red hair, then **160** did not.

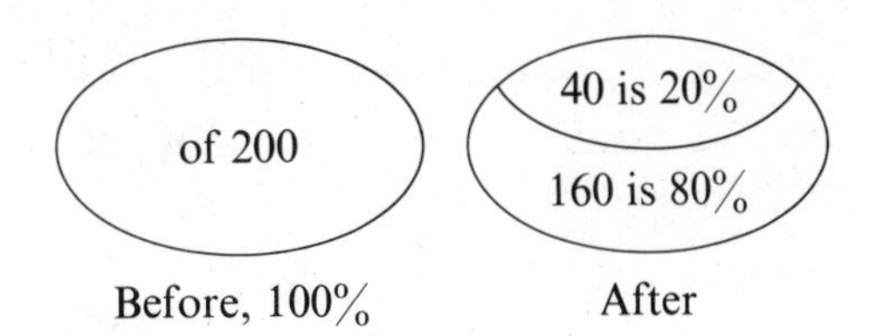

example 94.A.2 When the fog lifted, the number of frogs that could be seen was 140 percent greater than before. If 40 frogs could be seen before, how many could be seen after the fog lifted?

solution We must be careful. A 140 percent increase means 240 percent total.

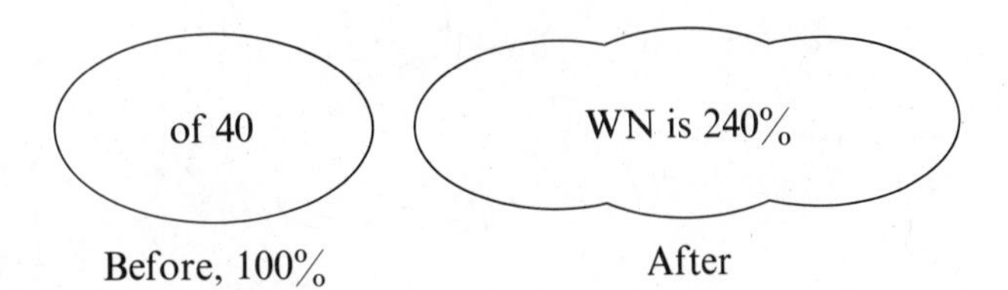

Now we can state the question as:

What number is 240 percent of 40?

Now we use the percent equation to solve.

$$\frac{P}{100} \cdot \text{of} = \text{is} \qquad \text{equation}$$

$$\frac{240}{100} \times 40 = WN \qquad \text{substituted}$$

$$\frac{9600}{100} = WN \qquad \text{multiplied}$$

$$\mathbf{96} = WN \qquad \text{divided}$$

Thus, we find that 96 is 240 percent of 40. Now we can complete the diagram.

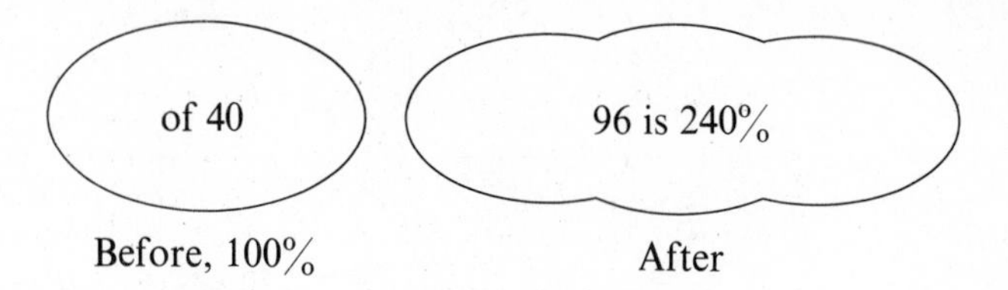

example 94.A.3 Forty percent of the trees were in bloom. If 1200 trees were not in bloom, how many trees were in bloom?

solution First we draw the diagrams and use the numbers that were given.

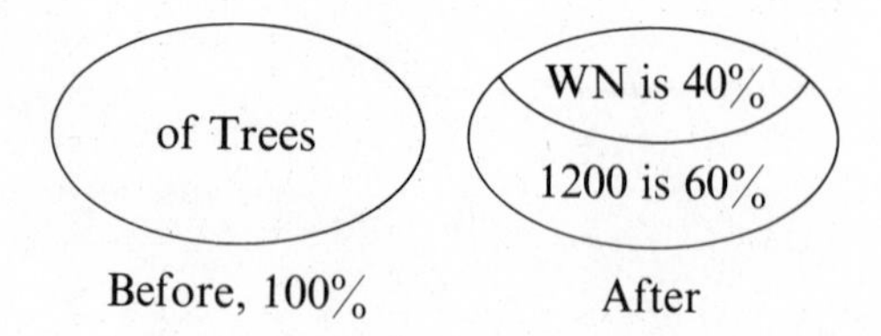

If 40 percent were in bloom, then 60 percent were not in bloom. We can state the problem as

1200 is 60 percent of the trees

Now we use the percent equation to solve.

$$\frac{P}{100} \cdot \text{of} = \text{is} \qquad \text{equation}$$

$$\frac{60}{100} \cdot \text{trees} = 1200 \qquad \text{substituted}$$

$$\frac{100}{60} \cdot \frac{60}{100} \cdot \text{trees} = 1200 \cdot \frac{100}{60} \qquad \text{multiplied by } \frac{100}{60}$$

$$\text{trees} = 2000 \qquad \text{simplified}$$

Now we draw the final diagram.

of 2000

800 is 40%

1200 is 60%

Before, 100%

After

If there were 2000 trees in all and 1200 were not in bloom, then **800** were in bloom.

problem set 94

*1. Twenty percent of the clowns had red hair. If 40 clowns had red hair, how many did not have red hair?

*2. When the fog lifted, the number of frogs that could be seen was 140 percent greater than before. If 40 frogs could be seen before, how many could be seen after the fog lifted?

*3. Forty percent of the trees were in bloom. If 1200 trees were not in bloom, how many trees were in bloom?

4. When all the extras had been counted, the number rose from 40 to 130. What percent increase was this?

5. What base 10 number does 101101 (base 2) represent?

6. Write 102 (base 10) using base 2 numerals.

7. Write 43 (base 10) using base 2 numerals.

8. What number is 130 percent of 230? Draw a diagram that depicts the problem.

9. If 70 is increased by 60 percent, what is the resulting number? Draw a diagram that depicts the problem.

10. What percent of 60 is 72? Draw a diagram that depicts the problem.

11. Find the perimeter of the following figure. Dimensions are in inches.

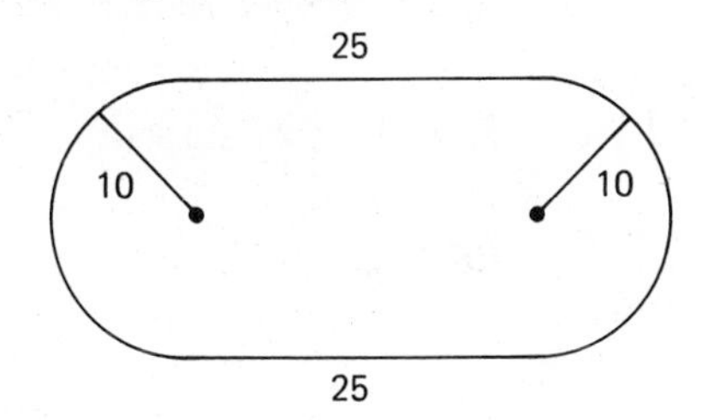

12. If $4x - 3 = 9$, what is the value of $\frac{1}{3}x + 2$?

13. If $3x + 5 = 17$, what is the value of $\frac{1}{2}x + 4$?

Use the distributive property to multiply.

14. $2a(3a + b + c)$

15. $3ab(a + b + c + d)$

Simplify by adding like terms.

16. $a^2 + b^2 + 3aa + 2bb - 3ab + 2ab$

17. $a^3 + 2aab + 2a^2b + 3b^2a + 3bba$

Solve:

18. $3x - 4 = 2x + 12$

19. $-4x - 5 = 3x + 4$

20. $-6x - 1 = -2x + 4$

21. $\dfrac{-\frac{2}{3}}{3\frac{1}{2}} = \dfrac{\frac{1}{4}}{x}$

Simplify:

22. $-3^2 - [-(2)]$

23. $-[-(-3)] + (-2)^3$

24. $ab^2ba^2b^3$

25. $2^2 - 2^3[2^2(3 - 2^2) + 2^2(3 - 1)]$

26. $xy^2x^3y^2x$

27. $m^3n^2m^2mn^3$

28. $\dfrac{2^3 - 2^2(3^2 - 2^3) + 2^2}{5(2^3 - 3^2)}$

29. Use the method of cut and try to estimate $\sqrt[3]{400}$ to the nearest whole number.

30. Evaluate: $ab + b$ if $a = -1$, $b = -3$

LESSON 95 Exponentials with negative bases

95.A exponentials

Beginners often make mistakes when they evaluate exponentials whose bases are negative numbers. We are aware of the difference in the meaning of the expressions

$$-2^2 \quad \text{and} \quad (-2)^2$$

The expression on the left has a value of -4. The expression on the right has a value of $+4$. When evaluating exponentials with variable bases, it is helpful to use parentheses when the letter is to be replaced with a negative number.

example 95.A.1 Evaluate: (a) a^2 (b) $-a^2$ if $a = -2$

solution (a) We first write

$$(\ \ )^2$$

Then we write -2 inside the parentheses and simplify:

$$(-2)^2 = \mathbf{4}$$

(b) This time we begin by writing

$$-(\ \ )^2$$

Then we write -2 inside the parentheses and simplify.

$$-(-2)^2 = \mathbf{-4}$$

example 95.A.2 Simplify: $a^2 - ab^2 - b^2$ if $a = -2$ and $b = -3$

solution Parentheses are not absolutely necessary, but we will use them to help us with the second term. We substitute and get

$$4 - (-2)(-3)^2 - 9$$

We simplify this and get

$$4 - (-18) - 9 \qquad \text{multiplied}$$

$$4 + 18 - 9 \qquad \text{simplified}$$

$$\mathbf{13} \qquad \text{added}$$

problem set 95

1. Forty percent of the students mispronounced the word *chic*. If 1440 knew how to pronounce this word, how many mispronounced it?

2. Twenty percent of the students were pertinacious in their pursuit of knowledge. If 400 students were not in this category, how many students were there in all?

3. The number that paid the piper rose 140 percent in 1 month. If 8640 paid the piper this month, how many paid him last month?

4. The ratio of sycophants to bullies was 14 to 1. If there were 750 in all, how many were sycophants?

5. What base 10 number does 110110 (base 2) represent?

6. Write 59 (base 10) using base 2 numerals.

7. What base 10 number does 11110 (base 2) represent?

8. What number is 170 percent of 120? Draw a diagram that depicts the problem.

9. If 80 is increased by 60 percent, what is the resulting number? Draw a diagram that depicts the problem.

10. What percent of 70 is 77? Draw a diagram that depicts the problem.

11. Find the area of the following figure. Dimensions are in feet.

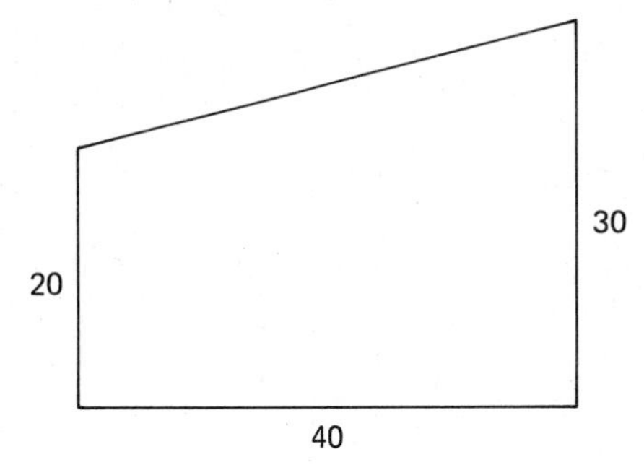

12. If $3x - 6 = 12$, what is the value of $\frac{1}{6}x + 3$?

13. If $4x - 3 = 21$, what is the value of $\frac{1}{2}x + 2$?

Use the distributive property to multiply.

14. $2bc(a + b + c)$

15. $ab(a + b + ab)$

Simplify by adding like terms:

16. $a^3 + 2abb + 3aa^2 + 2ab^2$

17. $yx^2 + xyx + 3y^2x + 2yxy$

Solve:

18. $4x - 3 = -x + 6$

19. $-5x - 2 = -x + 7$

20. $-3x + 5 = -2x - 3$

21. $\dfrac{-\frac{1}{3}}{2\frac{1}{3}} = \dfrac{\frac{1}{2}}{x}$

Simplify:

22. $-2^3 - [-(-3)]$

23. $-[-(-6)] + (-2)(-3)$

24. $2^3 - 2^2[2(3 - 2^2) + 2^2(2 \cdot 3 - 2^2)]$

25. $aba^2b^2a^3$

26. $xy^3x^2yy^2x$

27. $\dfrac{2^2 - 2^3(3^2 - 2^2) + 2^3}{6(5 - 2^2)}$

28. Use the method of cut and try to estimate $\sqrt[5]{250}$ to the nearest whole number.

Evaluate:

*29. $a^2 - ab^2 - b^2$ if $a = -2$ and $b = -3$

30. $a^3 + b^3$ if $a = -1$ and $b = -2$

LESSON 96 Multiple-term equations

96.A multiple-term equations

When equations have like terms, the first step in the solution is to combine like terms.

example 96.A.1 Solve: $4x + 2 + 3x - 2x = 12$

solution The first step is to combine like terms.

$$5x + 2 = 12 \qquad \text{combined like terms}$$

Now we finish by adding -2 to both sides and then dividing by 5.

$$\begin{array}{rcrl} 5x + 2 & = & 12 & \\ -2 & & -2 & \text{add } -2 \text{ to both sides} \\ \hline 5x & = & 10 & \\ \mathbf{x} & = & \mathbf{2} & \text{divided by 5} \end{array}$$

Check:

$$4(2) + 2 + 3(2) - 2(2) = 12 \longrightarrow 8 + 2 + 6 - 4 = 12 \longrightarrow 12 = 12 \qquad \text{check}$$

example 96.A.2 Solve: $3x - 2 - x - 4x = 5x + 12$

solution First we simplify the left side of the equation by adding like terms.

$$-2x - 2 = 5x + 12 \qquad \text{added like terms}$$

Next we decide to eliminate the variable on the left side, so we add $+2x$ to both sides.

$$\begin{array}{rcrl} -2x - 2 & = & 5x + 12 & \\ +2x & & +2x & \text{add } +2x \text{ to both sides} \\ \hline -2 & = & 7x + 12 & \end{array}$$

Now we finish by adding -12 to both sides and then dividing by 7.

$$\begin{array}{rcrl} -2 & = & 7x + 12 & \\ -12 & & -12 & \text{add } -12 \text{ to both sides} \\ \hline -14 & = & 7x & \\ \mathbf{-2} & = & \mathbf{x} & \text{divided} \end{array}$$

Check:

$$3(-2) - 2 - (-2) - 4(-2) = 5(-2) + 12$$

$$-6 - 2 + 2 + 8 = -10 + 12 \longrightarrow 2 = 2 \qquad \text{check}$$

problem set 96

1. Sixty percent of the boys did not like to euphemize. If 56 boys were not in this category, how many boys were there in all?

2. A 180 percent increase in 1 month could not be prevented. If the new total was 1120, what was the total before the increase?

3. When the rats invaded the town, some people ran and some stayed to fight the rats. The ratio of the runners to the fighters was 2 to 3. If 400 ran, how many people lived in the town before the rats came?

4. Four times a number was 24 greater than the product of the number and -2. What was the number?

5. What base 10 number does 111011 (base 2) represent?

6. Write 42 (base 10) using base 2 numerals.

7. If 65 is increased by 60 percent, what is the resulting number? Draw a diagram that depicts the problem.

8. What percent of 960 is 1392? Draw a diagram that depicts the problem.

9. Write $.0012\frac{3}{4}$ as a fraction of whole numbers.

10. Complete the following table. Begin by inserting the reference numbers.

Fraction	Decimal	Percent
$\frac{7}{20}$	(a)	(b)

11. Find the least common multiple of 24, 30, and 36.

12. If $6x + 4 = 28$, what is the value of $\frac{1}{4}x + 3$?

13. If $7x - 6 = 22$, what is the value of $\frac{1}{2}x - 3$?

Use the distributive property to multiply.

14. $2abc(a + b + c)$

15. $ab(abc + 2ab - 2bc)$

Simplify by adding like terms.

16. $ab^2 + a^3b + a^2ab - 3abb$

17. $m^2n^3 - mn^2 + mmnn^2 + 6mnn$

Solve:

*18. $3x - 2 - x - 4x = 5x + 12$

19. $6\frac{1}{3}x - 2\frac{1}{3} = 3\frac{1}{6}$

20. $6x + 2 - 3x + 2x = 9x - 12$

21. Use the method of cut and try to estimate $\sqrt{33}$ to one decimal place.

Simplify:

22. $-2^3 - [-(-2)]$

23. $abb^2a^2bb^3$

24. $-[-(-2)] + (-3)(2) + (-1)^3$

25. $-2^3 - 2[2^3(2^2 - 3) + 2(2^3 - 3^2)]$

26. $xy^2y^3x^2y$

27. $\dfrac{2^3 - 2^2(2^3 - 3^2) + 3}{3(2^3 - 1)}$

28. $\frac{1}{2}\left(2\frac{1}{3}\cdot\frac{1}{4}-\frac{1}{2}\right)$

Evaluate:

29. $a^2 + b^2 + 2ab \qquad \text{if } a = -1, b = 2$

30. $a^3 + b^3 \qquad \text{if } a = -2, b = -2$

LESSON 97 Points, lines, and rays · Angles · Copying angles by construction

97.A points, lines, and rays

Here we show a series of dots, each one smaller than the one to its left.

If we continue drawing the dots with each one smaller than the one to its left, we would finally have a dot so small that it could not be seen without magnification. **This dot would still be larger than a mathematical point because a mathematical point is so small that it has no size at all.**

A curve is an endless connection of mathematical points. A line is a straight curve. Any two points on a line can be used to name a line. Because a line is made of mathematical points, a line has no width. When we draw a line with a pencil, the line we draw is a graph of the mathematical line and marks the location of the mathematical line. We sometimes put arrowheads on the ends of lines to emphasize that the lines continue without end in both directions.

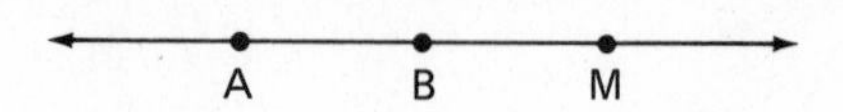

Any two points on a line can be used to name a line. We draw a bar above two letters to indicate that we are naming a line. Thus any of the following notations name the line shown.

$$\overline{AB} \qquad \overline{BA} \qquad \overline{AM} \qquad \overline{MA} \qquad \overline{MB} \qquad \overline{BM}$$

Many authors draw arrowheads on the ends of the overbar to emphasize that they are naming the whole line.

$$\overleftrightarrow{AB} \qquad \overleftrightarrow{BA} \qquad \overleftrightarrow{AM} \qquad \overleftrightarrow{MA} \qquad \overleftrightarrow{MB} \qquad \overleftrightarrow{BM}$$

These authors use the overbar with no arrowheads to name the **line segment** between the two points. They would use either

$$\overline{AB} \quad \text{or} \quad \overline{BA}$$

to name the line segment between points A and B.

P Q X

A **ray** is sometimes called a **half line.** A ray begins at one point and continues without end through the other point. A single-arrowhead overbar is often used to help name a ray. We could call the ray on the last page either

$$\overrightarrow{PQ} \quad \text{or} \quad \overrightarrow{PX}$$

97.B angles

An **angle** is formed by the intersection of two rays. Some people define an angle to be the opening between the two rays. Others say that the angle is the rays themselves. The **vertex** of an angle is the point of intersection of the two rays.

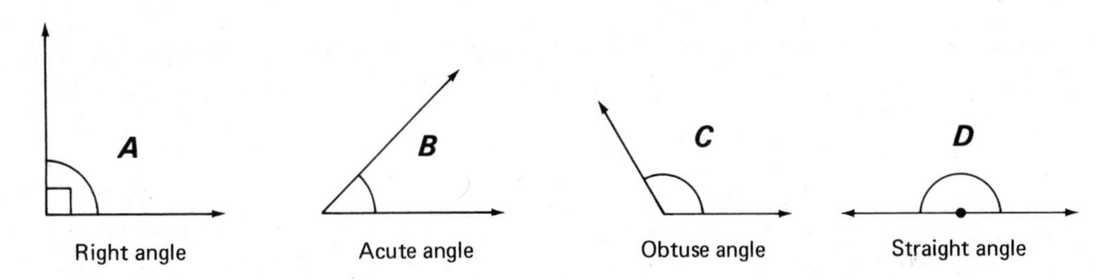

Angle A is formed by two perpendicular rays. We say the angle A is a **right angle.** The little square drawn in this angle tells us that the angle is a right angle, or a 90° angle. Angles smaller than right angles are called **acute angles,** as B; angles larger than right angles are called **obtuse angles,** as C. If the rays form a straight line, we call the angle a **straight angle.** A straight angle can be formed by two 90° angles.

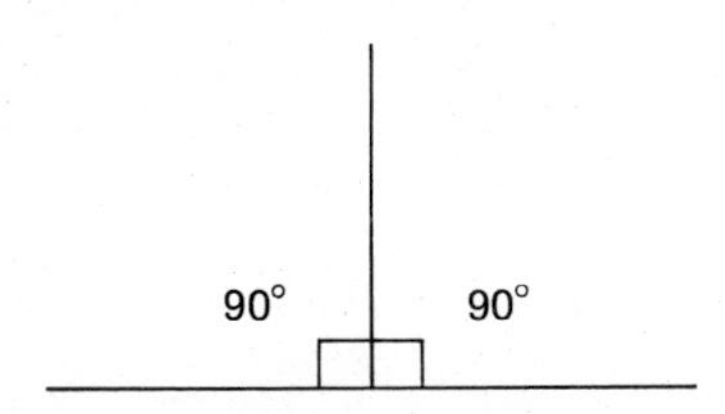

From this we see that a straight angle is a 180° angle.

97.C measuring angles

Sometimes we use the vertex and a point on each of the rays to name an angle

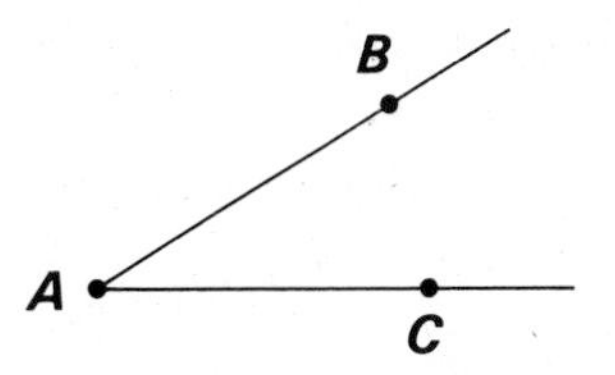

Using this notation, the angle shown here would be called angle BAC or angle CAB. We often use the symbol

$$\angle$$

to mean angle. Thus the angle above could be designated by writing

$$\angle BAC \quad \text{or} \quad \angle CAB$$

A protractor is often used to measure angles. The intersection of the rays is called the vertex of the angle and is placed under the dot or circle on the protractor, and one ray is aligned with the bottom line on the protractor, in the following figure.

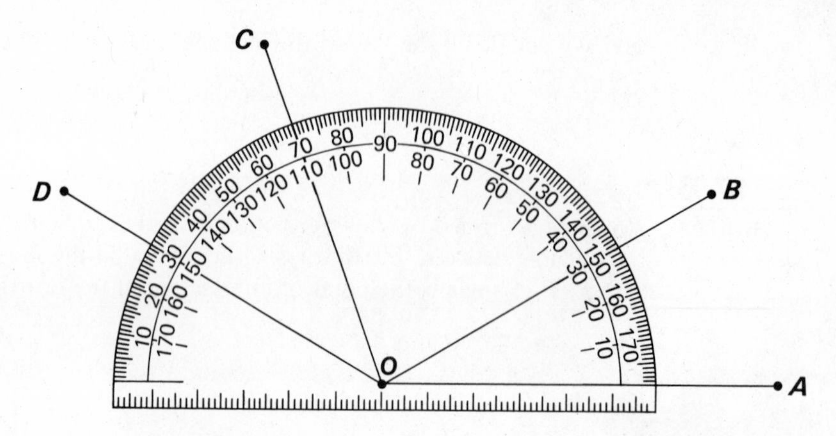

The measures of the three angles shown are

$$\angle AOB = 30^\circ \qquad \angle AOC = 110^\circ \qquad \angle AOD = 150^\circ$$

97.D copying angles by construction

An angle can be copied by using a ruler and a compass and five steps.

Given angle

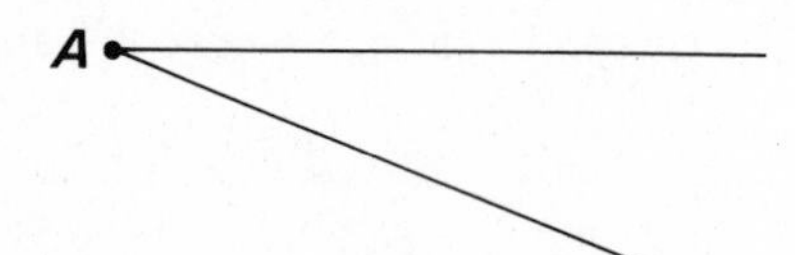

1. Draw one side.

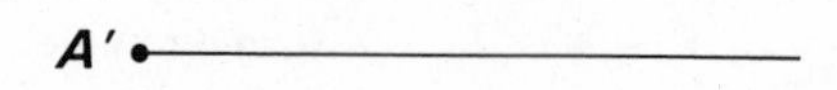

2. Draw arc on given angle.

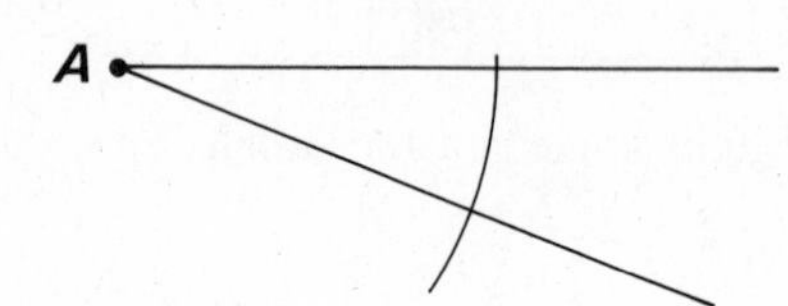

3. Draw same arc

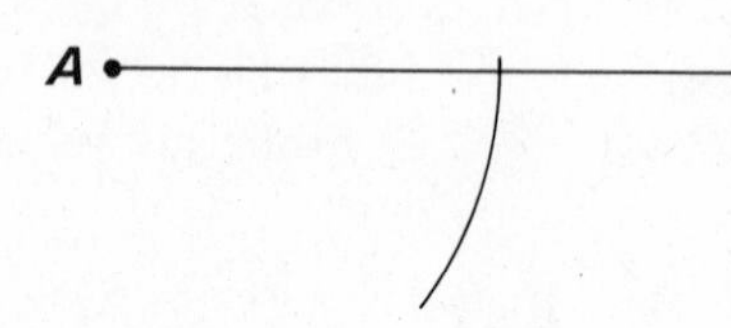

4. Draw arc on arc.

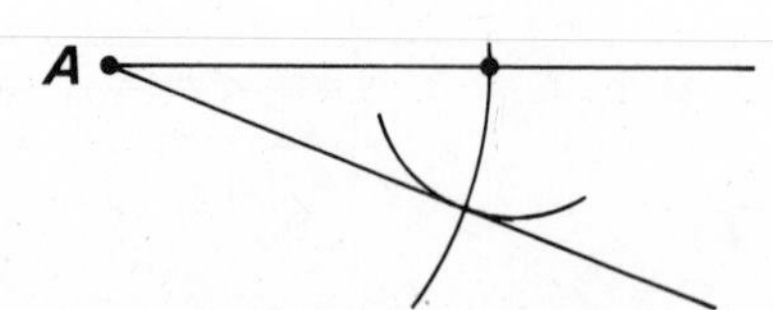

5. Draw same arc and draw ray.

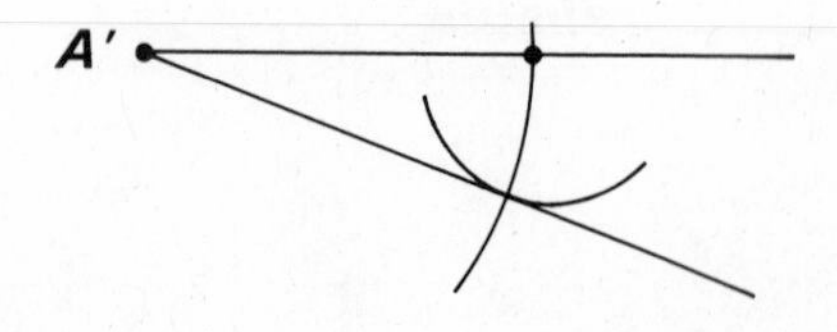

problem set 97

1. The wail of anguish turned into a paean of thanksgiving when the teacher curved the grades. Forty percent of the students were saved. If 1440 students were not saved, how many students were there in all?

2. The number that eschewed unauthorized assistance rose 260 percent in 1 month. If the number was 400 last month, what was the number this month?

3. The ratio of froward students to those who were conciliatory was 2 to 11. If 390 students attended the conference, how many were froward?

4. The sum of a number and 40 was 13 greater than 10 times the number. What was the number?

5. Use a protractor to draw a 60° angle. Then use a compass and straightedge to copy the angle.

6. What base 10 number does 11001 (base 2) represent?

7. Write 75 (base 10) using base 2 numerals.

8. If 66 is increased by 50 percent, what is the resulting number? Draw a diagram that depicts the problem.

9. What percent of 620 is 837? Draw a diagram that depicts the problem.

10. Find the volume in cubic feet of a solid whose base is shown and whose height is 2 yards. Dimensions are in feet.

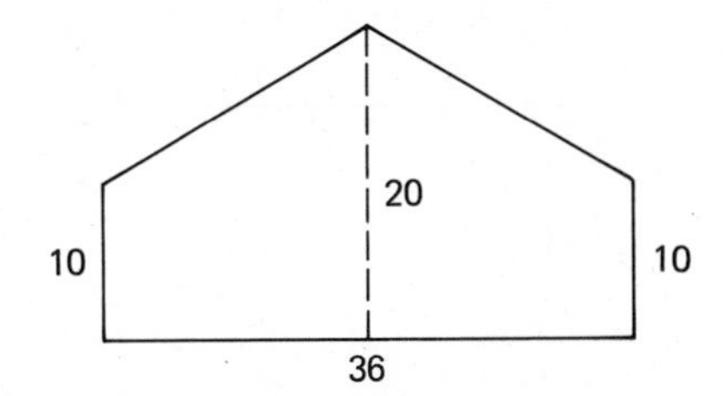

11. Convert 2 miles to inches.

12. If $3x + 6 = 24$, what is the value of $\frac{1}{3}x - 2$?

13. If $7x + 6 = 27$, what is the value of $2x - 3$?

Use the distributive property to multiply.

14. $2ab(b + a + c)$

15. $mn(mn + n + m)$

Simplify by adding like terms.

16. $ab^2 + a^2b + 3abb + 2aab - a^2$

17. $xy^3 + x^2y + xyy^2 + xxy$

Solve:

18. $2x + 3 - 3x - x = 6x + 12$

19. $2\frac{1}{4}x - 2\frac{1}{2} = 3\frac{1}{4}$

20. $3x + 2x - 4x - 5 = 6x + 6$

21. $\dfrac{\frac{1}{3}}{\frac{9}{2}} = \dfrac{\frac{1}{4}}{x}$

Simplify:

22. $-[-(-3)] + (-2)(3)$

23. $-[-(-2)] + (-2)$

24. $ab^2b^3a^2b$

25. $xy^3y^2x^2x^3$

26. $\dfrac{2^4 - 2^3(2^3 - 4) + 2}{2^3(3^2 - 2^3)}$

27. $\frac{1}{3}\left(3\frac{1}{3} \cdot \frac{1}{3} - \frac{7}{9}\right)$

Evaluate:

28. $c^2 - 2ab$ if $a = -2, b = -1, c = -3$

29. $a + b^3$ if $a = -1, b = -2$

30. Use the method of cut and try to estimate $\sqrt[3]{500}$ to the nearest whole number.

LESSON 98 English volume conversions

98.A English volume conversions

There are quite a few cubic inches in a cubic foot. We can see this clearly by looking at this drawing of a cubic foot.

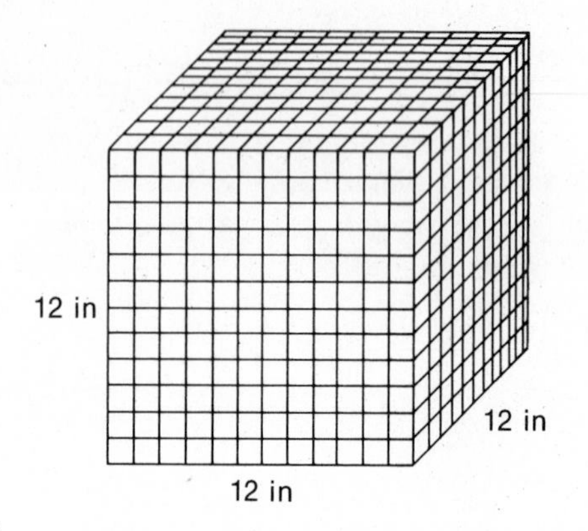

We see that there are

$$12 \times 12 = 144$$

144 cubes in the front face; and the cubes are 12 deep so there are a total of

$$144 \times 12 = 1728 \text{ cubes}$$

We will use three unit multipliers to make volume conversions. We will concentrate on the process rather than on numerical answers. Thus the answers will not be worked out all the way.

example 98.A.1 Convert 14 cubic feet to cubic inches.

solution We will write cubic feet (ft^3) as ft · ft · ft so we can see how we use the three unit multipliers.

$$14\,\cancel{\text{ft}}\cdot\cancel{\text{ft}}\cdot\cancel{\text{ft}} \times \frac{12\text{ in}}{1\,\cancel{\text{ft}}} \times \frac{12\text{ in}}{1\,\cancel{\text{ft}}} \times \frac{12\text{ in}}{1\,\cancel{\text{ft}}} = \mathbf{14(12)(12)(12)\ in^3}$$

A pocket calculator can be used to do the multiplication if a numerical answer is necessary.

example 98.A.2 Convert 140 cubic yards (yd^3) to cubic inches (in^3).

solution We will go from yd^3 to ft^3 to in^3.

$$140\,\cancel{\text{yd}^3} \times \frac{3\,\cancel{\text{ft}}}{1\,\cancel{\text{yd}}} \times \frac{3\,\cancel{\text{ft}}}{1\,\cancel{\text{yd}}} \times \frac{3\,\cancel{\text{ft}}}{1\,\cancel{\text{yd}}} \times \frac{12\text{ in}}{1\,\cancel{\text{ft}}} \times \frac{12\text{ in}}{1\,\cancel{\text{ft}}} \times \frac{12\text{ in}}{1\,\cancel{\text{ft}}} = \mathbf{140(3)(3)(3)(12)(12)(12)\ in^3}$$

example 98.A.3 Convert 50,163,529 cubic inches to cubic miles (mi^3).

solution We will go from in^3 to ft^3 to mi^3.

$$50{,}163{,}529\,\cancel{\text{in}^3} \times \frac{1\,\cancel{\text{ft}}}{12\,\cancel{\text{in}}} \times \frac{1\,\cancel{\text{ft}}}{12\,\cancel{\text{in}}} \times \frac{1\,\cancel{\text{ft}}}{12\,\cancel{\text{in}}} \times \frac{1\text{ mi}}{5280\,\cancel{\text{ft}}} \times \frac{1\text{ mi}}{5280\,\cancel{\text{ft}}} \times \frac{1\text{ mi}}{5280\,\cancel{\text{ft}}}$$

$$= \mathbf{\frac{50{,}163{,}529}{(12)(12)(12)(5280)(5280)(5280)}\ mi^3}$$

problem set 98

1. Bucolic themes dominated the landscapes as 70 percent of the paintings were bucolic in nature. If 2000 paintings were exhibited, how many were not bucolic?
2. Ninety-six percent of the newcomers caviled every time an event was announced. If 140 did not cavil, how many newcomers were there?
3. The ratio of aesthetes to philistines was 3 to 14. If 1700 attended the art show, how many were philistines?
4. The sum of a number and 40 is 5 less than the product of the number and 6. What is the number?
5. Use a protractor to draw a 30° angle. Then use a compass and a straightedge to copy the angle.
6. What base 10 number does 101011 (base 2) represent?
7. Write 28 (base 10) using base 2 numerals.
8. If 125 is increased by 60 percent, what is the resulting number? Draw a diagram that depicts the problem.
9. Find the volume in cubic centimeters of a solid whose base is shown and whose height is 3 meters. Dimensions are in centimeters.

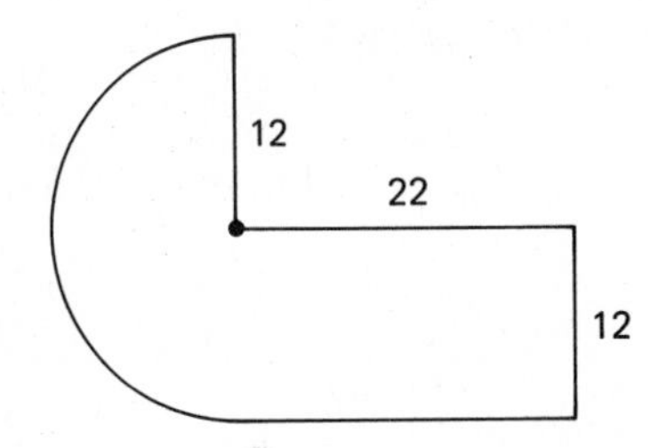

*10. Convert 50,163,529 cubic inches to cubic miles.

11. Convert 10 cubic feet to cubic inches.

12. If $3x - 5 = 22$, what is the value of $\frac{1}{3}x + \frac{1}{2}$?

13. If $16x + 21 = 53$, what is the value of $\frac{1}{3}x + \frac{1}{2}$?

14. Use the distributive property to multiply: $3ac(ab + ac + bc)$

15. Use the cut-and-try method to estimate $\sqrt[4]{700}$ to the nearest whole number.

Simplify by adding like terms.

16. $ab^2 + abb + 2a^2ab + 2abb - baa^2$
17. $-xy^3 + x^2xyy^2 + x^3y^3 - 3xyy^2$

Solve:

18. $3x + 4 - 2x - x = 4x - 5$
19. $3\frac{1}{2}x - 1\frac{3}{4} = -\frac{3}{2}$
20. $6 + x - 2x + 3x - 4x = 3x + 5$
21. $\dfrac{-\frac{1}{4}}{\frac{1}{3}} = \dfrac{\frac{16}{9}}{x}$

Simplify:

22. $(-3)(-2) + (-3)^2$
23. $aba^2b^2a^3b$
24. $(-3)^2(2) - 3$
25. $xy^3x^2xxy^3$

26. $\dfrac{-(-3)^2 - 2^3(2^3 - 2 \cdot 6) + 3}{2^2(2^3 - 4)}$

27. $\dfrac{1}{4}\left(2\dfrac{1}{3} \cdot \dfrac{1}{4} - \dfrac{7}{12}\right)$

Evaluate:

28. $-b^2 + a + b$ if $a = -2, b = -1$

29. $-b^3 + 2ab$ if $a = -1, b = -2$

30. Write 6213825.32 in words.

LESSON 99 Perpendicular bisectors · Angle bisectors

99.A perpendicular bisectors

The word **bisect** means to divide into two equal parts. A **perpendicular bisector** of a line segment is a line that is perpendicular to the line segment at the **midpoint** of the line segment. To construct a perpendicular bisector of the line segment $\overline{MN}$, use the compass to draw from point M two arcs of equal radii. The radii should be a little longer than half of the line segment (a). Then in (b) we draw the same arcs from point N.

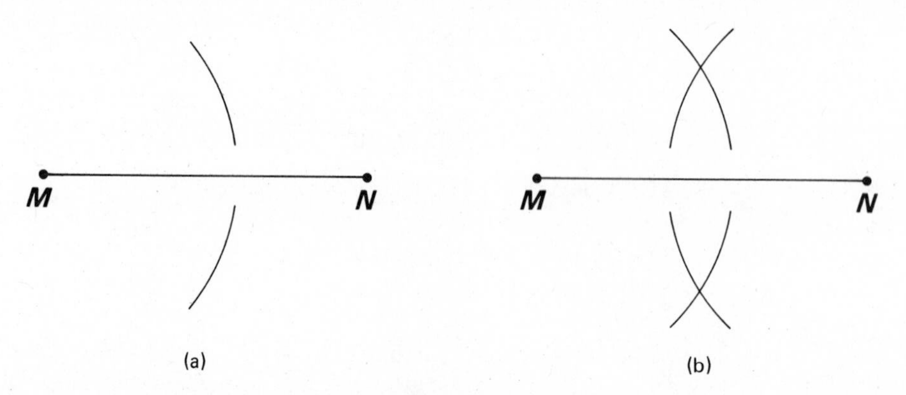

(a) (b)

The points where the arcs cross are **equidistant** (the same distance) from points M and N. Any point on the line that connects these intersections will also be equidistant from points M and N. A line drawn through these points is the perpendicular bisector of $\overline{MN}$.

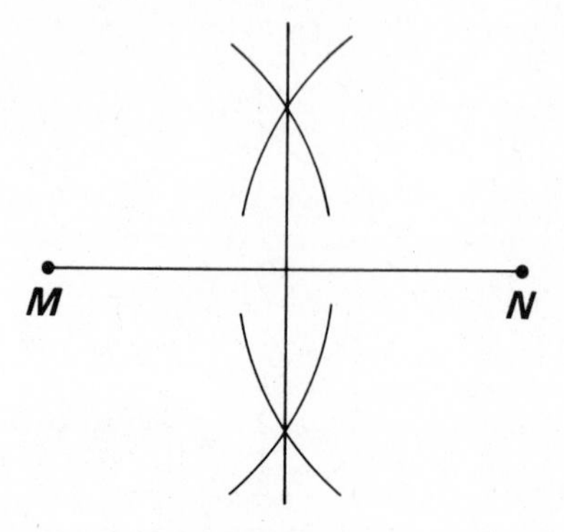

example 99.A.1 Construct a right triangle whose base is 4 centimeters and whose height is 3 centimeters.

solution The only way to construct a perpendicular is to bisect a line segment. Thus, we draw a line segment 8 centimeters long and bisect it.

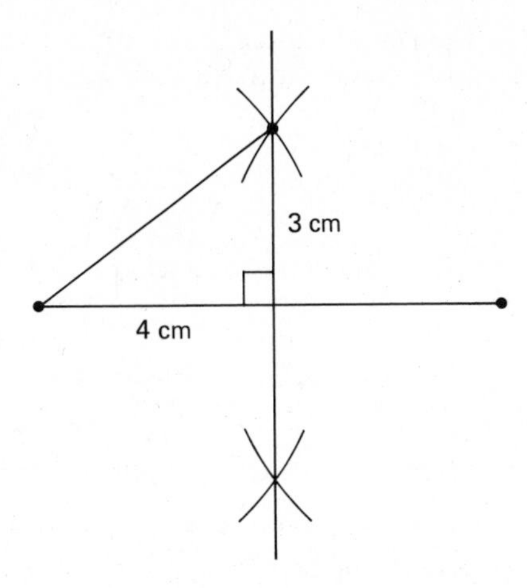

The last step was to measure 3 centimeters on the bisector and then draw the triangle.

99.B angle bisectors

An angle bisector divides an angle into two equal parts. A similar technique is used to bisect an angle as is used to bisect a line segment. To bisect the angle shown in (a)

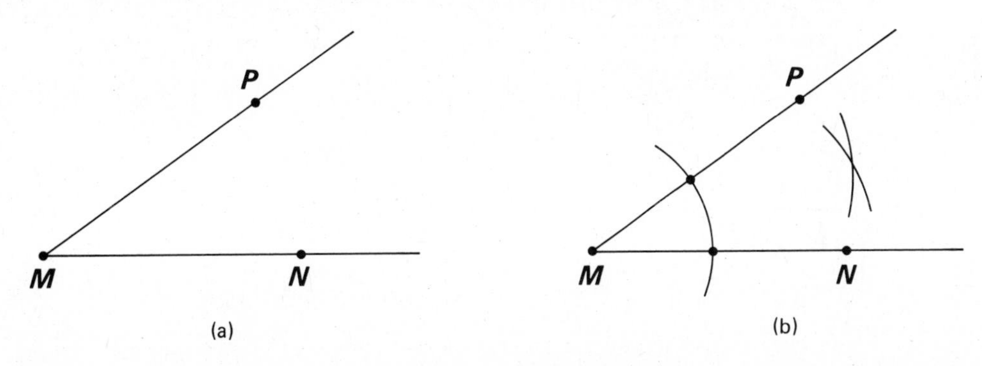

(a) (b)

we first draw an arc that intersects both rays (b). Then arcs of equal radii are drawn from each of the points of intersection. The point where these arcs cross lies on the angle bisector.

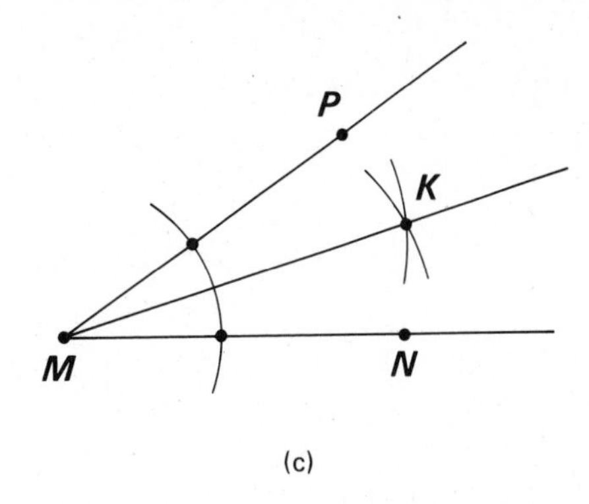

(c)

$$\angle NMK = \frac{1}{2} \angle NMP \quad \text{and} \quad \angle PMK = \frac{1}{2} \angle PMN$$

problem set 99

1. Sixty percent recoiled in horror when the beast emerged. If 1400 did not recoil in horror, how many saw the beast emerge?

2. The ratio of the affable to the truculent was 9 to 4. If 260 had forced their way inside, how many were affable?

3. Thank goodness that only 2 percent thought that pugnacity was synonymous with manliness. If 400 had this view, how many were there in all?

4. The product of a number and 4 is 24 greater than the product of the same number and 8. What is the number?

5. Draw a 4-centimeter line segment with a ruler and construct the perpendicular bisector of the line segment.

6. Use a protractor to draw a 60° angle. Then use a straightedge and a compass to construct the bisector of the angle.

7. Use a protractor to draw a 45° angle. Then use a compass and a straightedge to copy the angle.

8. What base 10 number does 1011011 (base 2) represent?

9. Write 25 (base 10) using base 2 numerals.

10. If 130 is increased by 70 percent, what is the resulting number? Draw a diagram that depicts the problem.

11. Find the volume in cubic inches of a solid whose base is shown and whose height is 1 foot. Dimensions are in inches.

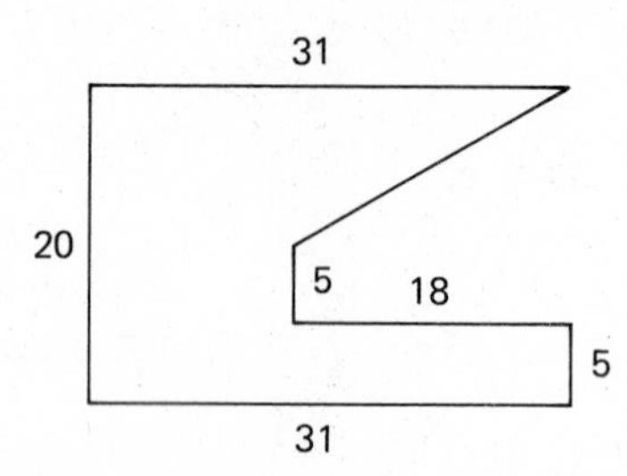

12. Convert 161 cubic yards to cubic inches.

13. Convert 612,312,561 cubic inches to cubic miles.

14. If $8x - 16 = 32$, what is the value of $2x - 4$?

Use the distributive property to multiply:

15. $4ab(b + c - ab)$

16. $-mn(n + m - mn)$

Simplify by adding like terms:

17. $ac^2 + a^2b + 2acc + 3aba - 2ac^2$

18. $-xy^2 + xxy + 3xy^2 - 2yx^2$

Solve:

19. $4x - 3x + 2x + 6 = -2x - 6$

20. $-6\frac{2}{3}x + \frac{2}{3} = -\frac{3}{2}$

21. $\dfrac{-\frac{1}{3}}{\frac{2}{9}} = \dfrac{3}{x}$

Simplify:

22. $-[-(-3^2)] + (-2)(-4) + (-2)^3$

23. $2^3 - 2^2[-2(-3 + 2^2)(2^3 - 2^2) + 1]$

24. $6ab^2a^2bb^2$

25. $2xy^2y^3x^2x$

26. $\dfrac{-(-3)^2 + 2^2(2^3 - 3^2) + 2}{2^2(2^3 - 4)}$

27. $\dfrac{1}{3}\left(2\dfrac{1}{4} \cdot \dfrac{1}{3} - \dfrac{5}{12}\right)$

Evaluate:

28. $2a^2 + 3a^3$ if $a = -1$

29. $2ab + b^2$ if $a = -1, b = 1$

30. Graph: $x \nleq 3$

LESSON 100 Base 8

100.A base 8

Any whole number greater than 1 can be used as a base for a number system. Base 2 numbers are used by engineers to design computers. Another important number system is the base 8 system. Base 10 numbers use 10 digits. Base 2 numbers use 2 digits, and as one might guess, base 8 numbers use 8 digits.

0, 1, 2, 3, 4, 5, 6, 7

The right-most place in base 8 has a place value of 1. The next place has a place value of 8 times 1, or 8. The next place has a place value of 8 times 8 times 1, or 64. Each place has a place value 8 times the value of the place to its right.

BASE 8 PLACE VALUES

etc.	4096	512	64	8	1

example 100.A.1 What base 10 number is 4032 (base 8)?

solution First we write our place value table for base 8.

BASE 8 PLACE VALUES

etc.	4096	512	64	8	1
		4	0	3	2

We see that 4032 base 8 means

4 × 512	for	2048
3 × 8	for	24
2 × 1	for	2
		2074

So 4032 (base 8) equals **2074** (base 10).

example 100.A.2 Write 1321 (base 10) as a base 8 numeral.

solution First we write the base 8 place-value table.

etc.	4096	512	64	8	1

There are no 4096s in 1321, but it seems that we have two 512s.

$$\begin{array}{r} 512 \\ \times\ 2 \\ \hline 1024 \end{array} \qquad \begin{array}{r} 1321 \\ -\ 1024 \\ \hline 297 \end{array}$$

Now we find that there are four 64s in 297.

$$\begin{array}{r} 64 \\ \times\ 4 \\ \hline 256 \end{array} \qquad \begin{array}{r} 297 \\ -\ 256 \\ \hline 41 \end{array}$$

and 41 has five 8s and 1 left over.

4096	512	64	8	1
	2	4	5	1

Thus, the number 1321 base 10 can be made up of two 512s, four 64s, five 8s, and one 1, so it can be written

2451 (base 8)

problem set 100

1. Twenty percent of the little people suffered from agoraphobia, and they would not go into the clearing. If 1600 did not suffer from agoraphobia, how many little people lived in the forest?

2. The ratio of the number of loners to the number who were gregarious was 3 to 14. If 15,300 lived in the valley, how many were gregarious?

3. The product of a number and -2 was increased by 7. This result was 27 greater than the product of the number and 3. What was the number?

4. Alfonso traveled 100 miles in 4 hours. Then he increased his speed by 5 miles per hour. How long did it take him to travel the last 120 miles?

5. Draw a 3-centimeter line segment with a ruler and construct the perpendicular bisector of the line segment.

6. Use a protractor to draw a 30° angle. Then use a straightedge and a compass to construct the bisector of the angle.

7. Use a protractor to draw an 80° angle. Then use a compass and a straightedge to copy the angle.

*8. What base 10 number is 4032 (base 8)?

*9. Write 1321 (base 10) as a base 8 numeral.

10. What base 10 number does 100110 (base 2) represent?

11. If 120 is increased by 80 percent, what is the resulting number? Draw a diagram that depicts the problem.

12. What percent of 50 is 60? Draw a diagram that depicts the problem.

13. Convert 12 cubic miles to cubic inches.

14. Convert 62,115 cubic inches to cubic yards.

15. If $16x - 12 = 116$, what is $4x - 3$?

Use the distributive property to multiply:

16. $2ac(ab + a - b)$

17. $-am(a^2 + m - am)$

18. Simplify by adding like terms: $3xy^2 - 2xy + 3xyy + 3xy - y^2x$

Solve:

19. $3x - 2x + 6 - 2x = x + 7$

20. $-2\frac{1}{3}x - \frac{3}{4} = \frac{1}{2}$

21. $\frac{-\frac{1}{3}}{\frac{4}{9}} = \frac{\frac{1}{4}}{x}$

Simplify:

22. $-[-(-4)^2] + (-3)(4) + (-2)^3$

23. $2^2 + 2^3[-2(-3 + 2^2)(2^2 - 1) + 2]$

24. $6a^2bab^2b^3$

25. $\frac{-(-2)^2 + 3^2(2^2 - 5) + 3}{2(2^3 - 4)}$

26. $\frac{1}{4}\left(2\frac{1}{3} \cdot \frac{1}{4} - \frac{7}{12}\right)$

Evaluate:

27. $a^3 - 2a$ if $a = -2$

28. $a^2 + 3ab^2$ if $a = -1, b = -3$

29. Graph: $x \ngtr 2$

30. Use the cut-and-try method to estimate $\sqrt{55}$ to one decimal place.

LESSON 101 Metric volume

101.A metric volume

A cubic meter is a relatively large unit of volume which equals approximately 264.17 gallons. A cubic centimeter is a relatively small unit of volume. It takes about 16.39 cubic centimeters to equal a cubic inch. The metric system has a unit of volume that is about the size of a quart—the liter, which equals 1000 cubic centimeters. A liter equals about 1.06 quarts. So when we see the word **liter,** we can think **"big quart."**

$$1 \text{ liter} = 1 \text{ big quart}$$

Because 1000 cubic centimeters makes a liter, we say that a cubic centimeter is a milliliter, which is one-thousandth of a liter.

$$1 \text{ milliliter} = 1 \text{ cubic centimeter}$$

Only one unit multiplier is required to convert between cubic centimeters and liters. More unit multipliers are normally required for any other volume conversions.

example 101.A.1 Convert 1,451,600 milliliters (ml) to liters.

solution Only one unit multiplier is required to make this conversion.

$$1{,}451{,}600 \cancel{\text{ml}} \times \frac{1 \text{ liter}}{1000 \cancel{\text{ml}}} = \mathbf{1451.6 \text{ liters}}$$

example 101.A.2 Convert 40 cubic meters (m^3) to cubic centimeters (cm^3).

solution We need three unit multipliers.

$$40\cancel{m^3} \times \frac{100\text{ cm}}{\cancel{m}} \times \frac{100\text{ cm}}{\cancel{m}} \times \frac{100\text{ cm}}{\cancel{m}} = 40(100)(100)(100)\text{ cm}^3$$

$$= \mathbf{40{,}000{,}000\ cm^3}$$

example 101.A.3 Convert 1400 cubic meters to liters.

solution We will go from m^3 to cm^3 to liters.

$$1400\cancel{m^3} \times \frac{100\cancel{cm}}{\cancel{m}} \times \frac{100\cancel{cm}}{\cancel{m}} \times \frac{100\cancel{cm}}{\cancel{m}} \times \frac{1\text{ liter}}{1000\cancel{cm^3}} = \frac{1400(100)(100)(100)}{1000}\text{ liters}$$

$$= \mathbf{1{,}400{,}000\ liters}$$

problem set 101

1. Sixty percent of the students were garrulous. If 2000 students were not garrulous, how many were garrulous?
2. The number of loquacious and the number of taciturn were in the ratio of 7 to 3. If 750 students were in the school, how many were taciturn?
3. Five was added to the product of a number and 7. The result was 3 less than the product of the same number and 5. What was the number?
4. The caravan traveled 40 miles in 8 hours. Then they doubled their speed. How long did it take the caravan to travel the last 60 miles?
5. Draw a 4-centimeter line segment with a ruler and construct the perpendicular bisector of the line segment.
6. Use a protractor to draw a 40° angle. Then use a straightedge and a compass to construct the bisector of the angle.
7. Use a protractor to draw a 70° angle. Then use a compass and a straightedge to copy the angle.
8. What base 10 number does 6025 (base 8) represent?
9. Write 3725 (base 10) using base 8 numerals.
10. What base 10 number does 101001 (base 2) represent?
11. What number is 240 percent of 375? Draw a diagram that depicts the problem.
12. What percent of 80 is 92? Draw a diagram that depicts the problem.

*13. Convert 1400 cubic meters to liters.

14. Convert 6,462,712 milliliters to liters.
15. Convert 44 cubic meters to cubic centimeters.
16. If $13x - 12 = 40$, what is $\frac{1}{2}x + 4$?
17. Use the distributive property to multiply: $2ab(ab + a^2 - b^2 + a)$
18. Simplify by adding like terms: $2xyy + 3xyx - 2x^2y + 5xy^2 - 3yxy$
19. Find the volume in cubic feet of a solid whose base is shown and whose height is 2 yards. Dimensions are in feet.

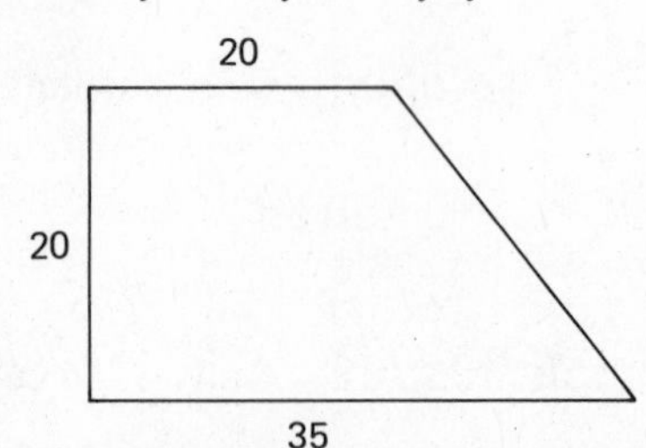

Solve:

20. $2x + 3x - 6 + 4x = 2x - 22$

21. $-1\frac{1}{3}x - \frac{5}{6} = \frac{2}{3}$

22. $\dfrac{-\frac{2}{5}}{\frac{4}{25}} = \dfrac{-\frac{1}{2}}{x}$

Simplify:

23. $-[-(-2)^2] + (-3)(-2)$

24. $2^2 + 2^3[-3(-2 + 2^2)(2^3 - 2^2) + 3^2]$

25. $5a^3bab^2a^2b^3$

26. $\dfrac{-(-2)^2 + 2^3(3^2 - 2^3) + 2^2}{2^2(3^2 + 1)}$

27. $\frac{1}{5}\left(2\frac{1}{2} \cdot \frac{1}{6} - \frac{1}{3} \cdot \frac{5}{4}\right)$

Evaluate:

28. $-ab + 2a^3$ if $a = -1, b = -2$

29. $-a^3b + b$ if $a = -2, b = -1$

30. Graph: $x \not\leq -1$

LESSON 102 Adding in base 10 and base 2

102.A adding in base 10

We review our procedure for adding in base 10 by noting that we split the sum of a column into two parts. One part is a whole number times the base, and the other part is the remainder. We record the remainder and carry the whole number to the next column.

$$\begin{array}{rl} 53\ 9 & \\ 84\ 4 & \\ 63\ 8 & \\ \hline \textcircled{21} & = 2(10) + 1 \end{array} \qquad \begin{array}{r} \overset{2}{5}39 \\ 844 \\ 638 \\ \hline 1 \end{array}$$

The sum of the first column is 21. This is 2 times the base (10), with 1 left over. We record the 1 and carry the 2 to the second column.

$$\begin{array}{rl} 5\ \overset{2}{3}\ 9 & \\ 8\ 4\ 4 & \\ 6\ 3\ 8 & \\ \hline \textcircled{12}\ 1 & \quad 12 = 1(10) + 2 \end{array} \quad \longrightarrow \quad \begin{array}{r} \overset{1\,2}{539} \\ 844 \\ 638 \\ \hline 21 \end{array}$$

When we sum the second column, we get 12. This is 1 times the base (10), with 2 left over. We record the 2 and carry the 1 to the next column.

$$\begin{array}{rl} \overset{1}{5}\ \overset{2}{3}9 & \\ 8\ 44 & \\ 6\ 38 & \\ \hline \textcircled{20}\ 21 & \quad 20 = 2(10) + 0 \end{array} \quad \longrightarrow \quad \begin{array}{r} \overset{2\,1\,2}{539} \\ 844 \\ 638 \\ \hline 2021 \end{array}$$

The total of the third column is 20. This is 2 times the base, with 0 left over. We record the 0 and carry the 2.

102.B adding in base 2

We will use the same procedure to add in base 2. We will split the sum of a column into two parts. One is a whole number times the base, and the other part is the remainder. We record the remainder and carry the whole number.

example 102.B.1 Add 1111 (base 2) + 1011 (base 2) + 1101 (base 2) + 1101 (base 2).

solution First we write the numbers in base 10 and add.

$$15 + 11 + 13 + 13 = 52 \text{ (base 10)}$$

Now let's add in base 2. We record the numbers vertically and add the first column. We will always think in base 10. First we add the right column. We will write the sum in base 10 and circle it.

```
 1111
 1011
 1101
 1101
 ----
    ④ = 2(2) + 0
```

The sum of the first column in base 10 is 4. This is 2 times the base (2), with a remainder of 0. We record the 0 and carry 2 to the next column and add this column.

```
     2
 1 1 1 1
 1 0 1 1
 1 1 0 1
 1 1 0 1
 -------
     ④0      4 = 2(2) + 0
```

This sum is 2 times the base, with a remainder of 0. We record 0 and carry 2. Then we sum the third column.

```
   2 2
 1 1 1 1
 1 0 1 1
 1 1 0 1
 1 1 0 1
 -------
   ⑤0 0      5 = 2(2) + 1
```

We get 5, which is 2 times the base, with a remainder of 1. We record 1 and carry the 2 and add the next column.

```
 2 2 2
 1 1 1 1
 1 0 1 1
 1 1 0 1
 1 1 0 1
 -------
⑥1 0 0      6 = 3(2) + 0
```

The number 6 is 3 times the base, with a remainder of 0.

```
3  2 2 2
   1 1 1 1
   1 0 1 1
   1 1 0 1
   1 1 0 1
 ---------
 ③0 1 0 0
```

But 3 is 1 times the base, with a remainder of 1. So our final sum is

110100

If we write a place-value table for base 2, we can find the number in base 10.

32	16	8	4	2	1
1	1	0	1	0	0

1 1 0 1 0 0 (base 2) = 32 + 16 + 4 = 52 (base 10)

Our sum was 52 when we added the numbers using base 10 numerals. This is our check.

example 102.B.2 Convert the base 2 numerals 1101 and 1011 to base 10 numerals and add in base 10. Then add the numbers in base 2 and convert the answer to base 10 to check.

solution We will use a place value table for base 2 to find the number in base 10.

8	4	2	1
1	1	0	1
1	0	1	1

1 1 0 1 = 8 + 4 + 1 = 13

1 0 1 1 = 8 + 2 + 1 = 11

24 (base 10)

Now we will add the numbers in base 2.

```
 1 1 1
  1 1 0 1
  1 0 1 1
 ---------
1 1 0 0 0
```

We use a place value table to write this in base 10.

16	8	4	2	1
1	1	0	0	0

1 1 0 0 0 = 16 + 8 = 24 (base 10) check

problem set 102

1. The freshman class was noted for being prolix. This was a little unfair because only 60 percent of the freshmen suffered from uncontrollable prolixity. If 800 freshmen were not prolix, how many were prolix?

2. The ratio of laconic answers to wordy answers was 1 to 17. If 7200 answers were given during the semester, how many were laconic?

3. The sum of twice a number and 14 was 32 less than the product of the number and 5. What was the number?

4. The first 50 spare mufflers cost a total of \$350. If the price of the next 50 mufflers was reduced by \$2 apiece, what was the total cost of these mufflers?

5. Draw a 6-centimeter line segment with a ruler and construct the perpendicular bisector of the line segment.

6. Use a protractor to draw a 50° angle. Then use a straightedge and compass to construct the bisector of the angle.

7. Use a protractor to draw a 40° angle. Then use a straightedge and a compass to duplicate the angle.

*8. Add 1111 (base 2) + 1011 (base 2) + 1101 (base 2) + 1101 (base 2).

9. Add 1001 (base 2) + 1101 (base 2) + 11001 (base 2) + 1111 (base 2).

10. Convert 111101 (base 2) to base 10.

11. Convert 101101 (base 2) to base 10.

12. What base 10 numeral does 6712 (base 8) represent?

13. What percent of 60 is 108? Draw a diagram that depicts the problem.

14. If 120 is increased by 60 percent, what is the resulting number? Draw a diagram that depicts the problem.

15. Convert 1200 cubic meters to liters.

16. Convert 16,215,321 milliliters to liters.

17. Use the distributive property to multiply: $3ab(a^2b + ab^2 - a - b)$

18. Simplify by adding like terms: $3xy^3 + 2x^3y - 5y^2xy - 6xyx^2 + 3xyyy$

19. Find the perimeter of the following figure. Dimensions are in feet.

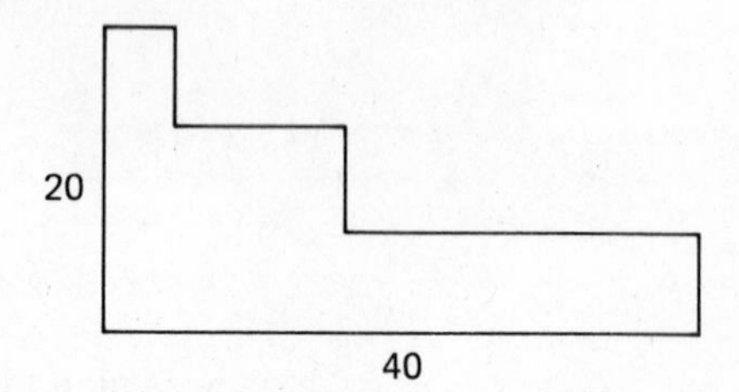

Solve:

20. $7x + 5 = -4x - 2x + 22$

21. $\frac{4}{3}x - \frac{2}{5} = 3\frac{1}{3}$

22. $\dfrac{\frac{1}{4}}{\frac{16}{3}} = \dfrac{\frac{1}{3}}{x}$

Simplify:

23. $-[-(-1)^3] + (-4)$

24. $2aa^2bab^2b^3$

25. $\dfrac{-(-2)^3 + 2^2(2^3 - 3^2) + 2^3}{2(2^2 - 1)}$

26. $\frac{1}{4}\left(3\frac{1}{3} \cdot 1\frac{1}{4} - \frac{5}{6} \cdot \frac{1}{2}\right)$

Evaluate:

27. $xy^2 + 2xy$ if $x = -1,\ y = -2$

28. $-ab^2 + b$ if $a = -2,\ b = -1$

29. Write $.0012\frac{3}{4}$ as a fraction of whole numbers.

30. Use the cut-and-try method to estimate $\sqrt[4]{600}$ to the nearest whole number.

LESSON 103 Probability

103.A probability

On the left we show a pair of dice. The singular of **dice** is **die**

so we call one of these a die. A die has six faces, each of which has one to six dots.

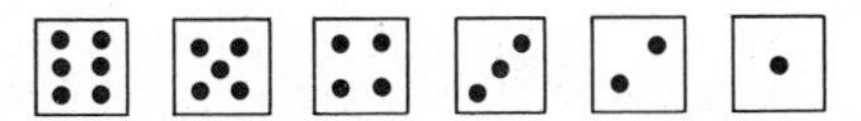

If we roll a die that has no flaws, it should come up 3 just as often as it comes up 5. Thus, we say that the probability of rolling a 3 is the same as the probability of rolling a 5. Since the die has six faces, if we rolled it a great number of times, we would expect each face to come up one-sixth of the time. We will use the symbol

$$P(\quad)$$

to designate the word **probability.** Thus, we would write that the probability of 3 equals $\frac{1}{6}$ by writing.

$$P(3) = \frac{1}{6}$$

The probability of an event is the number of ways the event can happen divided by the number of possible outcomes.

$$P(E) = \frac{\text{number of ways event can happen}}{\text{total number of possible outcomes}}$$

example 103.A.1 A single die is rolled. What is the probability that it will come up either 5 or 6?

solution There are 6 possible outcomes and 2 of them give us the desired result. Thus, the probability of rolling a 5 or 6 is

$$P(5 \text{ or } 6) = \frac{2}{6} = \mathbf{\frac{1}{3}}$$

example 103.A.2 A deck of cards contains 52 cards: 13 are spades, 13 are hearts, 13 are clubs, and 13 are diamonds. Jimmy tears up the ace of spades. Then he shuffles the deck and draws 1 card. What is the probability that the card is a spade?

solution There are 51 cards and 12 are spades. So

$$P(S) = \frac{12}{51} = \mathbf{\frac{4}{17}}$$

example 103.A.3 One die is red and the other die is green. Both are rolled. What is the probability of getting (a) a 7? (b) a 4?

solution We make a table that shows every possible outcome.

Green die

Red die	⚀	⚁	⚂	⚃	⚄	⚅
⚀	2	3	4	5	6	7
⚁	3	4	5	6	7	8
⚂	4	5	6	7	8	9
⚃	5	6	7	8	9	10
⚄	6	7	8	9	10	11
⚅	7	8	9	10	11	12

(a) The total of the dots is given by the numbers in the center of the table. Six of them are seven and there are 36 squares in all. So

$$P(7) = \frac{6}{36} = \mathbf{\frac{1}{6}}$$

(b) There are only 3 ways to get a 4. Thus, the probability of getting a 4 is

$$P(4) = \frac{3}{36} = \mathbf{\frac{1}{12}}$$

problem set 103

*1. A single die is rolled. What is the probability that it will come up either 5 or 6?

*2. A deck of cards contains 52 cards: 13 are spades, 13 are hearts, 13 are clubs, and 13 are diamonds. Jimmy tears up the ace of spades. Then he shuffles the deck and draws 1 card. What is the probability that the card is a spade?

*3. One die is red and the other die is green. Both are rolled. What is the probability of getting (a) a 7? (b) a 4?

4. Thirty percent of the airplanes in the show were biplanes. If 120 were biplanes, how many were not biplanes?

5. Draw a 5-centimeter line segment with a ruler and construct the perpendicular bisector of the line segment.

6. Use a protractor to draw an 80° angle. Then use a straightedge and a compass to construct the bisector of the angle.

7. Use a protractor to draw a 50° angle. Then use a straightedge and a compass to copy the angle.

8. Add 11011 (base 2) + 10111 (base 2) + 11010 (base 2) + 10011 (base 2).

9. Convert 10110 (base 2) to base 10.

10. Write 3228 (base 10) using base 8 numerals.

11. 45 is what percent of 225? Draw a diagram that depicts the problem.

12. Forty-two percent of what number is 126? Draw a diagram that depicts the problem.

13. Convert 52 cubic meters to cubic centimeters.

14. Convert 4100 cubic meters to liters.

15. If $12x - 61 = 83$, what is $\frac{1}{6}x - 7$?

16. Use the distributive property to multiply: $2ac(a + c - ac + a^2)$

17. Simplify by adding like terms: $2xyy + 3xyx - 6xy^2 + 4x^2y$

18. Find the volume in cubic meters of a solid whose base is shown and whose height is 3 centimeters. Dimensions are in centimeters.

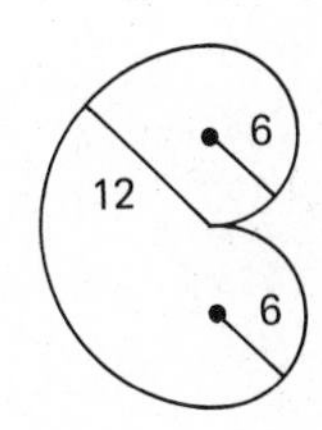

Solve:

19. $3x - 2x + 6 = 2x - 21$

20. $-2\frac{1}{3}x - \frac{5}{6} = \frac{1}{3}$

21. $\dfrac{-\frac{3}{4}}{\frac{9}{16}} = \dfrac{\frac{1}{3}}{x}$

Simplify:

22. $-[-(-1)^3] + (-4)$

23. $2^2aa^3ba^2b^3$

24. $\dfrac{-(-2)^3 + 2^2(2^2 - 2 \cdot 4)}{2^3(3^2 - 2^2)}$

25. $2\frac{1}{3}\left(2\frac{1}{2} \cdot \frac{1}{2} - \frac{1}{3} \cdot \frac{4}{5}\right)$

Evaluate:

26. $a^2b + ab^2$ if $a = -1, b = -2$

27. $a^2 - b^2$ if $a = -2, b = -3$

28. Graph: $x \nleq 3$

29. Express 10,080 as a product of primes.

30. Round off 62,978,134.32 to the nearest ten thousand.

LESSON 104 Numerals and numbers · Larger than and greater than

104.A numerals and numbers

A number is an idea. A numeral is the symbol we write to make us think of the idea. The three circles below have the quality of threeness, as do the three squares. The right diagram has this same quality of threeness even though the shapes shown are all different.

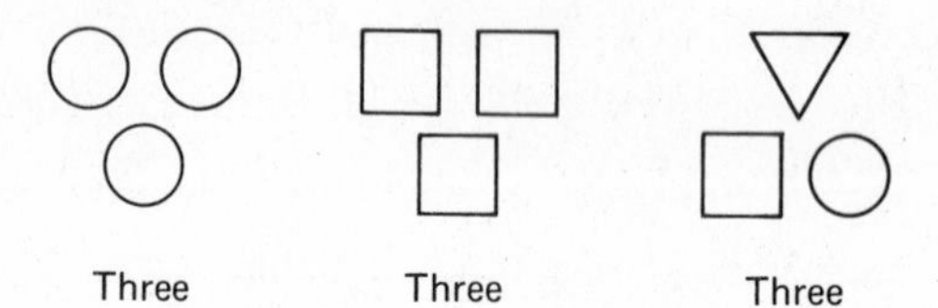

When we want to think of the **idea** of three, we often write the following symbol.

$$3$$

This symbol is not a number but is a numeral that makes us think of the number 3. The ancient Romans used the numeral

$$\text{III}$$

when they wanted to think of the number 3. There are many numerals for 3. All the following numerals have the value of 3.

$$\frac{30}{10}, \quad 2 + 1, \quad \frac{15}{5}, \quad 6 \div 2, \quad 14 - 11, \quad 1 + 1 + 1$$

Each of these numerals can also be called a **numerical expression. We see that the value of a numerical expression is the number that the expression represents. Since all these expressions represent the number 3, we say that each of them has a value of 3.** This distinction between a numeral and a number is often helpful. For instance, if we are asked to add the fractions

$$\frac{1}{4} + \frac{1}{2}$$

we cannot because the denominators are not the same. But we can add the numbers if we change the numeral for the second number to a numeral whose denominator is 4.

$$\frac{1}{4} + \frac{2}{4} = \frac{3}{4}$$

We did not change the number. We changed the numeral used to represent the number. We could do this because both

$$\frac{1}{2} \quad \text{and} \quad \frac{2}{4}$$

have the same value because they represent the same number.

From this we see that there is a very real difference between a number and a numeral. If we pay close attention to this difference, however, it often confuses rather than clarifies. So we will often use the word number when we really should say numeral. It is not a bad error especially if we remember that there is a difference between the two words and remember that:

A number is an idea

104.B larger than and greater than

Here we show the numbers 4 and 2.

4 2

We have written the number 4 larger than the number 2. Now we write the number 2 larger than the number 4.

4 2

We have been graphing numbers on the number line, and we have said that 4 is always greater than 2 because its graph is farther to the right on the number line.

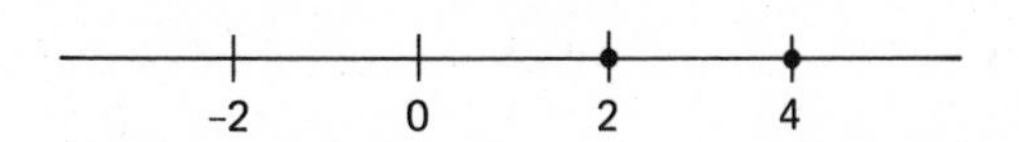

Note that we carefully use the words **greater than** and **less than** when we compare numbers. We do this because we want to avoid using the words **bigger than** and **larger than** because some people may think we are talking about the size of the numeral.

13 7

The number 13 is always greater than the number 7, in spite of the fact that above we show a numeral for 7 that is "larger than" the numeral used for 13.

problem set 104

1. One die is red and the other die is green. Both are rolled. What is the probability of getting (a) an 8? (b) a 9? (c) a 10?
2. The first three cards dealt from a well-shuffled deck of cards are spades. What is the probability that the next card dealt will be a spade?
3. Twenty percent of the flowers in the show were red. If 1400 flowers were not red, how many were red?
4. The ratio of ectomorphs to mesomorphs was 3 to 14. If there were 3400 ballplayers in all, how many were ectomorphs?
5. What is the difference between a numeral and a number?
6. Draw an 8-centimeter line segment with a ruler and construct the perpendicular bisector of the line segment.
7. Use a protractor to draw a 56° angle. Then use a straight-edge and a compass to construct the bisector of the angle.
8. Use a protractor to draw a 55° angle. Then use a straightedge and a compass to copy the angle.
9. Add 1011 (base 2) + 1001 (base 2) + 1000 (base 2) + 1100 (base 2).
10. Convert 111110 (base 2) to base 10.
11. What base 10 numeral does 3113 (base 8) represent?
12. Write 1039 (base 10) using base 8 numerals.
13. What percent of 70 is 112? Draw a diagram that depicts the problem.
14. If 130 is increased by 40 percent, what is the resulting number? Draw a diagram that depicts the problem.

15. Convert 1320 cubic meters to liters.

16. Convert 15,321,156 milliliters to liters.

17. Use the distributive property to multiply: $2ab(a + b - ab^2 + a^2b)$

18. Simplify by adding like terms: $2xyy^2 + 3x^2xy - 6yxy^2 + 7xyy^2$

19. Express in cubic meters the volume of a solid whose base is shown and whose sides are 3 meters high. Dimensions are in centimeters.

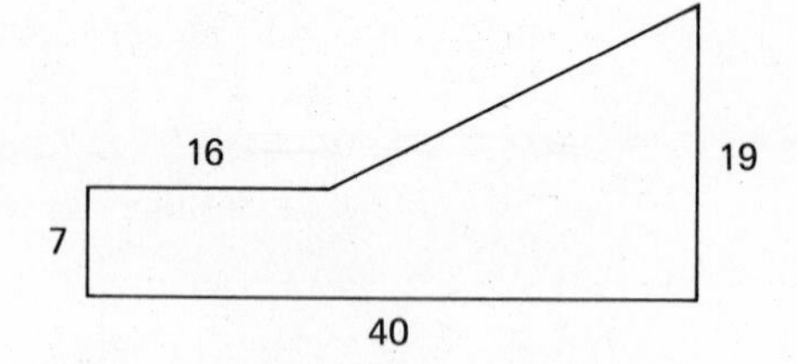

Solve:

20. $5x - 5 = -6x + 2x - 22$

21. $-\frac{2}{3}x - \frac{1}{5} = 1\frac{5}{6}$

22. $\dfrac{-\frac{1}{3}}{\frac{16}{9}} = \dfrac{-\frac{1}{3}}{x}$

Simplify:

23. $-[-(-2)^3] + (-2)$

24. $3a^2ba^3b^2b^3$

25. $\dfrac{-2^3 - 2^2(2^2 - 3) + 3}{3(2^3 - 1)}$

26. $\frac{1}{3}\left(2\frac{1}{2} \cdot \frac{1}{3} - \frac{1}{6} \cdot \frac{1}{2}\right)$

Evaluate:

27. $xy^2 + 2x^2y$ if $x = 1, y = -2$

28. Use the cut-and-try method to estimate $\sqrt{46}$ to one decimal place.

29. Write $.002\frac{3}{5}$ as a fraction of whole numbers.

30. Write 621321131.2 in words.

LESSON 105 Adding in base 8

105.A adding in base 8

Before we try addition in base 8 we will look again at addition in base 10.

$$\begin{array}{r} 41\ 7 \\ 53\ 8 \\ \underline{26\ 8} \\ \text{(23)} \end{array} \qquad 23 = 2(10) + 3 \qquad \begin{array}{r} 4\overset{2}{1}7 \\ 538 \\ \underline{268} \\ 3 \end{array}$$

When we added the first column, we got a sum of 23. This is 2 times the base, (10), with a remainder of 3. We record the remainder and carry the 2 to the next column. This is the same procedure that we will use for adding in base 8. But in base 8 the base will be 8.

example 105.A.1 Add 546 (base 8) + 52 (base 8) + 73 (base 8).

solution We record the numbers and add the first column, using base 10.

```
 54 6
  5 2
  7 3
 ----
  (11)     11 = 1(8) + 3
```

The sum is 11, which is 1 times the base, with a remainder of 3. We record the 3, carry the 1, and add the second column.

```
   1
 5 4 6
   5 2
   7 3
 -----
 (17) 3    17 = 2(8) + 1
```

The second column adds to 17, which is 2 times the base, with a remainder of 1. We record the remainder, carry the 2, and add the third column.

```
 2
 5 46
   52
   73
 ----
 (7)13
```

So 546 (base 8) + 52 (base 8) + 73 (base 8) = **713 (base 8).**

example 105.A.2 Add 42 (base 8) + 56 (base 8).

solution

```
  42
  56
  --
  (8)      8 = 1(8) + 0

  1
  4 2
  5 6
 ----
 (10) 0    10 = 1(8) + 2

  42
 156
 ---
 120
```

So 42 (base 8) + 56 (base 8) = **120 (base 8).**

problem set 105

1. There are 15 marbles in an urn. Five are red, 6 are black, and 4 are green. If one marble is drawn, what is the probability it will be either red or black?

2. Thirty percent of the natives were easily intimidated. If 450 natives were easily intimidated, how many natives were not easily intimidated?

3. The ratio of the squishy to the rock-hard was 7 to 13. If 1400 total were in the room, how many were squishy?

4. When the \$3400 was paid to the vendor, the students received 1700 items. If the vendor then reduced the price per item by 50 percent, how much did the students have to pay for the next 2000 items?

5. Draw an 8-centimeter line segment with a ruler and construct the perpendicular bisector of the line segment.

6. Use a protractor to draw a 66° angle. Then use a straightedge and a compass to construct the bisector of the angle.

7. Use a protractor to draw a 50° angle. Then use a straightedge and a compass to copy the angle.

8. Convert 111101 (base 2) to base 10.

9. Add 10110 (base 2) + 11001 (base 2) + 11101 (base 2).

*10. Add 42 (base 8) + 56 (base 8).

*11. Add 546 (base 8) + 52 (base 8) + 73 (base 8).

12. What percent of 95 is 152? Draw a diagram that depicts the problem.

13. If 320 is increased by 15 percent, what is the resulting number? Draw a diagram that depicts the problem.

14. Convert 62 cubic meters to milliliters.

15. Convert 15 cubic yards to cubic inches.

16. Convert 3 miles to inches.

17. Use the distributive property to multiply: $3ac(a + b - c + acb)$

18. Simplify by adding like terms: $2xxyyy - 6xy^2 + 3y^2x^2y + 3yxy$

19. Express in cubic centimeters the volume of a solid whose base is shown and whose sides are 2 meters high. Dimensions are in meters.

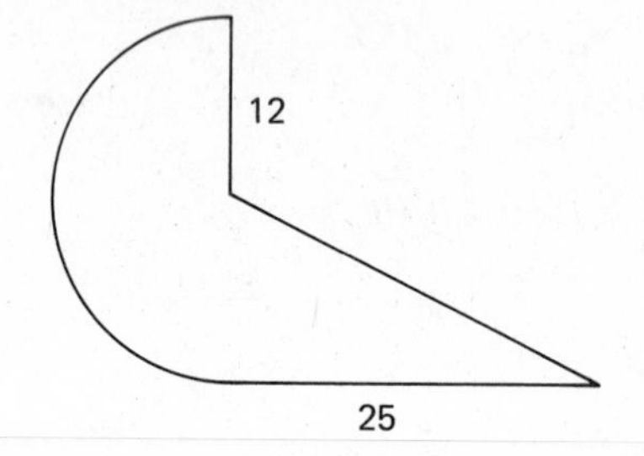

Solve:

20. $x + 5 = -3x + 2x - 6$

21. $-\frac{1}{6}x + 1\frac{1}{3} = -2\frac{5}{6}$

22. $\dfrac{-\frac{2}{3}}{\frac{4}{9}} = \dfrac{\frac{1}{2}}{-x}$

Simplify:

23. $-(-2)^3 + 2^3[-2(-3 + 5)(3 - 2^2) - 7]$

24. $2aba^2b^3a^2$

25. $\dfrac{-2^2 + 2^3(2^3 - 3^2) + 3}{3(2 \cdot 3 + 1)}$

26. $\frac{1}{2}\left(2\frac{1}{3} \cdot \frac{1}{4} - \frac{5}{6} \cdot \frac{1}{2}\right)$

Evaluate:

27. $xy^2 - x^2y + 2xy$ if $x = 2, y = -1$

28. $a - ba$ if $a = 1, b = -5$

29. Graph: $x \not> 1$

30. If $6x + 7 = 28$, what is $6x - 4$?

LESSON 106 Products of probabilities

106.A products of probabilities

The probability that two things will happen in a certain order is the product of the individual probabilities. A coin has two sides, a head and a tail

$$(H) \quad (T) \qquad P(H) = \frac{1}{2} \qquad P(T) = \frac{1}{2}$$

On a single toss, the probability of getting a head is one-half. **If a coin is tossed 4 times and comes up heads every time, the probability of a head on the next toss is still one-half because past events do not affect the probability of future events.**

However, the probability of two heads in two future tosses is the product of the probabilities.

$$P(HH) = P(H) \times P(H) = \frac{1}{2} \times \frac{1}{2} = \frac{1}{4}$$

The reason for this can be seen from the following. The four possible outcomes for two flips of the coin are

H	T		H	H		T	H		T	T

Only one of these outcomes contains two heads, so the probability of this outcome is one-fourth.

$$P(HH) = \frac{1}{4}$$

example 106.A.1 The first urn contains 4 blue marbles and 2 white marbles. The second urn contains 4 blue marbles and 11 white marbles. A marble is picked from each of the urns. What is the probability that both marbles are white?

solution The probability of getting a white marble from the first urn is two-sixths, and the probability of getting a white marble from the second urn is $\frac{11}{15}$.

$$P(W_1) = \frac{2}{6} \qquad P(W_2) = \frac{11}{15}$$

The probability of both marbles being white is the **product** of these probabilities.

$$P(W_1 W_2) = P(W_1)P(W_2) = \frac{2}{6} \times \frac{11}{15} = \mathbf{\frac{11}{45}}$$

example 106.A.2 A red die and a blue die are rolled. What is the probability of getting both a 6 on the red die and a 4 on the blue die?

solution The individual probabilities are:

$$P(6_R) = \frac{1}{6} \qquad P(4_B) = \frac{1}{6}$$

The probability of both occurring is the product of the probabilities.

$$P(6_R 4_B) = \frac{1}{6} \cdot \frac{1}{6} = \mathbf{\frac{1}{36}}$$

example 106.A.3 A fair coin is tossed five times. What is the probability of getting *HHTHH* in that order?

solution Each probability is $\frac{1}{2}$, so the probability of the five tosses being *HHTHH* is

$$P(HHTHH) = \left(\frac{1}{2}\right)\left(\frac{1}{2}\right)\left(\frac{1}{2}\right)\left(\frac{1}{2}\right)\left(\frac{1}{2}\right) = \mathbf{\frac{1}{32}}$$

problem set 106

*1. The first urn contains 4 blue marbles and 2 white marbles. The second urn contains 4 blue marbles and 11 white marbles. A marble is picked from each of the urns. What is the probability that both marbles are white?

*2. A red die and a blue die are rolled. What is the probability of getting both a 6 on the red die and a 4 on the blue die?

*3. A fair coin is tossed five times. What is the probability of getting *HHTHH* in that order?

4. The increase in the turnip crop was 260 percent. If last year's harvest was 230,000 tons, what was the harvest this year?

5. Draw a 6-centimeter line segment with a ruler and construct the perpendicular bisector of the line segment.

6. Use a protractor to draw a 50° angle. Then use a straightedge and a compass to construct the bisector of the angle.

7. Use a protractor to draw a 65° angle. Then use a straight edge and a compass to copy the angle.

8. Add 10111 (base 2) + 10011 (base 2) + 10001 (base 2).

9. Convert 10101 (base 2) to base 10.

10. Add 67 (base 8) + 76 (base 8).

11. Convert 217 (base 8) to base 10.

12. What percent of 90 is 162? Draw a diagram that depicts the problem.

13. Sixty percent of what number is 72? Draw a diagram that depicts the problem.

14. Convert 18,721,135 milliliters to liters.

15. Convert 181 cubic miles to cubic inches.

16. Use the distributive property to multiply: $2ac(ab + bc - c)$

17. Simplify by adding like terms: $2xyx + 3yxy^2 - 3x^2y + 5y^3x - 6yx^2$

18. Express in cubic feet the volume of a solid whose base is shown and whose sides are 2 yards high. Dimensions are in feet.

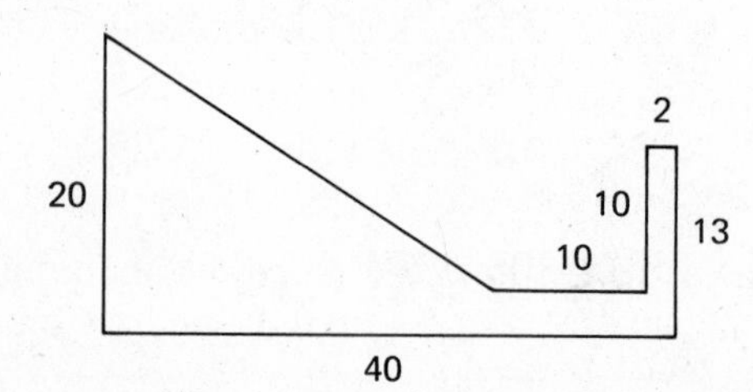

19. Find the surface area of a rectangular solid whose length, width, and height are 4 feet, 2 feet, and 10 feet, respectively.

Solve:

20. $3x - 3 + 6x - 2 = -8x + 5$

21. $\frac{1}{4}x + \frac{2}{3} = 3\frac{5}{12}$

22. $\frac{\frac{2}{3}}{\frac{4}{9}} = \frac{\frac{1}{3}}{x}$

Simplify:

23. $-[-(-1)^4] + (-3)(-2)$

24. $5aba^2b^3a^2$

25. $3\frac{1}{2} \times 2\frac{1}{3} \div \frac{1}{3} \div \frac{1}{2}$

26. $139.287 - 19.876$

Evaluate:

27. $-xy + y$ if $x = -2, y = -3$

28. $ab^2 - a^2b$ if $a = 2, b = -1$

29. If $3x - 12 = 12$, what is the value of $\frac{1}{4}x + 10$?

30. Use the cut-and-try method to estimate $\sqrt[3]{330}$ to the nearest whole number.

LESSON 107 *More construction*

107.A more construction

We can construct a triangle if we are given the lengths of all three sides.

example 107.A.1 Construct a triangle whose sides are 4 cm, 3 cm, and 2 cm.

solution We draw a line segment 4 cm long and from one end of the segment draw an arc whose radius is 3 cm.

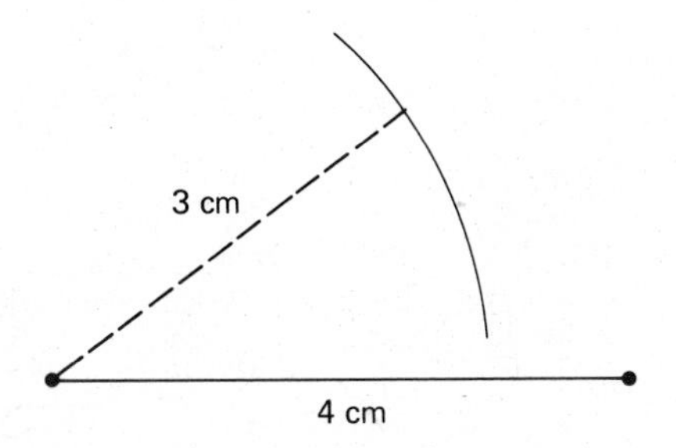

Now from the other end of the line segment, we draw an arc with a 2-cm radius. The intersection of the arcs is the other vertex of the triangle.

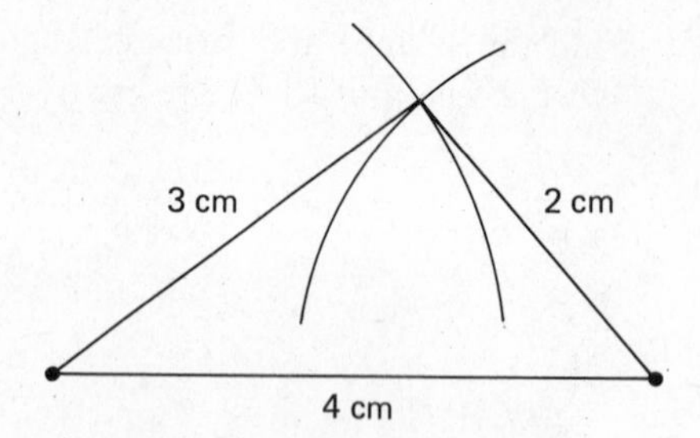

example 107.A.2 Given the line shown, construct a perpendicular to the line at point A.

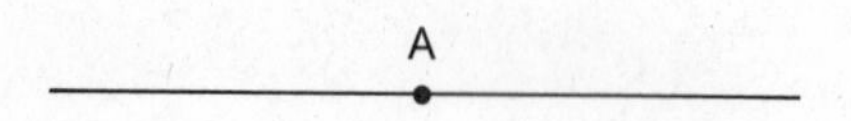

solution First we swing an arc in both directions from A. This arc will cut the line at points we call B and C. Then we construct the perpendicular bisector of $\overline{CB}$.

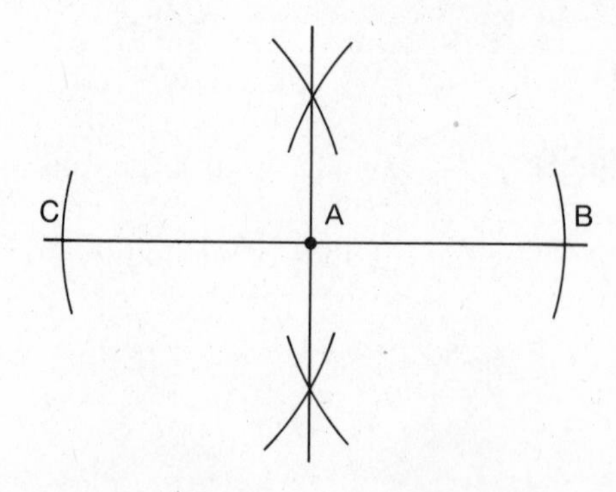

example 107.A.3 Draw a perpendicular to a line that passes through a point not on the line.

solution We draw a line and call the point P.

P

Next we swing a big arc from P. We call the points where this arc intersects the line points A and B. Then we construct the perpendicular bisector of $\overline{AB}$.

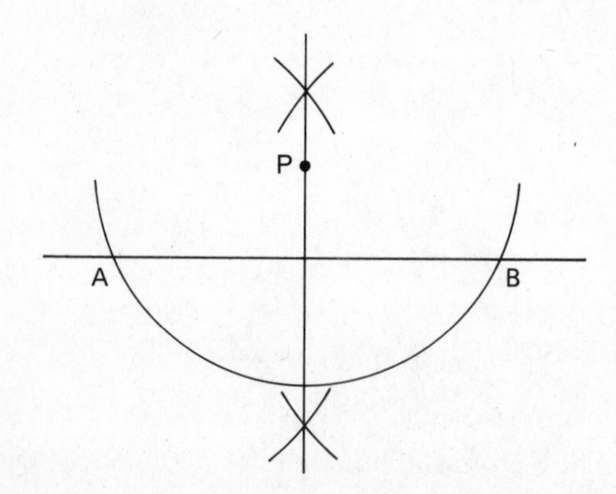

problem set 107

1. A single die is rolled twice. What is the probability that the first roll will be a 6 and the second roll will be a 2?

2. A pair of dice are rolled. What is the probability of rolling an 11?

3. As the official approached, the number that appeared to be busy increased 150 percent. If 1000 now appeared to be busy, how many appeared busy before the official approached?

4. The ratio of industrious to insouciant was 2 to 5. If 1400 were industrious, how many were insouciant?

*5. Construct a triangle whose sides are 4 cm, 3 cm, and 2 cm.

6. Use a ruler to draw a line segment 4 centimeters long. Construct a perpendicular to the line at a point 1 centimeter from the left endpoint.

7. Draw a line segment and a point outside the line. Construct a perpendicular to the line which passes through the point.

8. Use a protractor to draw a 36° angle. Then use a straightedge and a compass to construct the bisector of the angle.

9. Add 10101 (base 2) + 11011 (base 2).

10. Convert 1010 (base 2) to base 10.

11. Add 57 (base 8) + 74 (base 8).

12. What percent of 70 is 119? Draw a diagram that depicts the problem.

13. Seventy percent of what number is 210? Draw a diagram that depicts the problem.

14. Convert 1620 cubic meters to liters.

15. Convert 105 cubic miles to cubic yards.

16. Use the distributive property to multiply: $2ab(a + c + b^2 - ac)$

17. Simplify by adding like terms: $2yxxy^2 - 3xy^2x + 3y^2x^2y - 6xyxy$

18. Express in cubic meters the volume of a solid whose base is shown and whose sides are 4 centimeters high. Dimensions are in centimeters.

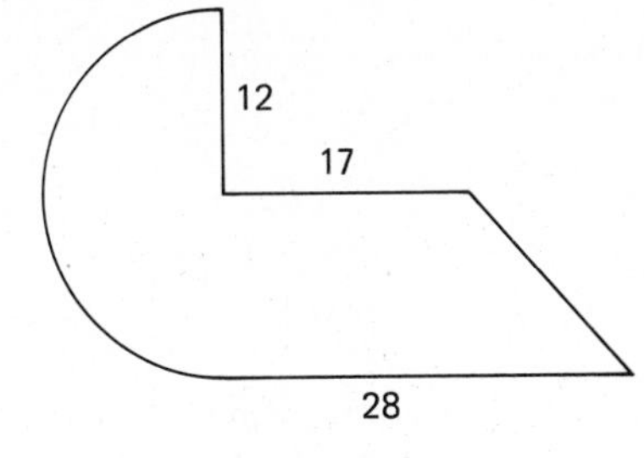

19. Find the surface area of the given prism. Dimensions are in inches.

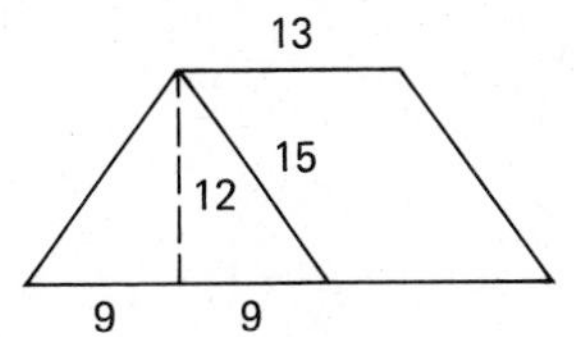

Solve:

20. $4x - 3 + 2x - 6 = -7x + 6$

21. $-1\frac{1}{3}x + \frac{1}{4} = \frac{5}{12}$

22. $\dfrac{-\frac{1}{6}}{\frac{5}{12}} = \dfrac{\frac{1}{2}}{x}$

Simplify:

23. $2(6 - 3 \cdot 4) + 3^2 - (-3)^2$

24. $4ab^2cb^3c^2a^3$

25. $\dfrac{-3^2 + (-2)^3 - 3(2 - 2^2)}{2(3^2 - 2^3)}$

26. $2\frac{1}{4} \times 1\frac{1}{3} \div \frac{1}{5} \div 1\frac{1}{3}$

Evaluate:

27. $xy^2 + 2xy$ if $x = -1, y = 2$

28. $xy^3 + x^2y$ if $x = -2, y = 3$

29. If $3x - 5 = 13$, what is $\frac{1}{3}x - 2$?

30. Write 13621451.4 in words.

LESSON 108 Negative exponents

108.A negative exponents

It is convenient to have another way to write the reciprocal of an exponential. Here we show an alternate way to write 5^2 and 1 over 5^2.

$$5^2 = \frac{1}{5^{-2}} \qquad \frac{1}{5^2} = 5^{-2}$$

On the left we moved the 5^2 from the top to the bottom and changed the sign of the exponent from plus to minus. On the right we moved the 5^2 from the bottom to the top and changed the sign of the exponent from plus to minus. We can state this rule by saying the following:

> **An exponential can be written in reciprocal form if the sign of the exponent is changed.**

example 108.A.1 Simplify: (a) $\dfrac{1}{4^{-3}}$ (b) 2^{-4}

solution (a) **To simplify, we must have positive exponents.** Therefore, we write $\frac{1}{4^{-3}}$ in reciprocal form and change the sign of the exponent.

$$\frac{1}{4^{-3}} = 4^3$$

Now 4^3 means $4 \cdot 4 \cdot 4$, or 64.

$$\frac{1}{4^{-3}} = 4^3 = \mathbf{64}$$

(b) **To simplify we must have positive exponents.** Therefore, we write 2^{-4} in reciprocal form as

$$\frac{1}{2^4}$$

Now 2^4 equals 16 so our answer is

$$\mathbf{\frac{1}{16}}$$

example 108.A.2 Simplify: (a) $\dfrac{1}{3^{-3}}$ (b) 3^{-2}

solution (a) First we write the reciprocal form of the expression.

$$\frac{1}{3^{-3}} = 3^3$$

And this can be easily simplified.

$$\frac{1}{3^{-3}} = 3^3 = \mathbf{27}$$

(b) Again we write the expression in reciprocal form.

$$3^{-2} = \frac{1}{3^2}$$

And this can be easily simplified.

$$3^{-2} = \frac{1}{3^2} = \mathbf{\frac{1}{9}}$$

problem set 108

1. A fair coin comes up heads twice in a row. If it is flipped again, what is the probability that it will be heads again?
2. An urn contains 2 white marbles, 3 green marbles, and 5 red marbles. A marble is drawn and then replaced. Then a second marble is drawn. What is the probability that the first marble will be white and the second will be green?
3. Forty percent of the items in the showcase were covered with mildew. If there were 200 items in the showcase, how many were not covered with mildew?
4. Zorba asked for 340 percent more than the man was willing to give. If the man was willing to give 400 drachmas, how many drachmas did Zorba ask for?
5. Construct a triangle whose sides are 3 cm, 4 cm, and 5 cm.
6. Use a ruler to draw a line segment 5 centimeters long. Construct a perpendicular to the line at a point 2 centimeters from the left endpoint.
7. Draw a line segment and a point outside the line. Construct a perpendicular to the line which passes through the point.
8. Use a protractor to draw a 110° angle. Then use a straightedge and a compass to construct the bisector of the angle.
9. Add 10111 (base 2) + 10110 (base 2).
10. Add 26 (base 8) + 57 (base 8).

11. Convert 617 (base 8) to base 10.

12. What percent of 60 is 69? Draw a diagram that depicts the problem.

13. If 120 is increased by 25 percent, what is the resulting number? Draw a diagram that depicts the problem.

14. Convert 1425 cubic meters to liters.

15. Convert 26 cubic yards to cubic inches.

16. Use the distributive property to multiply: $3ab(ab + b - cab)$

17. Simplify by adding like terms: $3xyx^2 - 2xy^2 + 4x^3y - 3xyy$

18. Express in cubic centimeters the volume of a solid whose base is shown and whose sides are 1 meter high. Dimensions are in meters.

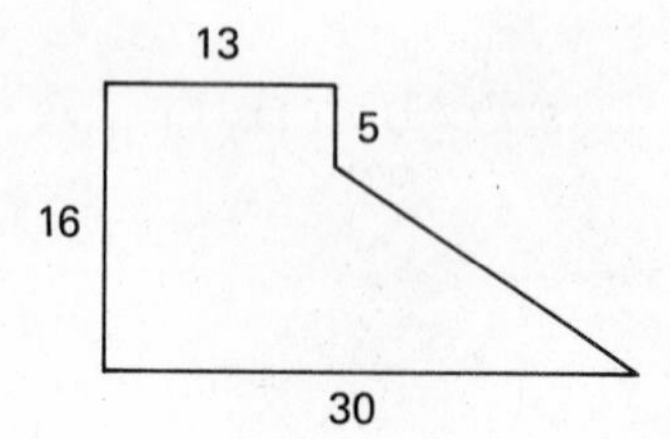

Solve:

19. $3x - 3 + 6x = -5x + 6$

20. $-1\frac{2}{3}x - \frac{1}{6} = 1\frac{2}{3}$

21. $\dfrac{-\frac{3}{4}}{\frac{9}{16}} = \dfrac{-\frac{1}{2}}{x}$

Simplify:

22. *(a) $\dfrac{1}{3^{-3}}$ (b) 2^{-3} (c) $\dfrac{1}{2^{-3}}$ (d) 3^{-3}

23. $-(-2)^2 - 3(3 - 2 \cdot 2^2)$

24. $5ab^2a^2ba^3b$

25. $\dfrac{-2^3 + (-3)^2 - 4(2 - 3^2)}{3(2^3 - 3)}$

26. $\frac{1}{3}\left(1\frac{1}{3} \cdot \frac{3}{4} - \frac{1}{6} \cdot \frac{3}{2}\right)$

Evaluate:

27. $xy^2 - xy$ if $x = -2, y = -3$

28. $a^2b - a^3b^2$ if $a = -1, b = -3$

29. If $6x - 3 = 21$, what is $\frac{1}{2}x - 12$?

30. Graph: $x \not\geq 3$

LESSON 109 Rectangular coordinates

109.A rectangular coordinates

Here we show two number lines. The lines are perpendicular and intersect at the origin of both lines. We call the vertical line the y axis and call the horizontal line the x axis.

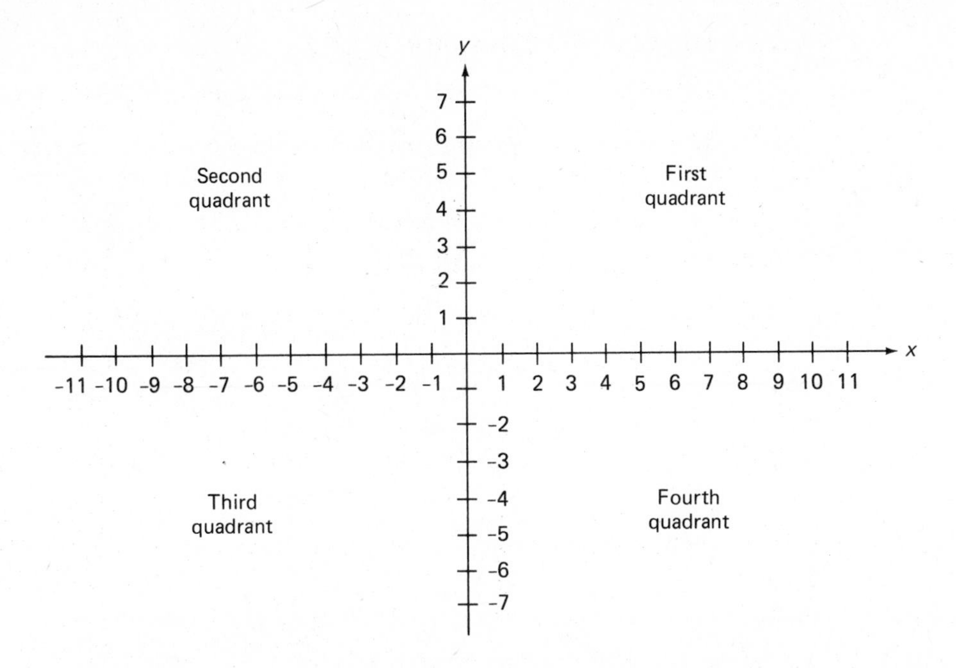

The number lines (axes) divide the plane into four quarters, or **quadrants.** The upper right quadrant is called the first quadrant. The others follow in a counterclockwise direction.

Any point in the plane can be located by telling how far it is to the right or left of the y axis and how far up or down it is from the x axis. Two numbers are associated with every point in the plane. The first number tells how far to the right (+) or to the left (−) the point is from the y axis. This number is called the ***x* coordinate** of the point. The second number tells how far above (+) or below (−) the point is from the x axis. This is called the ***y* coordinate** of the point.

example 109.A.1 Four points are graphed on this rectangular coordinate system. What are the x and y coordinates of the points?

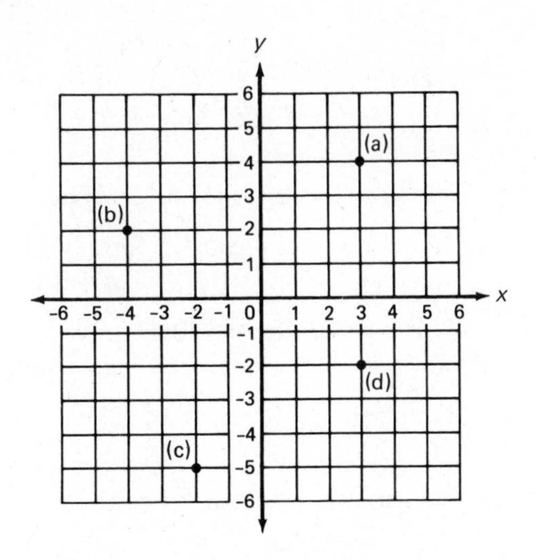

solution Point (a) is 3 units to the right and 4 units up.

$$\text{(a)}\quad \boldsymbol{x = 3, y = 4}$$

Point (b) is 4 units to the left and 2 units up.

$$\text{(b)}\quad \boldsymbol{x = -4, y = 2}$$

Point (c) is 2 units to the left and 5 units down.

$$\text{(c)} \quad x = -2, y = -5$$

Point (d) is 3 units to the right and 2 units down.

$$\text{(d)} \quad x = 3, y = -2$$

example 109.A.2 Graph the following points: (a) (4, 2) (b) (4, −3) (c) (−4, −3)

solution The first number is always the x coordinate and the second number is always the y coordinate. Thus,

(a) is 4 units to the right and 2 units up

(b) is 4 units to the right and 3 units down

(c) is 4 units to the left and 3 units down

Now we graph the points.

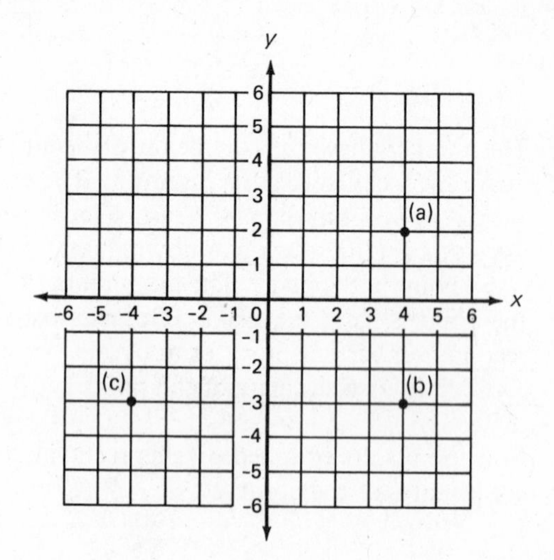

problem set 109

1. A fair coin is tossed three times. What is the probability that all three tosses will be heads?

2. A pair of dice is rolled twice. What is the probability of getting a 7 on both rolls?

3. It had never happened before, but this time the increase was 350 percent. If the total this time was 1800, what was the total last time?

4. The ratio of blue starlings to white egrets was 14 to 3. If 8500 birds had nested down by suppertime, how many were white egrets?

5. Graph the points on a rectangular coordinate system: (a) (4, 3) (b) (−2, 5) (c) (3, 6)

6. Construct a triangle whose sides are 4 cm, 3 cm, and 6 cm.

7. Use a ruler to draw a line segment 7 centimeters long. Construct a perpendicular to the line at a point 4 centimeters from the left endpoint.

8. Use a protractor to draw a 120° angle. Then use a straightedge and a compass to construct the bisector of the angle.

9. Draw a line segment and a point outside the line. Construct a perpendicular to the line which passes through the point.

10. Add 11011 (base 2) + 10111 (base 2).

11. Convert 461 (base 8) to base 10.

12. What percent of 80 is 108? Draw a diagram that depicts the problem.

13. Ninety percent of what number is 261? Draw a diagram that depicts the problem.

14. Convert 68,921,300 milliliters to liters.

15. Convert 2 cubic miles to cubic feet.

16. Use the distributive property to multiply: $2abc(a + b^2 + c - 1)$

17. Simplify by adding like terms: $2x^2y^2 - 3xyxy + 3x^3y - 2xy^2x$

18. Express in cubic inches the volume of a solid whose base is shown and whose sides are 2 feet high. Dimensions are in inches.

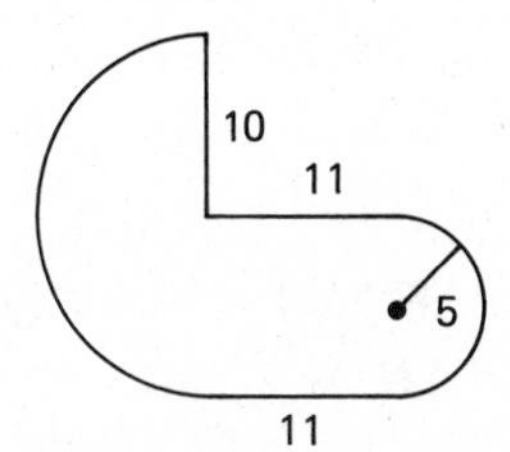

Solve:

19. $2x - 4 + 3x = x + 21$

20. $-2\frac{1}{4}x - \frac{1}{3} = 2\frac{1}{2}$

21. $\dfrac{-\frac{1}{2}}{\frac{3}{4}} = \dfrac{-\frac{1}{3}}{x}$

Simplify:

22. (a) $\dfrac{1}{3^{-2}}$ (b) 2^{-4} (c) $\dfrac{1}{2^{-4}}$ (d) 3^{-2}

23. $(-1)^3 - 2(2^3 - 2 \cdot 3)$

24. $4a^2bab^2a^3b$

25. $\dfrac{-2^3 + (-2)^2 - 3(3 - 3^2)}{2(2^3 - 3^2)}$

26. $\frac{1}{2}\left(1\frac{2}{3} \cdot \frac{1}{4} - \frac{1}{2} \cdot \frac{5}{6}\right)$

Evaluate:

27. $-2xy + 2x^2$ if $x = -1$, $y = 2$

28. $mn - mn^2$ if $m = -2$, $n = -1$

29. If $4x - 21 = 3$, what is $\frac{1}{4}x - 3$?

30. Find the least common multiple of 6, 9, and 16.

LESSON 110 Advanced equations

110.A advanced equations

The equations that we have discussed thus far have required adding like terms on only one side of the equation. Often we simplify the expressions on both sides of the equals sign as the first step.

example 110.A.1 Solve: $6x - 5 + x - 2 = 2x - 3 + x$

solution We begin by adding like terms to simplify both sides of the equation.

$$7x - 7 = 3x - 3$$

Now we eliminate the $3x$ by adding $-3x$ to both sides.

$$\begin{array}{rcll} 7x - 7 &=& 3x - 3 & \text{added like terms} \\ -3x & & -3x & \text{add } -3x \text{ to both sides} \\ \hline 4x - 7 &=& -3 & \\ +7 & & +7 & \text{add } +7 \text{ to both sides} \\ \hline 4x &=& 4 & \\ x &=& \mathbf{1} & \text{divided both sides by 4} \end{array}$$

Check:

$$6(1) - 5 + (1) - 2 = 2(1) - 3 + (1) \longrightarrow 0 = 0 \quad \text{check}$$

example 110.A.2 Solve: $3x - 2 - x = 4 + x + 3$

solution Again we begin by adding like terms on both sides. Then we solve.

$$\begin{array}{rcll} 2x - 2 &=& x + 7 & \text{added like terms} \\ -x & & -x & \text{add } -x \text{ to both sides} \\ \hline x - 2 &=& 7 & \\ +2 & & +2 & \text{add } +2 \text{ to both sides} \\ \hline x &=& \mathbf{9} & \end{array}$$

Check:

$$3(9) - 2 - (9) = 4 + (9) + 3 \longrightarrow 27 - 2 - 9 = 4 + 9 + 3$$
$$\longrightarrow 16 = 16 \quad \text{check}$$

problem set 110

1. The urn contained 20 marbles: 10 were black, 5 were red, and 5 were green. One marble was drawn and then put back. Then a second marble was drawn. What is the probability that both marbles were black?

2. In Problem 1 what is the probability that the first marble was black and the second marble was red?

3. Ten percent of those present were able to concentrate on what the speaker was saying. If 600 were concentrating, how many were not concentrating?

4. The ratio of the weight of canned goods to the weight of bulk commodities was 4 to 11. If there were 300,000 tons stored in the warehouse, what was the weight of the bulk commodities?

5. Graph the points on a rectangular coordinate system: (a) $(3, -2)$ (b) $(-1, 4)$ (c) $(-2, -3)$

6. Construct a triangle whose sides are 5 cm, 6 cm, and 4 cm.

7. Draw a line segment and a point outside the line. Construct a perpendicular to the line which passes through the point.

8. Convert 10101 (base 2) to base 10.

9. Convert 521 (base 8) to base 10.

10. What percent of 60 is 99? Draw a diagram that depicts the problem.

11. If 130 is increased by 140 percent, what is the resulting number? Draw a diagram that depicts the problem.

12. Convert 256 cubic meters to cubic centimeters.

13. Convert 12 cubic miles to cubic inches.

14. Use the distributive property to multiply: $2ac(ab + 2ac - a)$

15. Simplify by adding like terms: $2x^2y^2 - 3x^2y^3 - 3xyxy + 7x^2y^2y$

16. If $3x - 5 = 13$, what is $\frac{1}{3}x - 7$?

17. Find the volume of a solid whose base is shown and whose sides are 3 feet high. Dimensions are in feet.

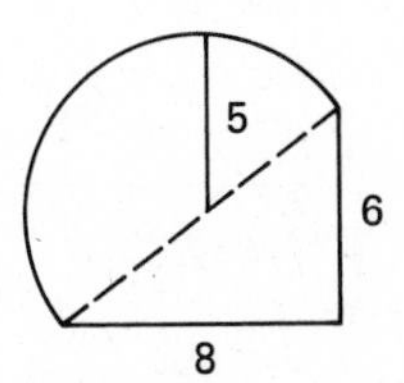

Solve:

18. $3x - 6 + 4x = -3x + 16 - x$

19. $-3\frac{1}{3}x - \frac{1}{6} = \frac{7}{18}$

20. $\dfrac{-\frac{1}{4}}{\frac{8}{9}} = \dfrac{-\frac{1}{4}}{x}$

***21.** $6x - 5 + x - 2 = 2x - 3 + x$

***22.** $3x - 2 - x = 4 + x + 3$

Simplify:

23. (a) $\dfrac{1}{2^{-3}}$ (b) 2^{-3} (c) $\dfrac{1}{2^{-1}}$ (d) 3^{-3}

24. $(-2)^3 - 3(2^3 - 3^2)$

25. $5a^2ba^2ba^3$

26. $\dfrac{-2^3 + (3)^2 - 2^2(2^3 - 3^2)}{3(3 \cdot 2 - 5)}$

27. $\frac{1}{3}\left(1\frac{1}{4} \cdot \frac{1}{3} - \frac{1}{6} \cdot \frac{1}{2}\right)$

Evaluate:

28. $xy^2 - x^3y^3$ if $x = -1$, $y = -2$

29. $mn + n^2$ if $m = -1$, $n = 5$

30. Graph: $x \nleq -2$

Appendix

Appendix

Index

Answers

problem set 1

1. (a) 60,000 (b) 900 (c) 3 **3.** 3,666,766 **5.** 51,000,027,520
7. 407,000,090,742,072
9. Fifty-one million, seven hundred twenty-three thousand, six hundred forty-two
11. Thirty-two billion, six hundred fifty-two
13. 1148 **15.** 12,395 **17.** 36,593 **19.** 1654 **21.** 1550 **23.** 34,693
25. 57,983 **27.** 31,531 **29.** 3945 **31.** 3356

problem set 2

1. 24,000 **3.** 83,752,910,000 **5.** 83,722,000 **7.** 777,727,757 **9.** 107,047,020
11. Seven hundred thirty-one million, two hundred eighty-four thousand, six
13. Nine billion, three million, one thousand, two hundred fifty-six **15.** 621 **17.** 1618
19. 13,215 **21.** 22,858 **23.** 219,258 **25.** 213,820 **27.** 1036 **29.** 1992

problem set 3

1. 35 **3.** 36 **5.** 288 **7.** 89 **9.** 4,150,000 **11.** 84,000,000 **13.** 225,223
15. 14,705,052 **17.** Seven hundred seven million, seventy thousand, seven hundred five
19. Nine billion, five hundred ten million, two hundred eighty-two thousand, one hundred five
21. 2294 **23.** 2083 **25.** 2230 **27.** 255,945 **29.** 27,396

problem set 4

1. 141,035 **3.** 655,400 **5.** 123,084 **7.** 290,745 **9.** 136 **11.** 1669
13. 51,783,600 **15.** 67,727 **17.** 207,075,903
19. Four million, fifty-one thousand, six hundred thirty-two
21. Ninety-eight million, sixty-one thousand, five hundred two
23. 372 **25.** 2253 **27.** 266,638 **29.** 8019

problem set 5

1. 6, R11 **3.** 1167, R36 **5.** 181, R2 **7.** 91,485 **9.** 323,584 **11.** 151,335
13. 666 **15.** 925 **17.** 720,000,000 **19.** 7333 **21.** 47,000,014
23. five million, twenty-one **25.** Thirty-nine thousand, two **27.** 4394
29. 2,685,261

problem set 6

1. 260.384 **3.** .02053 **5.** 1.2688 **7.** 5.157 **9.** 14.0168 **11.** 157, R13
13. 811, R1 **15.** 62, R4 **17.** 5,144,000 **19.** 773,757,797 **21.** .041,301
23. Four thousand, one hundred sixty-five and one hundred sixty-two ten-thousandths
25. Five hundred four thousand, three hundred twenty-seven and one million, five hundred ten thousand, five hundred twelve billionths
27. 949.4521 **29.** 5195.86

problem set 7

1. 4,716,231.4 **3.** .04132 **5.** .089744 **7.** .00128856 **9.** 1.3992

11. 36.353 **13.** 37.87 **15.** 392 **17.** 2100, R27 **19.** 91,600,000

21. 88,838,887 **23.** 702.00942

25. Nine million, fifty-six thousand, two hundred thirteen and fifty-seven thousand, three hundred twenty-eight hundred-millionths

27. 1164.762 **29.** 4915.524

problem set 8

1. 413.6268 **3.** 9.31521 **5.** .0165164 **7.** .00200304 **9.** .736 **11.** 388.086

13. 25.79 **15.** 1003.21 **17.** 4526.76 **19.** 42.123457 **21.** 47,000, 067,000.00417

23. Six thousand, one hundred eighty-four hundred-millionths

25. Seventy-five million, four thousand, two hundred thirteen and one thousand, six hundred fifty-two ten-millionths

27. 5652.06 **29.** 350.6492

problem set 9

1. 92 inches **3.** 31.64215 **5.** 417.3652 **7.** .07584 **9.** .0185262 **11.** 1.642

13. 2.8529 **15.** 436.1838 **17.** 668 **19.** 5705.48 **21.** 61.3737378

23. 742,000,537.010948

25. One hundred twenty-eight thousand, six hundred forty-seven hundred-millionths

27. Fifty-one thousand, seven hundred eighty-six and seven hundred eighty-five hundred-thousandths

29. 11,556.876

problem set 10

1. (a) 300; 4888; 9132; 72,654 (b) 300; 9132; 72,654 (c) 235, 300 (d) 300

3. 112 meters **5.** 40,265 **7.** .078848 **9.** 1255.42 **11.** .799 **13.** 1.8147

15. 396.8 **17.** 383.70 **19.** 6392.54 **21.** 4300 **23.** 762,000,442.12792

25. One hundred twenty thousand, one hundred twenty-three billionths

27. Five million, one hundred sixty-two and eight ten-thousandths **29.** 2986.47

problem set 11

1. 1500 **3.** King, .404 **5.** 112 inches **7.** 13.2116 **9.** .123631 **11.** 36.4011

13. 49.41 **15.** 61.8088 **17.** 1.079 **19.** 6.74 **21.** 3.56 **23.** 223.09

25. 1,625,000,250,025.123

27. Two hundred twenty-three million, ninety-two thousand, eight hundred seventy

29. 7707.8

problem set 12

1. 10,979.04098 inches **3.** 95

5. (a) 302; 9172; 3132; 62,120 (b) 3132 (c) 625; 62,120 (d) 62,120

7. 112 feet **9.** 36,821.1 **11.** 5651.47

13. (a) $2 \cdot 2 \cdot 3 \cdot 7$ (b) $2 \cdot 2 \cdot 3 \cdot 3 \cdot 5 \cdot 7$ (c) $2 \cdot 2 \cdot 2 \cdot 3 \cdot 3 \cdot 3 \cdot 5 \cdot 5$ **15.** 1.6119

17. .33 **19.** 3.05 **21.** 251.03 **23.** 4700 **25.** 961,313,000,025

27. Sixteen and five hundred sixty-two ten-thousandths **29.** 2898.29

problem set 13

1. 350,625 3. 13,942,000.000128

5. (a) 1020, 130, 1332, 132 (b) 1020, 125, 130, 185 (c) 1020, 1332, 132 (d) 1020, 130

7. 112 cm 9. 3,118,361.52 11. 5036.31

13. (a) $2 \cdot 2 \cdot 2 \cdot 3 \cdot 3 \cdot 5$ (b) $2 \cdot 2 \cdot 2 \cdot 2 \cdot 3 \cdot 3 \cdot 5$ (c) $2 \cdot 2 \cdot 2 \cdot 2 \cdot 2 \cdot 3 \cdot 3 \cdot 5$

15. 1.4816 17. 1931.098 19. 19.36 21. 259.27 23. 1231.626

25. 321,617,212.231

27. Six hundred thirteen and one hundred sixty-two thousandths 29. 728.15

problem set 14

1. 440 3. .00001197

5. (a) 120, 1620 (b) 120, 135, 1620 (c) 120, 122, 1332, 1620 (d) 120, 135, 1332, 1620

7. 150 km 9. (a) $\frac{2}{7}$ (b) $\frac{3}{7}$ (c) $\frac{1}{9}$ 11. 11,361.21 13. 144.64

15. (a) $2 \cdot 2 \cdot 2 \cdot 3 \cdot 3 \cdot 5 \cdot 5$ (b) $2 \cdot 2 \cdot 3 \cdot 3 \cdot 5 \cdot 5$ (c) $2 \cdot 3 \cdot 3 \cdot 5 \cdot 5$ 17. 175.482

19. 1052.96 21. 36,544.8 23. 509.34 25. 20,100,000,000

27. Eleven thousand, one hundred twenty-three and one hundred twenty-one thousandths

29. 1332.48

problem set 15

1. 14,907,987 3. 189,903 5. (a) 2133, 312, 630 (b) 212, 312, 610, 630

7. 168 ft 9. .03 11. (a) $\frac{7}{6}$ (b) $\frac{2}{3}$ (c) $\frac{2}{3}$ 13. 12,361.311 15. 70.848

17. 999.15 19. .63 21. 272.34 23. 2.51 25. 42.617618

27. One million, eight hundred seventy-six thousand, two hundred eleven and thirty-two hundredths 29. 7842

problem set 16

1. 272 3. 79,027 5. (a) 650, 625, 15, 20, 30 (b) 650, 20, 30 7. 162 yd

9. .06 11. $\frac{6}{7}$ 13. $\frac{1}{3}$ 15. 686.56 17. 167,318.38 19. 193.41

21. 151.08 23. $2 \cdot 2 \cdot 2 \cdot 3 \cdot 3 \cdot 3 \cdot 5$ 25. 87,621.321789

27. One hundred seventy-two and three hundred twelve thousandths 29. 32,347.12

problem set 17

1. 982 3. 103,173 5. 238 7. 40 cm^2 9. .18 11. .92 13. $2 \cdot 3 \cdot 5$

15. $2 \cdot 2 \cdot 3 \cdot 3 \cdot 5 \cdot 7$ 17. $\frac{3}{4}$ 19. 538.5141 21. 62,538.76 23. 67.4643

25. 90.52 27. 2.0707071 29. 1418.04

problem set 18

1. 71,266 3. 44,102,079 5. 29, 31, 37, 41, 43, 47 7. 570 cm^2 9. $\frac{2}{5}$

11. $\frac{5}{7}$ 13. .94 15. .85 17. $2 \cdot 2 \cdot 2 \cdot 3 \cdot 3 \cdot 3 \cdot 5 \cdot 7$ 19. 8826.43

21. 181.2783 23. 3.18 25. 9.86960 27. 7562.47 29. 777,777

problem set 19

1. Charles: 975.2157, Mary: 975.0137; Mary's guess was closer **3.** 13,292 **5.** 37

7. $\frac{4}{15}$ **9.** $\frac{14}{27}$ **11.** 600 yd^2 **13.** $\frac{2}{3}$ **15.** $\frac{16}{45}$ **17.** .29

19. $2 \cdot 2 \cdot 2 \cdot 2 \cdot 2 \cdot 3 \cdot 3 \cdot 5 \cdot 7$ **21.** 79.488 **23.** 593.448 **25.** 19.78

27. Nine million, six hundred ninety-nine thousand, six hundred ninety **29.** 545.6

problem set 20

1. 36,825 **3.** 114,012 **5.** 9, 18, 27, 36, 45 **7.** $\frac{5}{24}$ **9.** $\frac{1}{4}$ **11.** 484 m^2

13. $\frac{3}{4}$ **15.** $\frac{3}{5}$ **17.** .46 **19.** $2 \cdot 2 \cdot 2 \cdot 3 \cdot 3 \cdot 5 \cdot 5 \cdot 7$ **21.** 109,670.4

23. 8.4041 **25.** 3.14

27. One hundred eleven million, five hundred forty-six thousand, four hundred thirty-five

29. 197.49631

problem set 21

1. 1,527,474,973.0173 **3.** 1713 **5.** 31, 37, 41, 43, 47 **7.** 12.25 **9.** $\frac{3}{8}$ **11.** $\frac{1}{5}$

13. 380 cm^2 **15.** $\frac{5}{6}$ **17.** .65 **19.** $2 \cdot 2 \cdot 2 \cdot 3 \cdot 3 \cdot 5 \cdot 7$ **21.** 112.179

23. 121.89 **25.** 7.18 **27.** 34.7182 **29.** .0098762

problem set 22

1. 3,576,999.998662 **3.** 13 hr **5.** $1189\frac{1}{3}$ **7.** $\frac{280}{297}$ **9.** 4 **11.** 357 ft^2

13. $\frac{16}{45}$ **15.** $\frac{15}{16}$ **17.** .27 **19.** $2 \cdot 2 \cdot 2 \cdot 2 \cdot 5 \cdot 5 \cdot 5 \cdot 7$ **21.** 60.547

23. 6852.449 **25.** 3.77 **27.** .30 **29.** 123,713.6

problem set 23

1. 27,049.4995 **3.** 79 **5.** (a) 61, 67 (b) 63 **7.** 6 ft **9.** 504 inches

11. $\frac{6}{5}$ **13.** 170 yd **15.** $\frac{45}{64}$ **17.** $\frac{1}{4}$ **19.** .89 **21.** $2 \cdot 3 \cdot 3 \cdot 3 \cdot 5 \cdot 5 \cdot 5$

23. 691.04 **25.** 61,059.377 **27.** 23,937.67

29. Sixty-seven million, two hundred eleven thousand, three hundred sixty-one and seventy-two hundredths

problem set 24

1. 6065 **3.** 43 **5.** (a) 23, 29 (b) 24, 28 **7.** 5.28 meters **9.** 17 ft

11. $\frac{2}{3}$ **13.** 1 **15.** 520 in^2 **17.** $\frac{3}{4}$ **19.** .21 **21.** $2 \cdot 2 \cdot 2 \cdot 2 \cdot 2 \cdot 3 \cdot 5$

23. 7465.7 **25.** 12.829 **27.** 68.06 **29.** 3,817,300

problem set 25

1. 54,285 **3.** 3284 **5.** (a) 41, 43, 47 (b) 42, 45, 48 **7.** 4631 cm

9. .416 km **11.** 1 **13.** 128 ft **15.** $\frac{13}{24}$ **17.** $\frac{25}{72}$ **19.** .35

21. $2 \cdot 3 \cdot 3 \cdot 5 \cdot 7$ **23.** 17.3536 **25.** 17.87 **27.** 3403.5

29. One hundred eleven million, three hundred twenty-one thousand, six hundred fifty-four and seven-tenths

problem set 26

1. 4.075 m **3.** 9.625 **5.** 1525 **7.** 1.3615 m **9.** 1.899 km **11.** 1

13. 575 m^2 **15.** $\frac{6}{7}$ **17.** $\frac{9}{16}$ **19.** .65 **21.** $2 \cdot 3 \cdot 5 \cdot 5 \cdot 7$ **23.** 1.8931

25. 711.999 **27.** 181.15

29. Six million, two hundred eleven thousand, three hundred fifty-seven and five-tenths

problem set 27

1. 2095 **3.** 296,002 **5.** 1229 **7.** 243 feet **9.** 489,900 cm **11.** 2

13. 140 ft **15.** 400 ft^2 **17.** $\frac{7}{9}$ **19.** .54 **21.** $2 \cdot 2 \cdot 3 \cdot 3 \cdot 3 \cdot 3$ **23.** 1713.88

25. 18,517.7228 **27.** 15,283 **29.** 781.10564

problem set 28

1. .03753 **3.** Thirty-three million, seven hundred forty-five thousand, twenty-six **5.** $2\frac{4}{5}$

7.

9. $\frac{67}{9}$ **11.** 1.9272 meters **13.** $\frac{1}{6}$ **15.** $\frac{4}{3}$ **17.** 800 cm **19.** $\frac{7}{8}$ **21.** .70

23. $1440 = 2 \cdot 2 \cdot 2 \cdot 2 \cdot 2 \cdot 3 \cdot 3 \cdot 5$ **25.** 116.6508 **27.** 260,376.67

29. Seventy-eight million, two hundred fifty-six thousand, one hundred thirteen and seven-tenths

problem set 29

1. 40,000.001078 **3.** 38 **5.** 4300 **7.** 25 **9.** $4\frac{1}{5}$ **11.** $\frac{59}{8}$ **13.** $\frac{62}{11}$

15. 95,040 ft **17.** $\frac{2}{15}$ **19.** $\frac{2}{3}$ **21.** .70 **23.** $2 \cdot 2 \cdot 3 \cdot 5 \cdot 5 \cdot 7$ **25.** .0375

27. 52.593 **29.** 2831.82113

problem set 30

1. 300 **3.** 30 inches **5.** 46,710 **7.** 42 **9.** $4\frac{2}{5}$ **11.** $\frac{23}{3}$ **13.** $\frac{37}{8}$

15. 192.62 **17.** $\frac{2}{3}$ **19.** 9.6 meters **21.** $\frac{16}{45}$ **23.** .53 **25.** $2 \cdot 3 \cdot 5 \cdot 5 \cdot 11$

27. 45.5275 **29.** 137840

31. Sixty-eight million, two hundred eleven thousand, three hundred fifty-one

problem set 31

1. \$50 **3.** 17 hr **5.** 1456 **7.** 121 **9.** $3\frac{1}{4}$ **11.** 40,500 **13.** 3400

15. 147,800 m **17.** $\frac{2}{3}$ **19.** 12,000 cm **21.** $\frac{7}{8}$ **23.** .83 **25.** 1.25154

27. 56.779 **29.** 415.629

problem set 32

1. 4420 **3.** 1027 **5.** 68,969 **7.** 1100 **9.** $5\frac{6}{7}$ **11.** 1800 **13.** 360

15. $\frac{14}{55}$ **17.** $\frac{4}{5}$ **19.** .0024081 **21.** 481.492 **23.** 425 ft^2 **25.** $\frac{52}{61}$ **27.** .83

29. 42,062,918,000

problem set 33

1. 6475 **3.** 82 **5.** \$91.60 **7.** 140 **9.** $5\frac{5}{7}$ **11.** 1080 **13.** $\frac{39}{40}$

15. 29 **17.** 2 **19.** .0000368 **21.** $\frac{10}{33}$ **23.** 450 ft^2

25. $2 \cdot 2 \cdot 2 \cdot 2 \cdot 2 \cdot 2 \cdot 5 \cdot 13$ **27.** .39 **29.** 99,540,000

problem set 34

1. 311 **3.** Graph **5.** 17 hr **7.** 40 **9.** $84\frac{1}{5}$ **11.** 840

13. $\frac{15}{16}$ **15.** 16 **17.** 14,295.24 **19.** 1730.99 **21.** 6.400136 **23.** 32

25. 345 ft^2 **27.** .6 **29.** 41,060,000

problem set 35

1. The second measurement was larger by .0033 **3.** \$20,000 **5.** 402 lb **7.** 20

9. $23\frac{1}{2}$ **11.** $\frac{25}{16}$ **13.** 43 **15.** .108378 **17.** 513.0011 **19.** 3 **21.** 4

23. 10 yd **25.** 1,520,640 inches **27.** 1.5 meters **29.** 5.55556

problem set 36

1. 236 lb **3.** \$997.80 **5.** \$120 **7.** 160 **9.** $7\frac{9}{11}$ **11.** $\frac{8}{15}$ **13.** 15

15. $\frac{43}{8}$ **17.** $695\frac{13}{20}$ **19.** 89,525 **21.** $\frac{7}{3}$ **23.** 24 **25.** 1.876258 km

27. 114 centimeters **29.** 659,000,000

problem set 37

1. \$480 **3.** 10 million **5.** 10 **7.** $6\frac{1}{5}$ **9.** 2520 **11.** $\frac{5}{8}$ **13.** $10\frac{7}{8}$

15. $751\frac{3}{5}$ **17.** $395\frac{11}{16}$ **19.** 575.2782 **21.** 674.056 **23.** 19

25. 633,600 inches **27.** 112 inches **29.** 1.052222

problem set 38

1. \$280 **3.** 6993 tons **5.** 10 **7.** $25\frac{3}{17}$ **9.** 28,200 **11.** $1\frac{19}{40}$ **13.** 21

15. $519\frac{1}{10}$ **17.** $491\frac{3}{4}$ **19.** 272.4 **21.** $\frac{3}{2}$ **23.** | 4 | 4 | 4 | 4 | 4 | **25.** 4

27. 725 cm^2 **29.** .74

problem set 39

1. 16 **3.** Graph **5.** 24 **7.** $24\frac{4}{13}$ **9.** 72 **11.** $7\frac{9}{10}$ **13.** 29

15. $682\frac{5}{8}$ **17.** $193\frac{11}{20}$ **19.** 116.23 **21.** $\frac{3}{2}$ **23.** 6 **25.** 9 **27.** 6.25611 km

29. .87

problem set 40

1. 473 **3.** 5040 **5.** 15 **7.** $13\frac{2}{3}$ **9.** 3780 **11.** $5\frac{1}{3}$ **13.** 19

15. $660\frac{7}{10}$ **17.** $395\frac{7}{8}$ **19.** 28,600 **21.** 4 **23.** 10 **25.** $\frac{4}{5}$ **27.** 34

29. (a) 41, 43 (b) 42

problem set 41

1. 600 units per hour **3.** 60 **5.** 9 **7.** $16\frac{2}{5}$ **9.** 840 **11.** $\frac{4}{3}$ **13.** 36

15. 35 **17.** 19 **19.** 286 **21.** 6601.91 **23.**

4	4	4	4	4	4

25. 44 **27.** 291 m^2 **29.** .79

problem set 42

1. Second guess, .066268 **3.** 88 **5.** 110 **7.** $\frac{89}{7}$ **9.** 1320 **11.** $\frac{14}{27}$

13. 12 **15.** $\frac{3}{4}$ **17.** $2\frac{31}{40}$ **19.** 9.7808 **21.** 8754.29 **23.**

2	2	2	2	2	2	2	2

25. 2 **27.** 1135 ft^2 **29.** 190,000,000

problem set 43

1. $32 **3.** Graph **5.** 60 **7.** $\frac{339}{8}$ **9.** 110 **11.** 500 **13.** $1\frac{1}{8}$

15. 123 **17.** $29\frac{17}{18}$ **19.** .028572 **21.** $\frac{7}{6}$ **23.** 0 **25.** 720 in^2 **27.** 140 cm

29. 18,000,000

problem set 44

1. $1600 **3.** 7,134,108 **5.** 40 **7.** $\frac{117}{16}$ **9.** $\frac{32}{7}$ **11.** 588 **13.** $\frac{3}{22}$

15. 10 **17.** $\frac{305}{8}$ **19.** $\frac{195}{176}$ **21.** $4\frac{5}{6}$ **23.** .02103 **25.** 34 **27.** 1127.5 ft^2

29. (a) 41, 43, 47 (b) 40, 45, 50

problem set 45

1. $1200 **3.** Graph **5.** 350 **7.** $\frac{17}{5}$ **9.** 12 **11.** 28 **13.** 105 **15.** 3

17. $7\frac{5}{6}$ **19.** 38 **21.** 6834.03 **23.** $\frac{361}{36}$ **25.** $\frac{46}{41}$ **27.** 1.04 m

29.

2	2	2	2	2	2	2	2	2	2	2	2

problem set 46

1. 132 **3.** 420 **5.** 240 **7.** 1400 **9.** 144 **11.** 6 **13.** 47 **15.** 34

17. $476\frac{5}{8}$ **19.** 1.7524 **21.** 63,568.33 **23.** 12 **25.** $\frac{152}{115}$ **27.** 589 ft^2

29.

3	3	3	3	3	3	3	3	3	3	3	3

problem set 47

1. 40 **3.** Graph **5.** 120 **7.** $12\frac{16}{25}$ **9.** 21 **11.** 45 **13.** 12 **15.** $1\frac{19}{32}$

17. 5 **19.** $105\frac{3}{10}$ **21.** .844 **23.** $\frac{38}{7}$ **25.** $2\frac{30}{47}$ **27.** 118 cm

29. | 5 | 5 | 5 | 5 | 5 |

problem set 48

1. $1600 **3.** 62,640 **5.** 600 **7.** $26\frac{5}{8}$ **9.** 16 **11.** $2\frac{7}{40}$ **13.** 27

15. $\frac{1}{16}$ **17.** $7\frac{3}{16}$ **19.** 2.5296 **21.** 65,455 **23.** $\frac{81}{95}$ **25.** 243 **27.** 604.5 m^2

29. | 4 | 4 | 4 | 4 | 4 |

problem set 49

1. 1600 **3.** Graph **5.** $\frac{7}{2}$ **7.** 72 **9.** $\frac{213}{31}$ **11.** 18 **13.** 32 **15.** $\frac{3}{4}$

17. $121\frac{9}{14}$ **19.** 8 **21.** 202.097 **23.** $1\frac{21}{115}$ **25.** 27 **27.** 6000 ft^3

29. | 5 | 5 | 5 | 5 | 5 | 5 |

problem set 50

1. 85,251 **3.** 250 **5.** 1350 m^2 **7.** 105 **9.** 72 **11.** 24 **13.** $\frac{29}{10}$

15. 46 **17.** $\frac{11}{12}$ **19.** $3\frac{19}{30}$ **21.** .26912 **23.** 541.67 **25.** $2\frac{12}{17}$ **27.** 8

29. | 3 | 3 | 3 | 3 | 3 | 3 | 3 | 3 | 3 |

problem set 51

1. 5643 **3.** $2940 **5.** 276 ft^3 **7.** 378 **9.** $\frac{7}{8}$ **11.** $8\frac{11}{15}$ **13.** $2\frac{5}{6}$

15. 27 **17.** $4\frac{3}{4}$ **19.** $3\frac{7}{20}$ **21.** $22\frac{9}{14}$ **23.** 44.079 **25.** $\frac{308}{51}$ **27.** 35

29. .0216218 km

problem set 52

1. 40 **3.** $100,000 **5.** $\frac{131}{80,000}$ **7.** 4200 yd^3 **9.** 180 **11.** 39 **13.** 96

15. 15 **17.** $\frac{5}{4}$ **19.** 14 **21.** $11\frac{1}{2}$ **23.** 1097.388 **25.** $\frac{203}{48}$ **27.** 72

29. 22,500 in

problem set 53

1. 155,982,014 **3.** 7700 **5.** 120 **7.** $\frac{29}{2500}$ **9.** 516 m^2 **11.** $\frac{3}{8}$ **13.** $9\frac{1}{20}$

15. $\frac{25}{6}$ **17.** 20 **19.** $\frac{11}{12}$ **21.** 0 **23.** .23582 **25.** 2797.639 **27.** $\frac{34}{45}$

29. 26

problem set 54

1. \$36,000 **3.** 698,042 **5.** 120 **7.** $\frac{119}{5000}$ **9.** 5200 cm^3 **11.** 84 **13.** $\frac{182}{23}$

15. $\frac{23}{14}$ **17.** 18 **19.** $\frac{19}{9}$ **21.** 2.5649 **23.** 3840.6 **25.** $\frac{29}{32}$ **27.** $\frac{1144}{135}$

29. 18

problem set 55

1. \$4524 **3.** 997.5 lb **5.** (a) 62 (b) 1488 **7.** $\frac{13}{800}$ **9.** $\frac{15}{19}$ **11.** 44

13. 360 **15.** 30 **17.** 66 **19.** $1\frac{3}{5}$ **21.** 1.5924 **23.** 45,020 **25.** $\frac{1}{6}$

27. $\frac{39}{8}$ **29.** 26

problem set 56

1. \$640 **3.** 1072 **5.** (a) 36 (b) 576 **7.** $\frac{69}{5000}$ **9.** $\frac{4}{5}$ **11.** 1040 ft^3

13. $27\frac{3}{7}$ **15.** $\frac{10}{3}$ **17.** $\frac{27}{28}$ **19.** 28 **21.** $4\frac{9}{40}$ **23.** 3.0736 **25.** 86,060

27. $\frac{13}{20}$ **29.** $\frac{1000}{147}$

problem set 57

1. 220,146 **3.** \$350 **5.** (a) $\frac{1}{4}$ (b) .25 (c) .0625 (d) 6.25 **7.** 60

9. .65 **11.** 180 **13.** $\frac{15}{16}$ **15.** $\frac{108}{5}$ **17.** 30 **19.** $\frac{37}{12}$ **21.** 17

23. 1249.829 **25.** $\frac{1}{10}$ **27.** $\frac{4}{7}$ **29.** 25

problem set 58

1. 213,222 **3.** 7538 lb **5.** (a) $\frac{4}{25}$ (b) .16 (c) .125 (d) 12.5 **7.** 80

9. .7 **11.** $\frac{95}{12}$ **13.** 360 **15.** $\frac{8}{3}$ **17.** $\frac{15}{4}$ **19.** 47 **21.** $\frac{20}{3}$ **23.** 1.2012

25. $20{,}383.\overline{6}$ **27.** $\frac{29}{10}$ **29.** $\frac{675}{22}$

problem set 59

1. (a) 80 yd/sec (b) 4800 yd **3.** 2 mi/hr

5. (a) $\frac{6}{25}$ (b) 24 (c) .6 (d) 60 **7.** 700 **9.** $.58\overline{3}$ **11.** $\frac{21}{4}$

13. 144 **15.** $\frac{5}{2}$ **17.** $\frac{25}{14}$ **19.** 48 **21.** $\frac{85}{12}$ **23.** 11.1924 **25.** 2602.14

27. $\frac{5}{48}$ **29.** $\frac{38}{3}$

problem set 60

1. 600 miles **3.** 8934 **5.** (a) $\frac{3}{25}$ (b) 12 (c) $.8\overline{3}$ (d) $83.\overline{3}$ **7.** 790

9. .875 **11.** $\frac{77}{20}$ **13.** 240 **15.** $\frac{54}{5}$ **17.** $\frac{45}{23}$ **19.** 31 **21.** $\frac{89}{12}$

23. .43368 **25.** $2937.1\overline{6}$ **27.** $\frac{4}{9}$ **29.** $\frac{100}{19}$

problem set 61

1. 20 sec **3.** 20 hr **5.** (a) $\frac{11}{50}$ (b) 22 (c) .84 (d) 84 **7.** 120

9. .83 **11.** $\frac{147}{40}$ **13.** 300 **15.** $\frac{160}{9}$ **17.** $\frac{203}{124}$ **19.** 68 **21.** $\frac{27}{4}$

23. 7.9443 **25.** 37,522 **27.** $\frac{7}{60}$ **29.** $\frac{5}{3}$

problem set 62

1. 70 minutes **3.** 20 hr **5.** (a) $\frac{4}{25}$ (b) 16 (c) .4 (d) 40 **7.** $.\overline{1}$

9. $\frac{125}{128}$ **11.** (a) 62.8 cm (b) 314 cm^2 **13.** 720 **15.** $\frac{48}{5}$ **17.** $\frac{49}{132}$

19. 35 **21.** $\frac{93}{20}$ **23.** .86982 **25.** 20,575 **27.** $\frac{91}{188}$ **29.** 176

problem set 63

1. 64 miles **3.** 124 miles **5.** (a) .6 (b) 60 **7.** $.\overline{6}$ **9.** $\frac{39}{68}$ **11.** 4368 in^3

13. $20\frac{1}{31}$ **15.** $\frac{25}{14}$ **17.** $\frac{113}{50}$ **19.** 66 **21.** $\frac{58}{21}$ **23.** .019602 **25.** 38,522

27. $\frac{35}{48}$ **29.** 55

problem set 64

1. 1344 **3.** $\frac{3}{8}$ **5.** (a) $\frac{9}{50}$ (b) 18 **7.** $.8\overline{3}$ **9.** $\frac{91}{12}$ **11.** 6600 cm^3

13. $458\frac{5}{7}$ **15.** $\frac{8}{3}$ **17.** $\frac{25}{38}$ **19.** 18 **21.** $\frac{61}{42}$ **23.** 2.3556 **25.** 250.87

27. $\frac{75}{17}$

29. Six hundred twenty-one million, seven hundred twenty-three thousand, one hundred thirty-one and seventy-two hundredths

problem set 65

1. 250 **3.** $\frac{1}{7}$ **5.** (a) .4 (b) 40 **7.** 240 **9.** $.\overline{3}$ **11.** 5950 cm^3

13. $\frac{114}{17}$ **15.** 4 **17.** $\frac{11}{26}$ **19.** 34 **21.** $\frac{281}{28}$ **23.** .22968 **25.** 2697

27. $\frac{817}{672}$

29. Six billion, nine hundred twenty-one million, three hundred twenty-one thousand, one hundred seventy-two and one tenth

problem set 66

1. 1050 **3.** 15 **5.** (a) $\frac{1}{4}$ (b) .25 **7.** 410 **9.** $.8\overline{3}$

11. (a) 91.4 ft (b) 557 ft^2 **13.** 180 **15.** $\frac{124}{11}$ **17.** $\frac{26}{21}$ **19.** $\frac{29}{9}$ **21.** 30

23. $9\frac{3}{4}$ **25.** 10,128.279 **27.** $\frac{67}{100}$ **29.** 40

problem set 67

1. 6 cents **3.** 3000 **5.** (a) $\frac{37}{100}$ (b) 37 **7.** 540 **9.** .9375

11. (a) 102.8 in (b) 714 in^2 **13.** 360 **15.** $\frac{35}{12}$ **17.** $\frac{207}{152}$ **19.** $\frac{115}{19}$

21. 82 **23.** $15\frac{17}{36}$ **25.** 10,824.3919 **27.** $1\frac{1}{2}$ **29.** 108

problem set 68

1. 4 **3.** 600 **5.** 1200

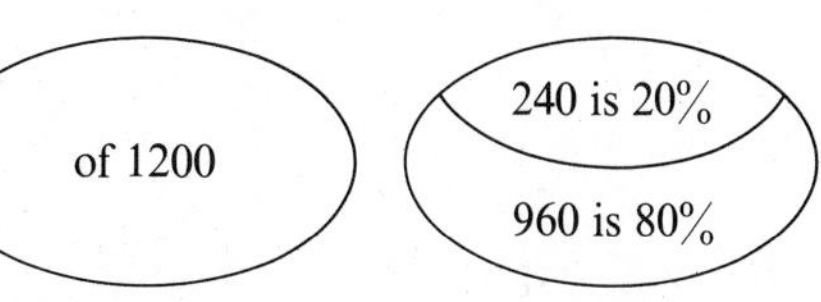

7. 63

of 210 | 63 is 30% | 147 is 70%

Before, 100% After

9. (a) $\frac{3}{10}$ (b) 30%

11. $\frac{129}{56}$ **13.** 854 ft^3 **15.** 360 **17.** $\frac{9}{4}$ **19.** $1\frac{35}{363}$ **21.** 64 **23.** $8\frac{5}{6}$

25. .032778 **27.** 40,292.5 **29.** $\frac{1378}{93}$

problem set 69

1. 25 cents, \$2.50 **3.** 3500 **5.** See lesson

7. 1200

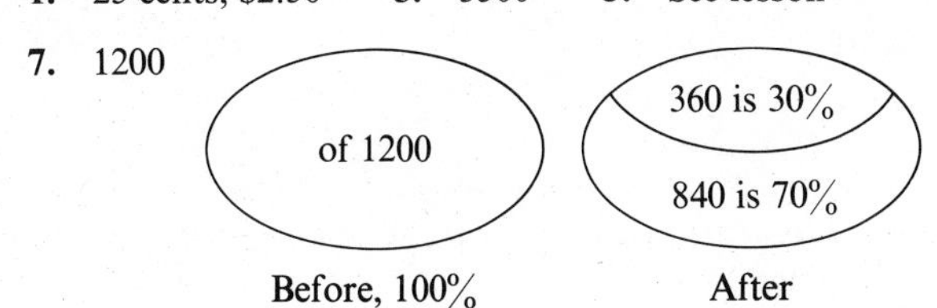

9. 20%

of 150 | 30 is 20% | 120 is 80%

Before, 100% After

11. $\frac{157}{100{,}000}$

13. (a) 88.4 in (b) 407 in^2 **15.** $\frac{443}{12}$ **17.** $\frac{28}{15}$ **19.** $\frac{11}{10}$ **21.** 111 **23.** $\frac{5}{3}$

25. .040551 **27.** 36,262 **29.** $\frac{1050}{17}$

problem set 70

1. 2160 **3.** 2200 **5.** See lesson

7. 2500

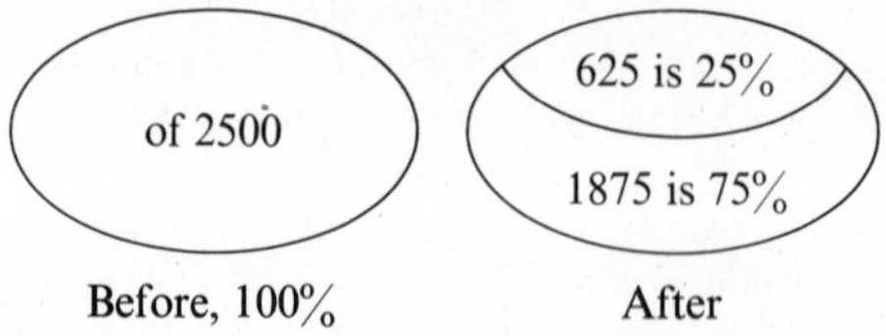

9. 40%

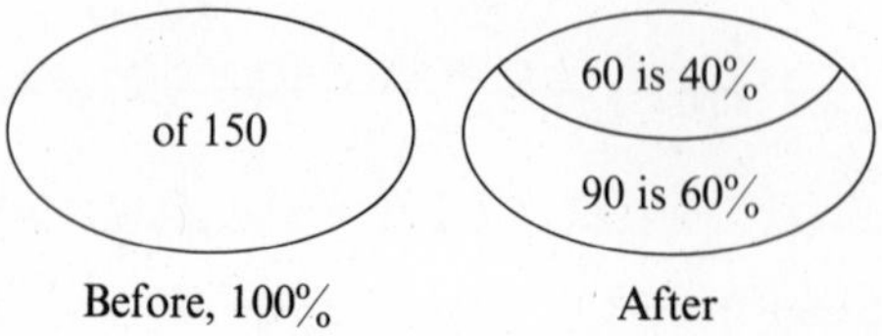

11. $\frac{37}{22{,}000}$

13. (a) 100 m (b) 400 m^2 **15.** $207\frac{1}{10}$ **17.** $\frac{57}{16}$ **19.** $\frac{41}{32}$ **21.** 76

23. $2\frac{15}{16}$ **25.** .060384 **27.** 32,030 **29.** $\frac{960}{169}$

problem set 71

1. 5390 **3.** 14,700 **5.** 168

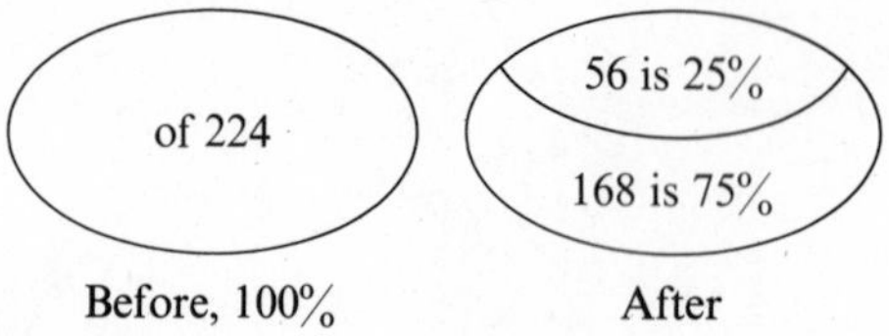

7. 410

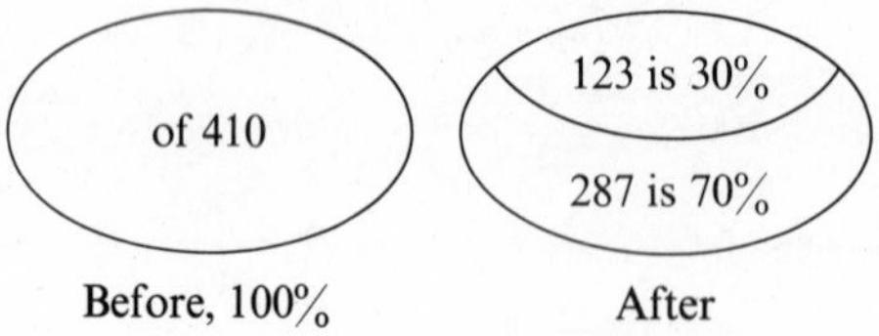

9. (a) .37 (b) 37%

11. .25 **13.** 89.25 m^3 **15.** 300 **17.** $\frac{15}{2}$ **19.** 23 **21.** 0 **23.** −10

25. .3996 **27.** 42,638

29. Six hundred twenty-five million, three hundred sixty-one thousand, eight hundred eleven and one hundredth

problem set 72

1. $\frac{3}{29}$ **3.** 3 hr **5.**

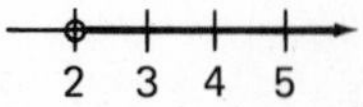

7. 288

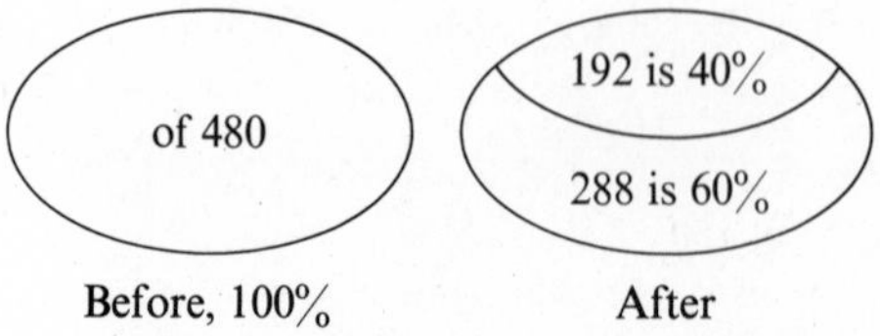

9. 20%

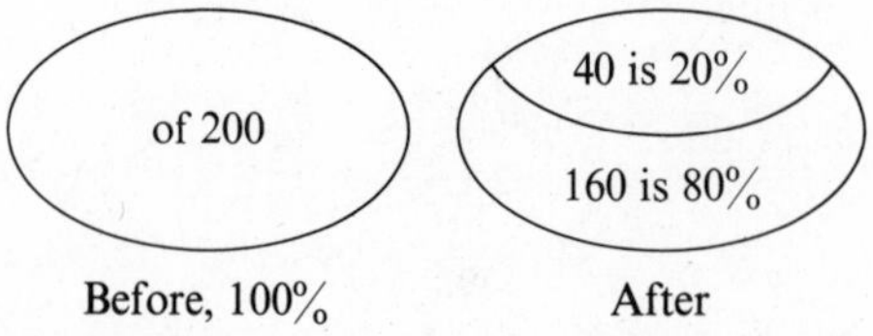

11. $\frac{7}{1100}$

13. (a) 85.68 ft (b) 401.04 ft^2 **15.** $\frac{89}{13}$ **17.** $\frac{176}{39}$ **19.** $\frac{15}{52}$ **21.** -10

23. -24 **25.** .22374 **27.** 36,385 **29.** 200,000,000

problem set 73

1. 240,000 **3.** 5 hr **5.**

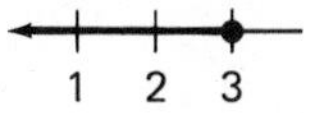

7. 496

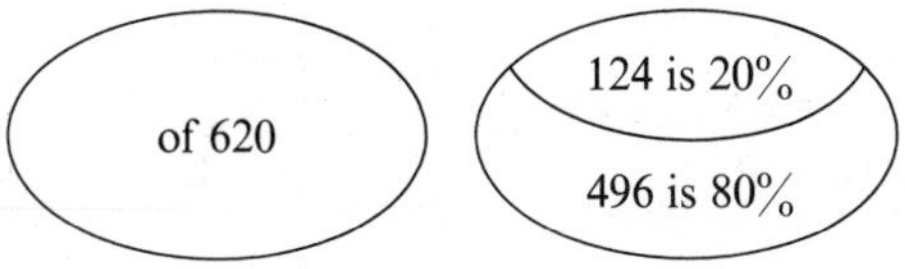

Before, 100% After

9. 20%

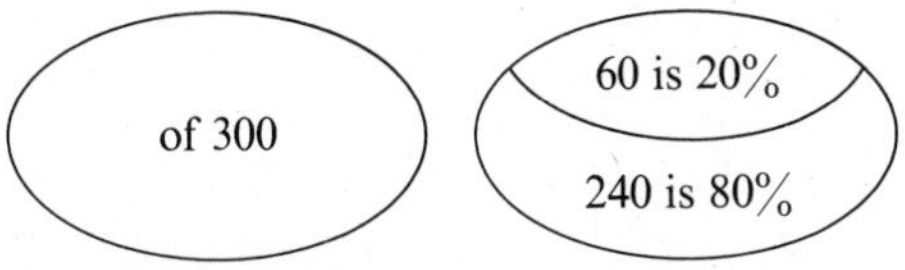

Before, 100% After

11. $\frac{39}{11{,}000}$ **13.** 4368 in^3

15. $231\frac{4}{7}$ **17.** $\frac{243}{80}$ **19.** $\frac{5}{11}$ **21.** 3 **23.** 0 **25.** $8\frac{17}{20}$ **27.** 3591.694

29. $\frac{11}{32}$

problem set 74

1. 990 **3.** 560 **5.**

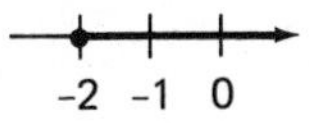

7. 112

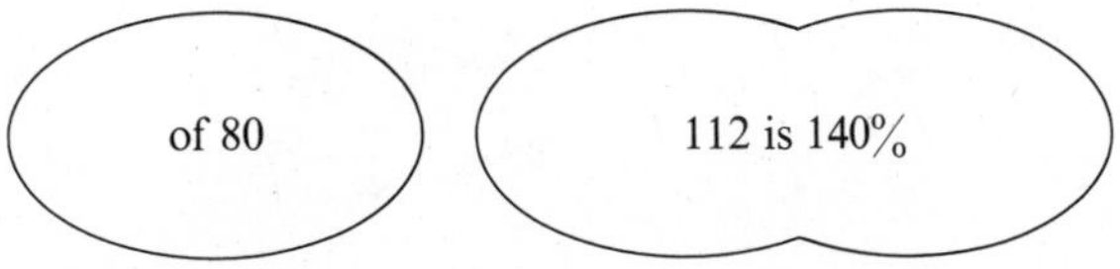

Before, 100% After

9. 450

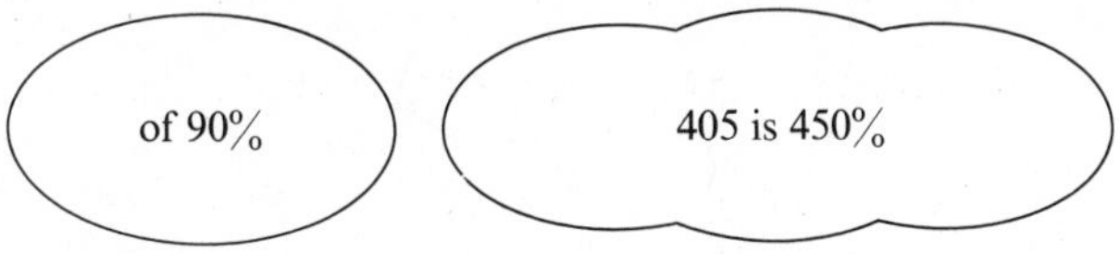

Before, 100% After

11. $\frac{47}{35{,}000}$

13. 2400 ft^3 **15.** $\frac{186}{29}$ **17.** $\frac{9}{14}$ **19.** $\frac{35}{38}$ **21.** 3 **23.** -10 **25.** 20

27. 1316.2509 **29.** 52

problem set 75

1. 2500 **3.** 1440 **5.**

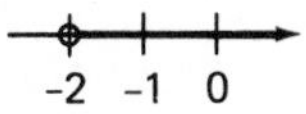

7. 108

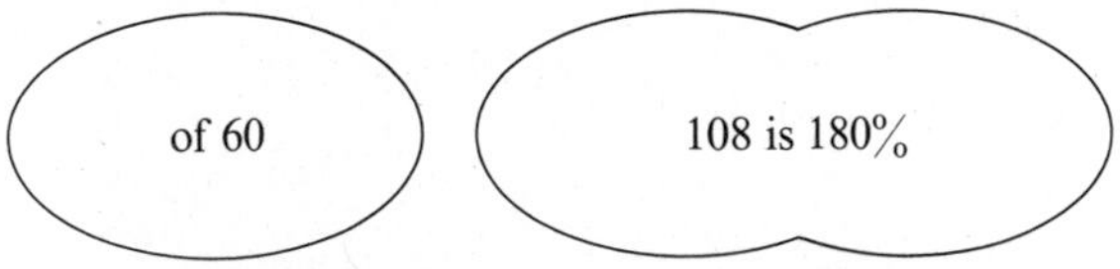

Before, 100% After

9. 250%

of 80 | 200 is 250%

Before, 100% | After

11. $\frac{283}{250{,}000}$

13. 41,000 cm^3 **15.** $2462\frac{1}{5}$ **17.** $\frac{3}{16}$ **19.** $\frac{81}{44}$

21. (a) 8 (b) 8 (c) -8 (d) -8

23. (a) 3 (b) 3 (c) -3 (d) -3 **25.** -32 **27.** $13\frac{1}{2}$ **29.** $2 \cdot 2 \cdot 5 \cdot 127$

problem set 76

1. 600 **3.** 385 mi **5.** -2 -1 0

7. 385

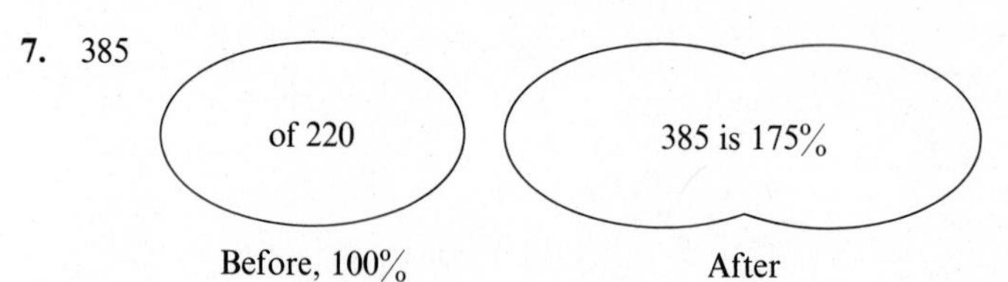

9. 175%

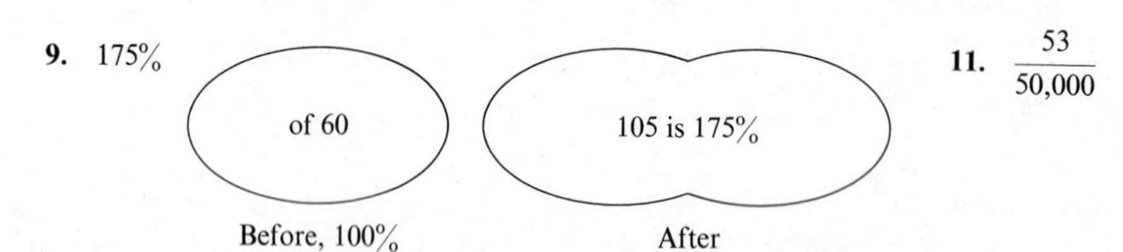

11. $\frac{53}{50{,}000}$

13. 150,000 cm^3 **15.** $\frac{179}{12}$ **17.** $\frac{7}{45}$ **19.** $\frac{31}{20}$ **21.** (a) -6 (b) -6 (c) 6

23. -3 **25.** -12 **27.** $\frac{197}{14}$ **29.** .0052

problem set 77

1. .690 **3.** 340 miles **5.** -3 -2 -1

7. 192

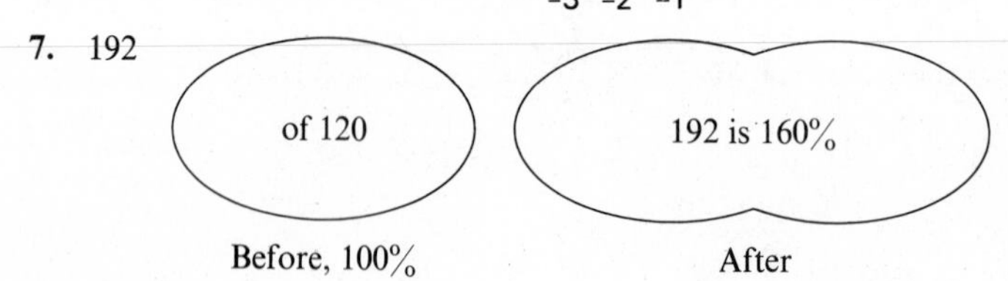

9. 250%

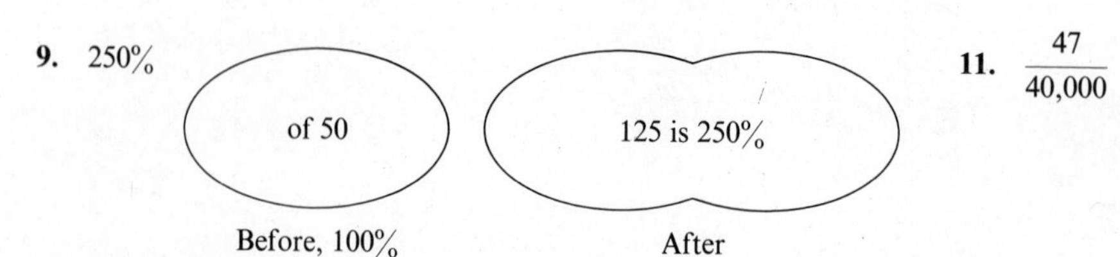

11. $\frac{47}{40{,}000}$

13. 52,500 m^3 **15.** $120\frac{11}{15}$ **17.** $\frac{5}{87}$ **19.** $\frac{23}{24}$

21. (a) -18 (b) 18 (c) -18 **23.** 3 **25.** -4 **27.** 0

29. One hundred thirty-six million, one hundred twenty-one thousand, one hundred thirty-four and five tenths

problem set 78

1. 4 hr **3.** 8000 **5.**

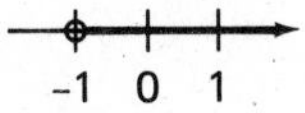

7. 196

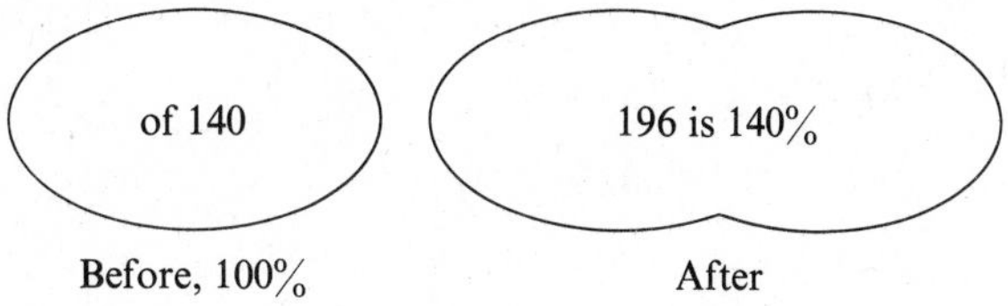

9. 140%

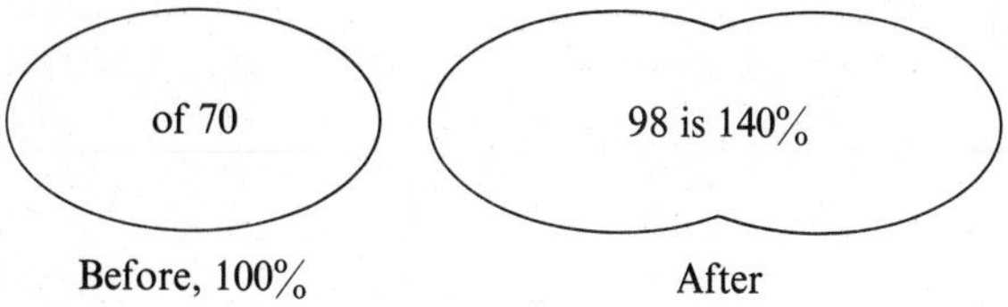

11. $\frac{19}{15,000}$

13. 1371 in^3 **15.** $113\frac{1}{12}$ **17.** $\frac{17}{40}$ **19.** $\frac{8}{7}$ **21.** (a) -18 (b) 10 (c) -7

23. -1 **25.** -8 **27.** -6 **29.** 31

problem set 79

1. $\frac{1}{8}$ **3.** 1000 **5.**

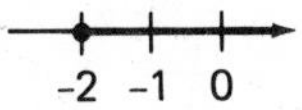

7. 288

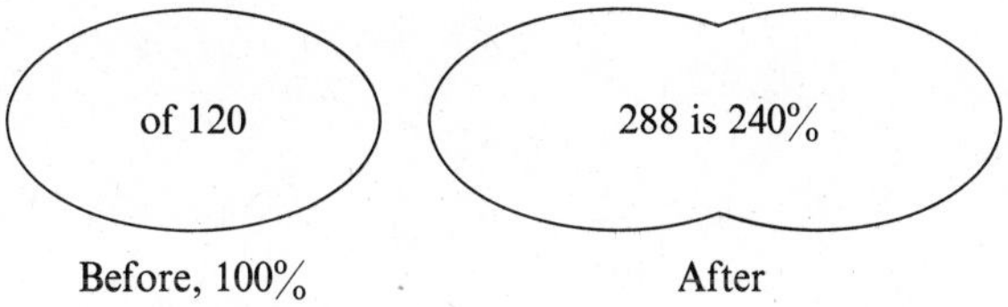

9. 160%

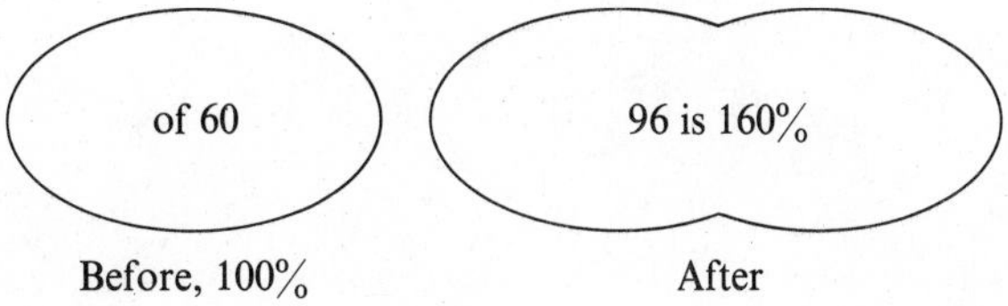

11. $\frac{1}{750}$

13. 42.2 m^3 **15.** 138,531,000 cm **17.** -2 **19.** $\frac{9}{40}$ **21.** 28

23. (a) -8 (b) -8 (c) 12 **25.** 51 **27.** 4 **29.** 19

problem set 80

1. $2400 **3.** 14,000

5. (a) $2N - 8$ (b) $-3(N + 5)$ (c) $-N - 4$ (d) $5(2N + 6)$

7. 646

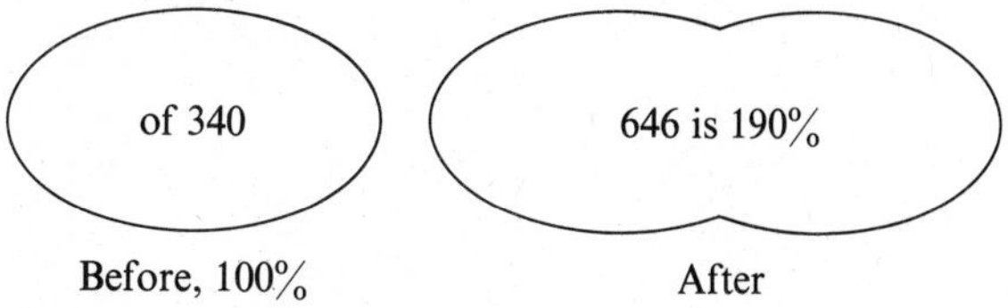

9. 160%

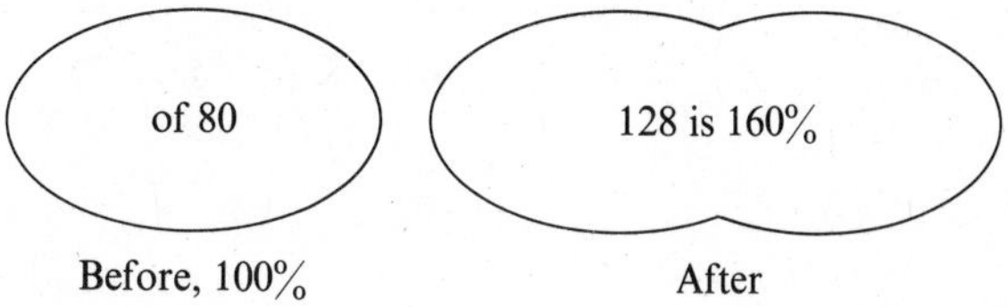

11. $\frac{77}{50,000}$

13. 17 yd **15.** $2 \cdot 2 \cdot 2 \cdot 2 \cdot 3 \cdot 3 \cdot 5$ **17.** $-\frac{15}{7}$ **19.** $-\frac{1}{2}$ **21.** $\frac{33}{4}$ **23.** 6

25. 155,302.5 **27.** .165494 **29.** -1

problem set 81

1. 56 inches **3.** \$50,400

5. (a) $4(3N - 6)$ (b) $-2(N + 5)$ (c) $-N - 25$ **7.** -3 -2 -1

9. 675

of 250 | 675 is 270%

Before, 100% | After

11. $\frac{93}{70{,}000}$

13. 126,720 inches **15.** $-\frac{5}{7}$ **17.** $\frac{11}{9}$ **19.** 40 **21.** (a) -6 (b) 4 (c) 6

23. -8 **25.** $\frac{49}{15}$ **27.** 653.0152 **29.** 11

problem set 82

1. 42 **3.** -3 **5.** 3 4 5 6 **7.** -2 -1 0

9. 675

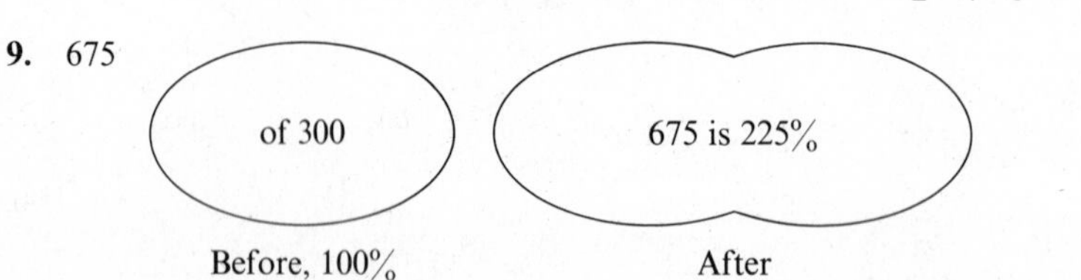

11. (a) $\frac{2}{5}$ (b) .4 **13.** 360 in^3 **15.** $-\frac{2}{3}$ **17.** $\frac{7}{19}$ **19.** 16

21. (a) -6 (b) -6 (c) 20 **23.** -16 **25.** $\frac{143}{5}$ **27.** 1169.0007 **29.** 1

problem set 83

1. 3 **3.** 3 **5.** 1 2 3

7. 805

9. 135%

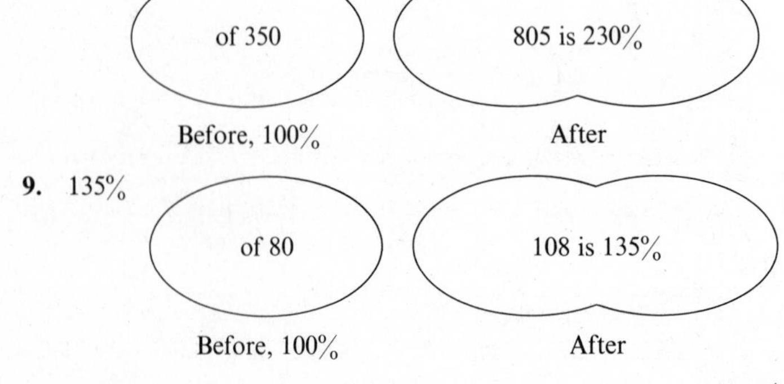

11. (a) $\frac{9}{25}$ (b) 36 **13.** 1042 yds^3 **15.** 6.3 **17.** $-\frac{1}{3}$ **19.** $-\frac{3}{2}$ **21.** 38

23. 32 **25.** $\frac{140}{9}$ **27.** $205{,}003.\overline{3}$ **29.** 5

problem set 84

1. 5 **3.** −5 **5.**

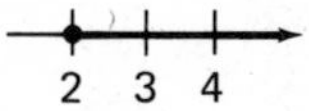

7. 286

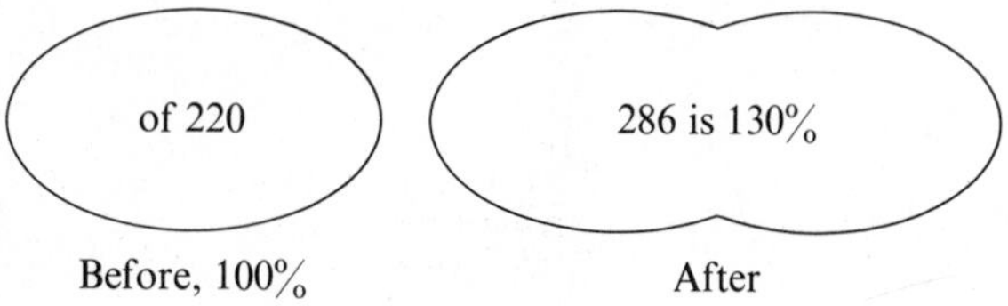

9. 160%

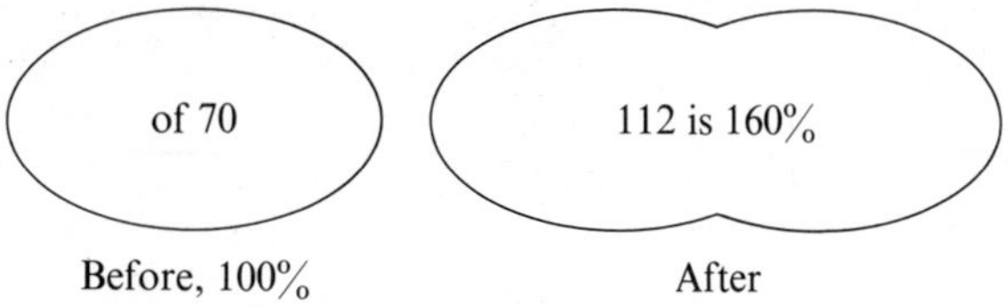

11. (a) $\frac{53}{100}$ (b) 53 **13.** 14,568 in^3 **15.** 1,268,700 cm **17.** $\frac{2}{3}$ **19.** $-\frac{51}{56}$

21. 26 **23.** −2 **25.** $\frac{3}{8}$ **27.** 30 **29.** 83

problem set 85

1. 20 **3.** 5 hr **5.**

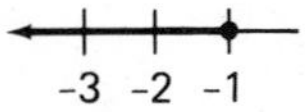

7. 384

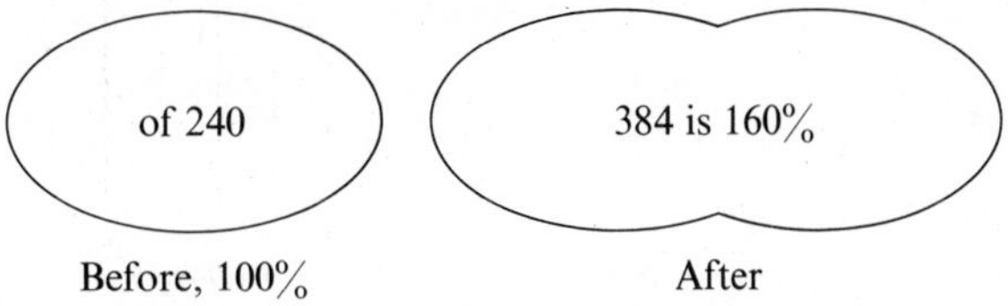

9. 160%

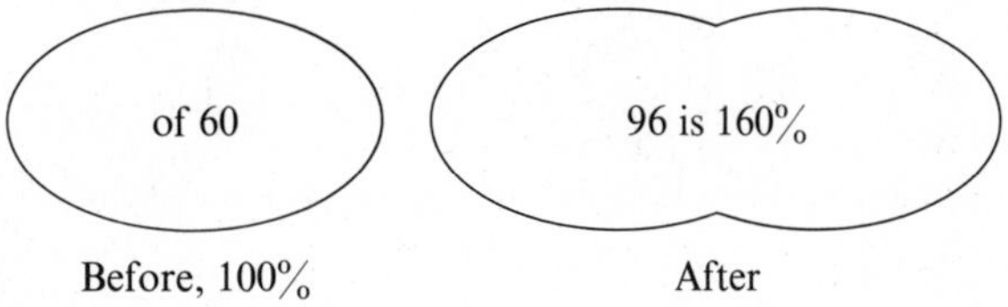

11. $\frac{71}{60{,}000}$ **13.** 3.3

15. $\frac{1}{2}$ **17.** $\frac{5}{3}$ **19.** $-\frac{14}{39}$ **21.** −24 **23.** $\frac{13}{3}$ **25.** $\frac{11}{72}$ **27.** $\frac{49}{19}$ **29.** 48

problem set 86

1. −20 **3.** $1400 **5.**

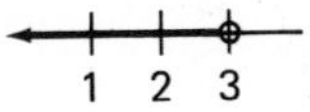

7. 621

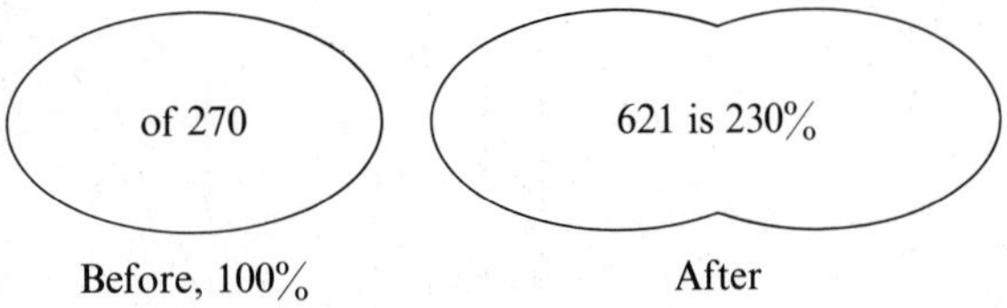

9. 145%

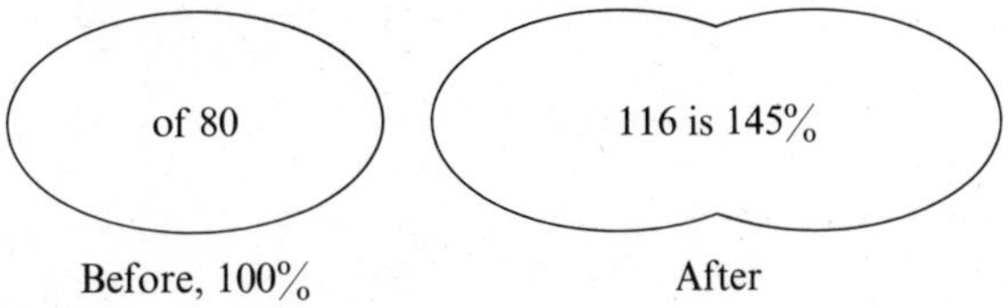

11. (a) .9 (b) 90 **13.** 5760 cm^3 **15.** .01287321 km **17.** 4 **19.** $-\frac{11}{4}$

21. $-\frac{3}{7}$ **23.** -30 **25.** 4 **27.** 69 **29.** -25

problem set 87

1. -64 **3.** 7 **5.** -1 0 1

7. 800

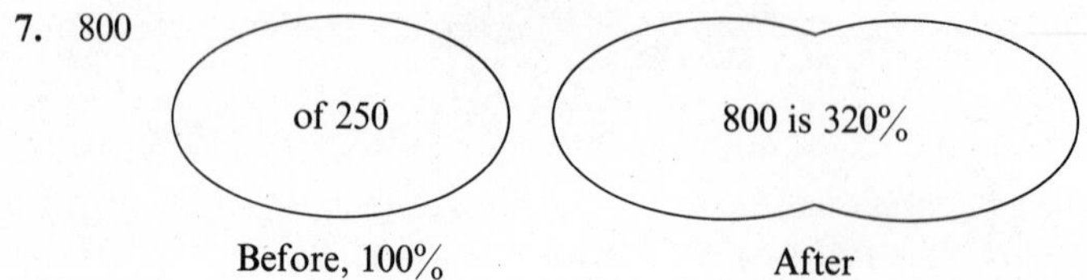

9. 180%

of 90 | 162 is 180%

Before, 100% After

11. $\frac{19}{5000}$

13. 120 **15.** 5280 yd **17.** $\frac{7}{6}$ **19.** $\frac{11}{3}$ **21.** $\frac{6}{7}$ **23.** 4 **25.** 1

27. 107 **29.** -1

problem set 88

1. 5 **3.** 720 **5.** -2 -1 0

7. 368

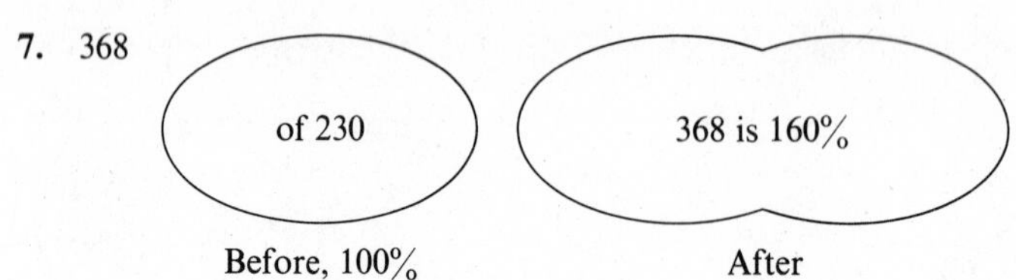

9. 185%

of 60 | 111 is 185%

Before, 100% After

11. $\frac{1}{80}$ **13.** 2.8

15. .07892321 km **17.** $\frac{7}{2}$ **19.** $-\frac{1}{5}$ **21.** $-\frac{9}{20}$ **23.** -52 **25.** 2 **27.** $\frac{9}{28}$

29. -18

problem set 89

1. 280 **3.** 11 **5.** -1 0 1

7. 195

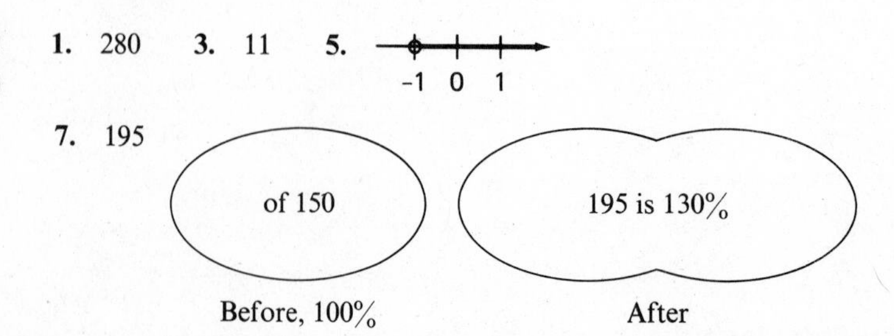

9. 25%

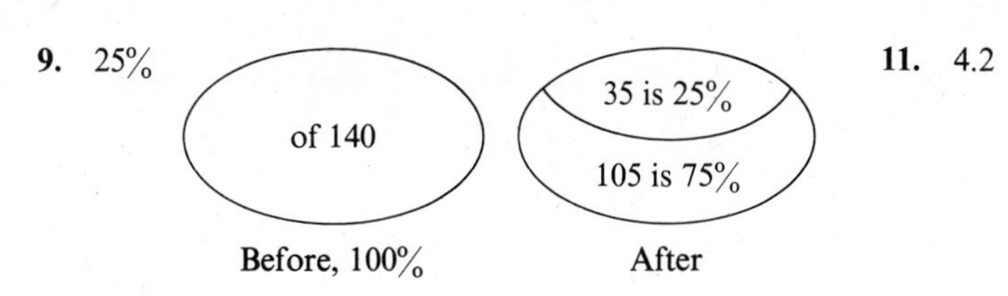

11. 4.2

13. (a) $\frac{17}{20}$ (b) .85 **15.** 6,213,112.1 cm **17.** 3 **19.** 1 **21.** $-\frac{6}{5}$

23. 4 **25.** -13 **27.** $\frac{17}{9}$ **29.** -14

problem set 90

1. 1400 **3.** $700 **5.**

-2 -1 0

7. 224

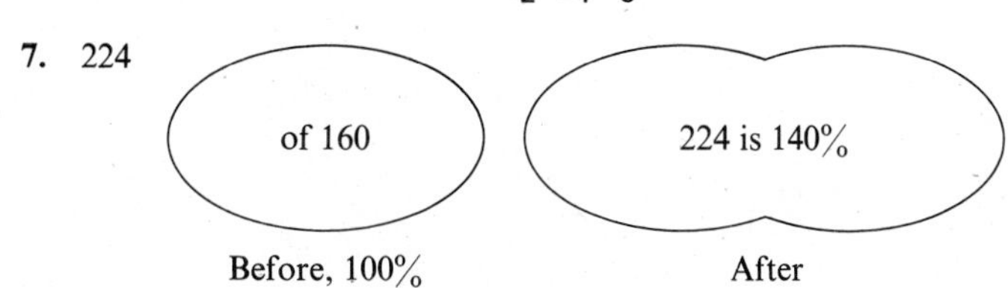

9. 600

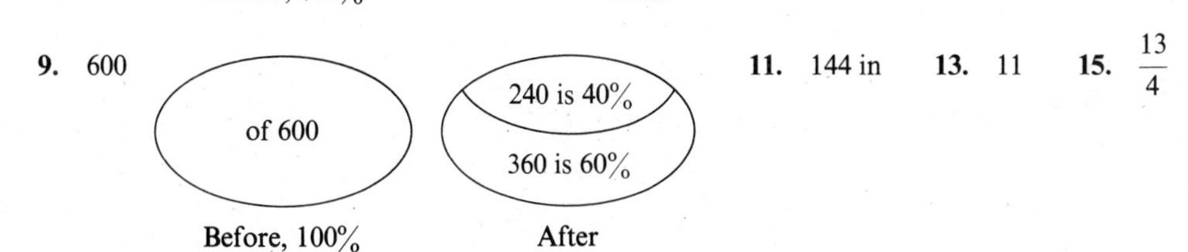

11. 144 in **13.** 11 **15.** $\frac{13}{4}$

17. $\frac{7}{5}$ **19.** $\frac{1}{5}$ **21.** -9 **23.** x^7y^{23} **25.** a^7m^8 **27.** -1 **29.** 196

problem set 91

1. 250 **3.** -4 **5.**

1 2 3

7. 240

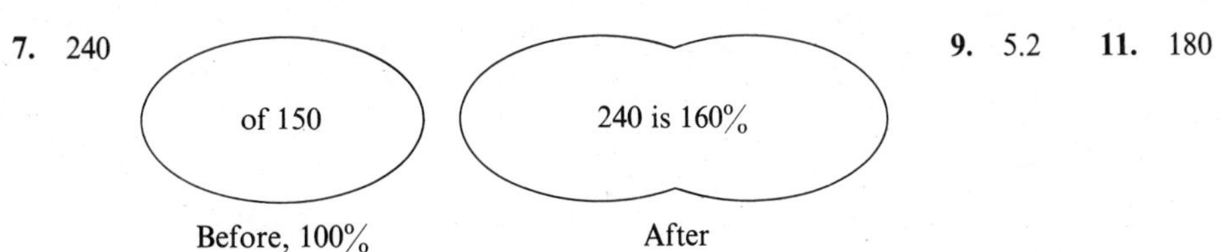

9. 5.2 **11.** 180

13. 28 **15.** $\frac{3}{2}$ **17.** $-\frac{2}{3}$ **19.** $4ab^2 - a + 3$ **21.** -17 **23.** 22 **25.** x^7y^9

27. $\frac{7}{24}$ **29.** $\frac{280}{3}$

problem set 92

1. 42 **3.** 6 hr **5.** 90

of 200
90 is 45%
110 is 55%
Before, 100%
After

7. 155%

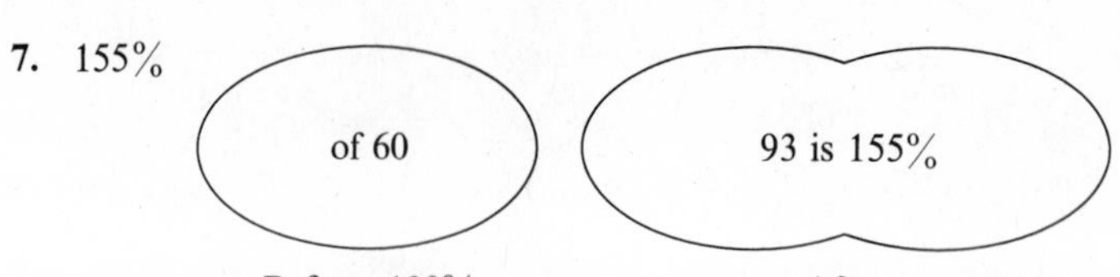

9. 736 cm^2 **11.** $\frac{37}{3}$

13. $2a^2b + 2ab^2$ **15.** $7xy + 9x + 3y + 5$ **17.** -2 **19.** 0

21. -1 **23.** -248 **25.** x^6y^7 **27.** -1 **29.** 6

problem set 93

1. 4250 **3.** 1308 **5.** 53 **7.** 110101

9. 105

of 60 | 105 is 175%

Before, 100% | After

11. 16,020 in^3

13. 16 **15.** $3a^2b + 3ab^2 - 12ab$ **17.** $3a^3 + b^3 + 6a^3b$ **19.** $-\frac{9}{8}$ **21.** $-\frac{25}{24}$

23. -10 **25.** a^4b^4 **27.** m^8n^5 **29.** 5

problem set 94

1. 160 **3.** 800 **5.** 45 **7.** 101011

9. 112

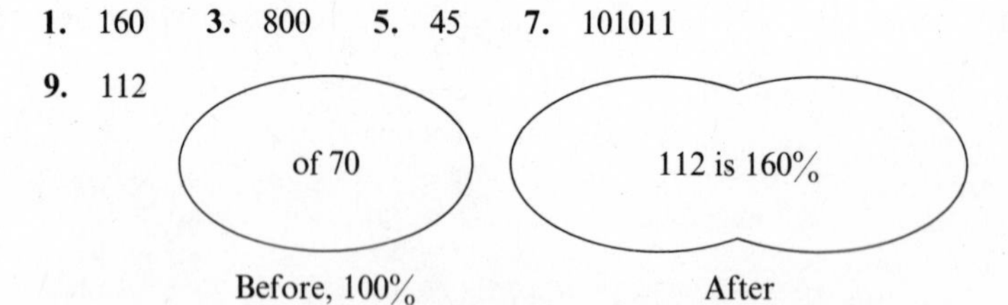

11. 112.8 inches

13. 6 **15.** $3a^2b + 3ab^2 + 3abc + 3abd$ **17.** $a^3 + 4a^2b + 6b^2a$ **19.** $-\frac{9}{7}$

21. $-\frac{21}{16}$ **23.** -11 **25.** -28 **27.** m^6n^5 **29.** 7

problem set 95

1. 960 **3.** 3600 **5.** 54 **7.** 30

9. 128

of 80 | 128 is 160%

Before, 100% | After

11. 1000 ft^2

13. 5 **15.** $a^2b + ab^2 + a^2b^2$ **17.** $2x^2y + 5y^2x$ **19.** $-\frac{9}{4}$ **21.** $-\frac{7}{2}$ **23.** 0

25. a^6b^3 **27.** $-\frac{14}{3}$ **29.** 13

problem set 96

1. 140 **3.** 1000 **5.** 59

7. 104

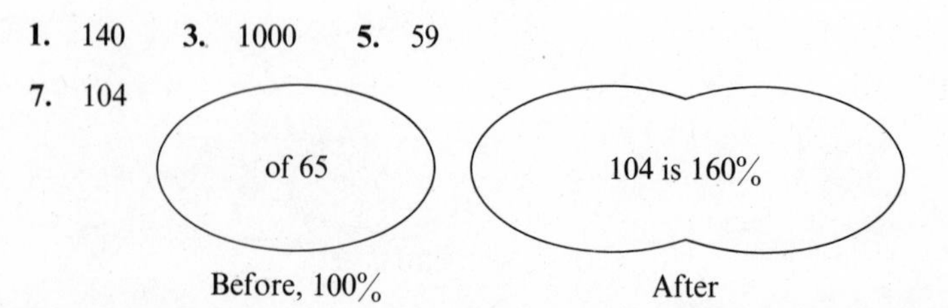

9. $\frac{51}{40,000}$ **11.** 360

13. -1 **15.** $a^2b^2c + 2a^2b^2 - 2ab^2c$ **17.** $2m^2n^3 + 5mn^2$ **19.** $\frac{33}{38}$ **21.** 5.7

23. a^3b^7 **25.** -20 **27.** $\frac{5}{7}$ **29.** 1

problem set 97

1. 2400 **3.** 60 **5.** Refer to Lesson 97 **7.** 1001011

9. 135% **11.** 126,720 inches

of 620

837 is 135%

Before, 100% After

13. 3 **15.** $m^2n^2 + mn^2 + m^2n$ **17.** $2xy^3 + 2x^2y$ **19.** $\frac{23}{9}$ **21.** $\frac{27}{8}$ **23.** -4

25. x^6y^5 **27.** $\frac{1}{9}$ **29.** -9

problem set 98

1. 600 **3.** 1400 **5.** Refer to Lesson 97 **7.** 11100 **9.** 147024 cm^3

11. 17,280 in^3 **13.** $\frac{7}{6}$ **15.** 5 **17.** $-4xy^3 + 2x^3y^3$ **19.** $\frac{1}{14}$ **21.** $-\frac{64}{27}$

23. a^6b^4 **25.** x^5y^6 **27.** 0 **29.** 12

problem set 99

1. 3500 **3.** 20,000 **5.** Refer to Lesson 99 **7.** Refer to Lesson 97 **9.** 11001

11. 5280 in^3 **13.** $\frac{612{,}312{,}561}{(12^3)(5280^3)}$ mi^3 **15.** $4ab^2 + 4abc - 4a^2b^2$ **17.** $ac^2 + 4a^2b$

19. $-\frac{12}{5}$ **21.** -2 **23.** 36 **25.** $2x^4y^5$ **27.** $\frac{1}{9}$ **29.** -1

problem set 100

1. 2000 **3.** -4 **5.** Refer to Lesson 99 **7.** Refer to Lesson 97 **9.** 2451 (base 8)

11. 216 **13.** $12(5280^3)(12^3)$ in^3

of 120

216 is 180%

Before, 100% After

15. 29 **17.** $-a^3m - am^2 + a^2m^2$ **19.** $-\frac{1}{2}$ **21.** $-\frac{1}{3}$ **23.** -28 **25.** $-\frac{5}{4}$

27. -4 **29.**

0 1 2

problem set 101

1. 3000 **3.** -4 **5.** Refer to Lesson 99 **7.** Refer to Lesson 97 **9.** 7215 (base 8)

11. 900 **13.** 1,400,000 liters

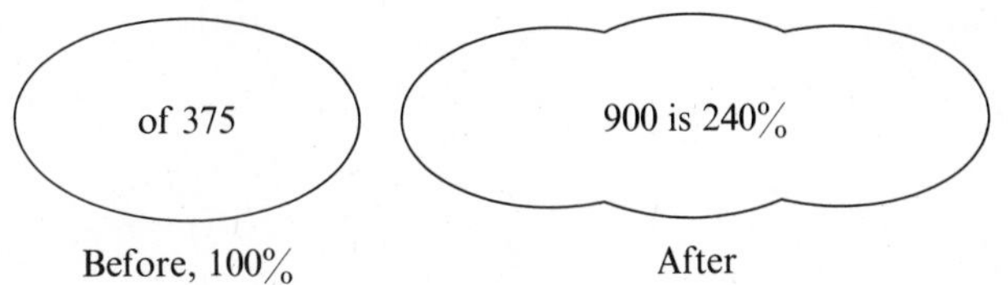

15. 44,000,000 cm^3 **17.** $2a^2b^2 + 2a^3b - 2ab^3 + 2a^2b$ **19.** 3300 ft^3 **21.** $-\frac{9}{8}$

23. 10 **25.** $5a^6b^6$ **27.** 0 **29.** -9

problem set 102

1. 1200 **3.** $\frac{46}{3}$ **5.** Refer to Lesson 99 **7.** Refer to Lesson 97 **9.** 111110

11. 45 **13.** 180%

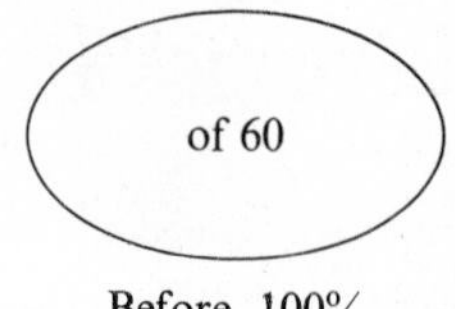

15. 1,200,000 liters **17.** $3a^3b^2 + 3a^2b^3 - 3a^2b - 3ab^2$ **19.** 120 ft **21.** $-\frac{14}{5}$

23. -5 **25.** 2 **27.** 0 **29.** $\frac{51}{40,000}$

problem set 103

1. $\frac{1}{3}$ **3.** (a) $\frac{1}{6}$ (b) $\frac{1}{12}$ **5.** Refer to Lesson 99 **7.** Refer to Lesson 97

9. 22 **11.** 20%

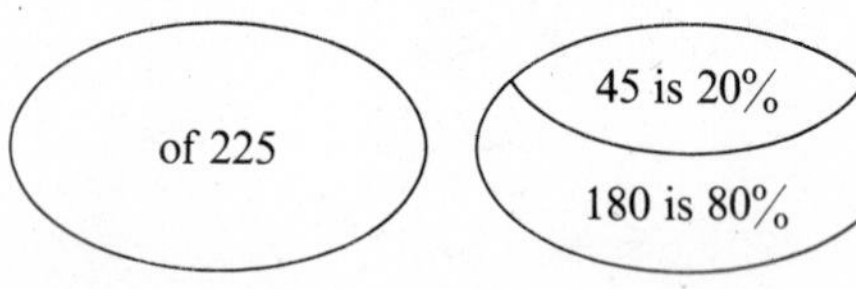

13. 52,000,000 cm^3

15. -5 **17.** $-4xy^2 + 7x^2y$ **19.** 27 **21.** $-\frac{1}{4}$ **23.** $4a^6b^4$ **25.** $\frac{413}{180}$

27. -5 **29.** $10,080 = 2^5 \cdot 3^2 \cdot 5 \cdot 7$

problem set 104

1. (a) $\frac{5}{36}$ (b) $\frac{1}{9}$ (c) $\frac{1}{12}$ **3.** 350

5. A number is an idea, and a numeral is the symbol we use to represent the number.

7. Refer to Lesson 99 **9.** 101000 **11.** 1611

13. 160%

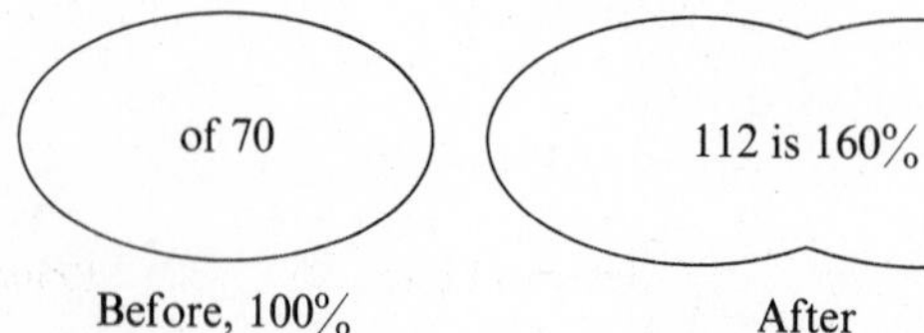

15. 1,320,000 liters

17. $2a^2b + 2ab^2 - 2a^2b^3 + 2a^3b^2$ **19.** .1272 m^3 **21.** $-\frac{61}{20}$ **23.** -10 **25.** $-\frac{3}{7}$

27. 0 **29.** $\frac{13}{5000}$

problem set 105

1. $\frac{11}{15}$ **3.** 490 **5.** Refer to Lesson 99 **7.** Refer to Lesson 97

9. 1001100 **11.** 713 (base 8)

13. 368

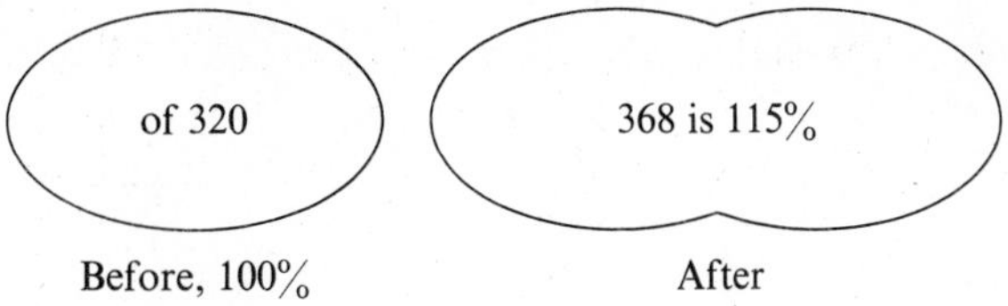

15. $15(3^3)(12^3)\text{ in}^3$

17. $3a^2c + 3abc - 3ac^2 + 3a^2bc^2$ **19.** $752{,}160{,}000\text{ cm}^3$ **21.** 25 **23.** -16

25. $-\frac{3}{7}$ **27.** 2 **29.**

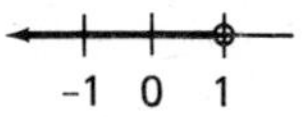

problem set 106

1. $\frac{11}{45}$ **3.** $\frac{1}{32}$ **5.** Refer to Lesson 99 **7.** Refer to Lesson 97 **9.** 21

11. 143 **13.** 120

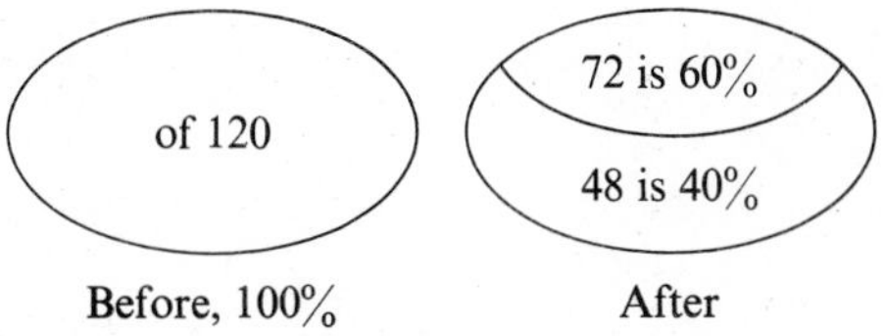

15. $181(5280^3)(12^3)\text{ in}^3$ **17.** $-7x^2y + 8xy^3$ **19.** 136 ft^2 **21.** -11 **23.** 7

25. 49 **27.** -9 **29.** 12

problem set 107

1. $\frac{1}{36}$ **3.** 400 **5.** Refer to Lesson 107 **7.** Refer to Lesson 107 **9.** 110000

11. 153 **13.** 300

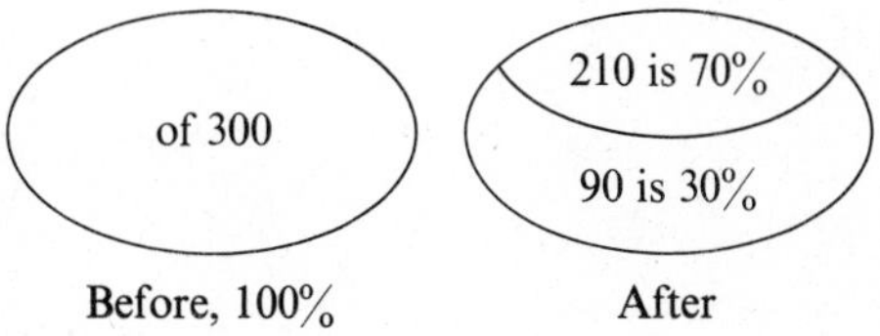

15. $\frac{105(5280)(5280)(5280)}{3(3)(3)}\text{ yd}^3$ **17.** $5x^2y^3 - 9x^2y^2$ **19.** 840 in^2 **21.** $-\frac{1}{8}$

23. -12 **25.** $-\frac{11}{2}$ **27.** -8 **29.** 0

problem set 108

1. $\frac{1}{2}$ **3.** 120 **5.** Refer to Lesson 107 **7.** Refer to Lesson 107 **9.** 101101

11. 399 **13.** 150

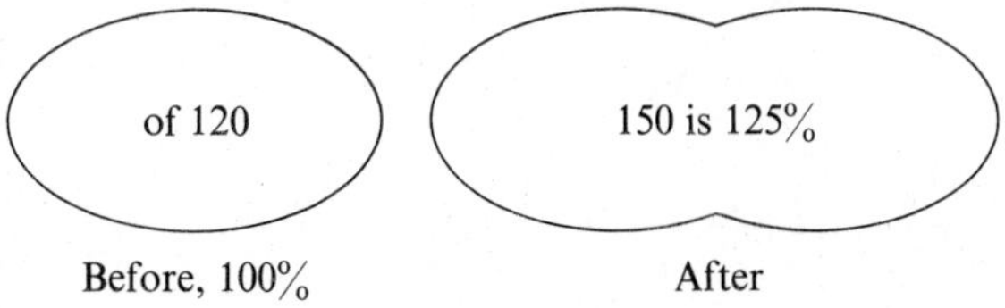

15. $26(3^3)(12^3)\text{ in}^3$ **17.** $7x^3y - 5xy^2$ **19.** $\frac{9}{14}$ **21.** $\frac{3}{8}$ **23.** 11 **25.** $\frac{29}{15}$

27. -24 **29.** -10

problem set 109

1. $\frac{1}{8}$ 3. 400 5.

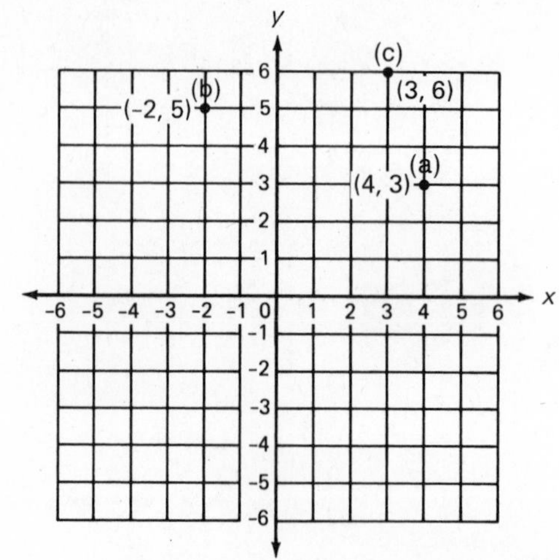

7. Refer to Lesson 107 9. Refer to Lesson 107 11. 305

13. 290

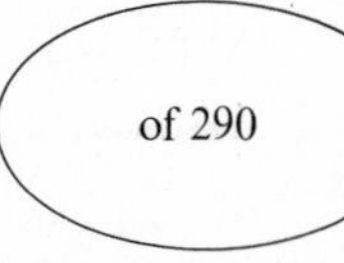

261 is 90%

29 is 10%

Before, 100% After

15. $2(5280^3)$ ft^3

17. $-3x^2y^2 + 3x^3y$ 19. $\frac{25}{4}$ 21. $\frac{1}{2}$ 23. -5 25. -7 27. 6 29. $-\frac{3}{2}$

problem set 110

1. $\frac{1}{4}$ 3. 5400 5.

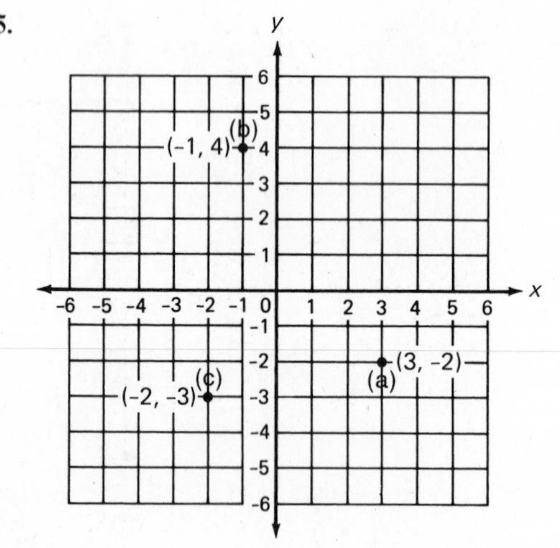

7. Refer to Lesson 107 9. 337

11. 312

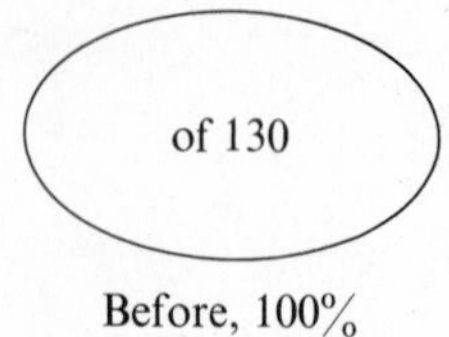

312 is 240%

Before, 100% After

13. $12(5280^3)(12^3)$ in^3 15. $-x^2y^2 + 4x^2y^3$ 17. 189.75 ft^3 19. $-\frac{1}{6}$ 21. 1

23. (a) 8 (b) $\frac{1}{8}$ (c) 2 (d) $\frac{1}{27}$ 25. $5a^7b^2$ 27. $\frac{1}{9}$ 29. 20